AGRICULTURAL PRODUCTIVITY
Measurement and Explanation

AGRICULTURAL PRODUCTIVITY
Measurement and Explanation

Susan M. Capalbo
John M. Antle
Editors

Resources for the Future
Washington, D.C.

© 1988 Resources for the Future

All rights reserved. No part of this publication may be reproduced by any means, either electronic or mechanical, without permission in writing from the publisher.

Printed in the United States of America

Published by Resources for the Future, Inc.
1616 P Street, N.W., Washington, D.C. 20036
Books from Resources for the Future are distributed worldwide by The Johns Hopkins University Press

Library of Congress Cataloging-in-Publication Data

Agricultural productivity: measurement and explanation/edited by Susan M. Capalbo, John M. Antle.
    p. cm.
Chiefly papers originally presented at a workshop, July 16–18, 1984.
"A product of RFF's Renewable Resources Division"—P. vi.
Bibliography: p.
Includes index.
ISBN 0-915707-37-3 (alk. paper)
    1. Agricultural productivity. 2. Agricultural productivity—United States. I. Capalbo, Susan M. II. Antle, John M. III. Resources for the Future. Renewable Resources Division. HD1415.A32 1987
338.1'6—dc19

                                        87-27224
                                              CIP

∞ The paper in this book meets the guidelines for permanence and durability of the Committee on Production Guidelines for Book Longevity of the Council on Library Resources.

DIRECTORS

Charles E. Bishop, *Chairman,* Anne P. Carter, William T. Creson, Henry L. Diamond, James R. Ellis, Lawrence E. Fouraker, Robert W. Fri, Jerry D. Geist, John H. Gibbons, Bohdan Hawrylyshyn, Thomas J. Klutznick, Frederic A. Krupp, Henry R. Linden, Richard W. Manderbach, Laurence I. Moss, William D. Ruckelshaus, Leopoldo Solís, Carl H. Stoltenberg, Barbara S. Uehling, Robert M. White, Macauley Whiting.

HONORARY DIRECTORS

Hugh L. Keenleyside, Edward S. Mason, William S. Paley, John W Vanderwilt

OFFICERS

Robert W. Fri, *President*
John F. Ahearne, *Vice President*
Edward F. Hand, *Secretary-Treasurer*

RESOURCES FOR THE FUTURE (RFF) is an independent nonprofit organization that advances research and public education in the development, conservation, and use of natural resources and in the quality of the environment. Established in 1952 with the cooperation of the Ford Foundation, it is supported by an endowment and by grants from foundations, government agencies, and corporations. Grants are accepted on the condition that RFF is solely responsible for the conduct of its research and the dissemination of its work to the public. The organization does not perform proprietary research.

RFF research is primarily social scientific, especially economic. It is concerned with the relationship of people to the natural environmental resources of land, water, and air; with the products and services derived from these basic resources; and with the effects of production and consumption on environmental quality and on human health and well-being. Grouped into five units—the Energy and Materials Division, the Quality of the Environment Division, the Renewable Resources Division, the National Center for Food and Agricultural Policy, and the Center for Risk Management—staff members pursue a wide variety of interests, including forest economics, natural gas policy, multiple use of public lands, mineral economics, air and water pollution, energy and national security, hazardous wastes, the economics of outer space, and climate resources. Resident staff members conduct most of the organization's work; a few others carry out research elsewhere under grants from RFF.

Resources for the Future takes responsibility for the selection of subjects for study and for the appointment of fellows, as well as for their freedom of inquiry. The views of RFF staff members and the interpretations and conclusions of RFF publications should not be attributed to Resources for the Future, its directors, or its officers. As an organization, RFF does not take positions on laws, policies, or events, nor does it lobby.

This book is a product of RFF's Renewable Resources Division and National Center for Food and Agricultural Policy. It was edited by Barbara Karni and designed by Martha Ann Bari. The index was prepared by Florence Robinson.

CONTENTS

Foreword Kenneth R. Farrell — xi
Acknowledgments — xv
Contributors — xvi

1 **Introduction and Overview** John M. Antle and Susan M. Capalbo — 1

   Overview of the Book   2
   Summary Observations   7

PART I   BACKGROUND — 15

2 **An Introduction to Recent Developments in Production Theory and Productivity Measurement** John M. Antle and Susan M. Capalbo — 17

   Basic Concepts in Modern Production Economics   18
   Approaches to Total Factor Productivity Measurement   48
   Measuring Interspatial and Intertemporal Total Factor Productivity   63
   Explaining Technological Change   66
   Dynamics   67
   Appendix 2-A.   The Translog Cost Model   74
   Appendix 2-B.   The Translog Profit Model   82
   Appendix 2-C.   TFP Measurement Using Exact Index Numbers   85

3 **A Review of the Evidence on Agricultural Productivity and Aggregate Technology** Susan M. Capalbo and Trang T. Vo — 96

   Historical Perspective: Pre-1970   97
   Historical Perspective: Post-1970   100
   Some Comparative Remarks   115
   Appendix 3-A.   U.S. Agricultural Data Base   125

4 The Statistical Base for Agricultural Productivity Research: A Review and Critique
C. Richard Shumway   138

A Description of the Productivity Data Base   139
Evaluation of the Data Base   144
Agency Response to Suggestions   152
Further Concerns   153

PART II MEASURING AGRICULTURAL PRODUCTIVITY AND TECHNICAL CHANGE   157

5 A Comparison of Econometric Models of U.S. Agricultural Productivity and Aggregate Technology   Susan M. Capalbo   159

Introduction   159
Full Static Equilibrium Models   162
Partial Static Equilibrium Models   171
Comparison of the Empirical Results   177
Conclusions and Implications   184

6 Productivity Measurement and the Distribution of the Fruits of Technological Progress: A Market Equilibrium Approach   Ramon E. Lopez   189

Competitive Market Equilibrium   191
Factor Biases of Technological Change   196
Price Effects of Technological Change: Food Prices Versus Land Prices   200
Application to U.S. Agriculture   202
Conclusion   205

7 Intertemporal and Interspatial Estimates of Agricultural Productivity
Michael Hazilla and Raymond J. Kopp   208

Introduction   208
Theoretical Models of Intertemporal and Interspatial Productivity   210
Data Construction Using Firm Enterprise Data System Budgets   218
Results   220
Conclusion   226

8 An Econometric Methodology for Multiple-Output Agricultural Technology: An Application of Endogenous Switching Models
Wallace E. Huffman 229

The Economic Model 231
The Econometric Model 232
A Procedure for Estimating the Econometric Model 239
Conclusion 241

PART III TOWARD EXPLAINING AGRICULTURAL PRODUCTIVITY 245

9 Induced Technical Change in Agriculture
Vernon W. Ruttan and Yujiro Hayami 247

Induced Innovation in the Theory of the Firm 248
Production and Productivity Growth 255
Factor Substitution Along the Metaproduction Function 262
A Test of the Induced Technical Change Hypothesis 266
Guiding Technological Change Along Alternative Paths 279
Appendix 9-A. Derivation of Equations for Measuring the Effect of Biased Technical Change 280
Appendix 9-B. Two-Level CES Production Function 281
Appendix 9-C. Factor Shares and Aggregate Input Price Indexes in the United States and Japan 285

10 Research, Extension, and U.S. Agricultural Productivity: A Statistical Decomposition Analysis   Robert E. Evenson 289

Methodology and Research Processes 290
The Evenson-Welch Study (1974) 304
Concluding Remarks 313

11 Endogenous Technology and the Measurement of Productivity   Yair Mundlak 316

The Choice of a Technique 316
Aggregation over Techniques 320
Identification and Estimation 323
Endogeneity of Technology 325
Empirical Implementation 328
Agenda for Future Research 329

12 Dynamics, Causality, and Agricultural Productivity   John M. Antle   332

Dynamics, Stochastics, and Aggregation   334
Causality in Production   344
The Dynamics of Innovation   347
Policy and Productivity   352
Empirical Evidence on the Dynamic Structure of
    U.S. Agricultural Production   356
Implications for Research   362

13 Incorporating Externalities into Agricultural Productivity Analysis   Sandra O. Archibald   366

Externalities and Regulation in Agriculture: Implications for
    Productivity   367
Modeling Agricultural Externalities   372
An Empirical Example: Accounting for Resistance Development
    in the Measurement of Pesticide Productivity   378
Conclusions   388
Appendix 13-A. Model Specification   389

Index   395

FOREWORD

Increased productivity has been a primary source of real economic growth in the United States throughout the twentieth century. Its benefits have been distributed widely within the U.S. economy in higher real per capita incomes, an increasingly varied array of goods and services, and higher standards of living for most Americans, and they have extended to other nations through expanded international trade and transfers of technology.

In agriculture, productivity growth has been persistent and substantial throughout the century, especially since the end of World War II. Gains in productivity have generated high net social returns, including an abundance of food and fiber at relatively low real cost to domestic consumers and an expanding volume of exports for consumption abroad. New and improved technologies have been substituted extensively for both land and human resources, making it possible to increase simultaneously the productivity of remaining resources and total output of the sector while freeing resources for production of goods and services elsewhere in the economy.

By the late 1970s, however, the capacity of the U.S. agricultural sector to sustain historical rates of productivity growth came increasingly into question in both the agricultural research and public policy communities. In the aftermath of rapid expansion in land use and total agricultural output, standard productivity indicators lent some credence to the popular hypothesis of future economic scarcity and slower productivity growth. Those advancing this hypothesis pointed to several potential constraints: possible real increases in the price of water and energy; the use of inherently less productive cropland as production moved to marginal croplands; environmental degradation and the likelihood of more stringent regulation to control nonpoint water pollution and to limit the use of productivity-enhancing pesticides and other chemical inputs; and the lagged potential effects of declining real public investment in agricultural research.

But specters of scarcity seem to have given way to expectations of chronic economic surplus. Although data for the 1980s must be interpreted cautiously, standard productivity measures suggest that both total factor and labor productivity in agriculture have continued at strong, positive rates. Disquieting evidence continues to accumulate about the externalities of agricultural production systems that undergirded high rates of past productivity growth, but the immediacy of public concern has receded from the level of the late 1970s. Optimists herald a new wave of technical change, the *sine qua non* of which is biotechnology and its promises of resource-saving and dramatic productivity-enhancing breakthroughs in both plant and animal sciences.

Opposing hypotheses of the course of future productivity growth should not be surprising. The future is uncertain and unpredictable, and the past is not necessarily a reliable indicator of what is to come. Nor are the spurious and transitory always easily differentiated from the legitimate and secular in observation of economic and social data. But although the course of productivity growth can neither be forecast with certainty nor planned or controlled with precision, we do know that the realized course will be of major social consequence.

In the broadest sense, societal standards of living can be enhanced only through gains in productivity as resources approach full utilization to meet evolving and ever-changing human needs. For agriculture, continued growth in productivity will remain important in enhancing efficiency in resource use and maintaining competitiveness in world markets. For consumers, increased productivity contributes to the maintenance and security of food supplies at reasonable cost. For agriculturally related business, the rate and patterns of future productivity growth are critical to long-range planning and investment. In the public sector, a wide range of policy actions may be needed—some as adjustments to realized patterns and rates of productivity growth, some as *a priori* actions to shape future patterns and rates of growth to social goals.

A substantial body of research results on agricultural productivity exists, but it lacks comprehensive assessments of the sources of growth and the public policy implications of possible future growth. With few exceptions, empirical research on productivity has focused on historical assessments of growth rates and measurement of structural indicators. This has been helpful in identifying the portions of output growth that cannot be explained by changes in tangible inputs, facilitating the formulation of working hypotheses with regard to major factors affecting growth patterns, and assessing the welfare implications of variations in productivity growth rates. Generally, however, that research has been highly aggregative and only partially successful in isolating the relative effects of a variety of possible determinants of productivity change. Furthermore, results of many of the applied productivity studies are dated.

Recent assessment of technical change and its implications for potential productivity growth and agriculture's productive capacity fail to incorporate adequately economic, institutional, and policy variables that influence or constrain application of technology. Many of the assessments have focused on a single or selected group of technologies, with little attention to input substitution possibilities or the adjustment process within the agricultural sector. Thus, although such assessments have generated a useful catalog of potential sources of future technical and productivity change, they have revealed little about the dynamics of that change.

Research is also needed to address a wide range of issues related to the social and economic consequences of productivity growth and related public policies. Critical is the need to assess and measure the costs of environmental externalities created by agricultural productivity-enhancing technologies and, conversely, the effects on agricultural productivity of alternative technologies and regulatory regimes to maintain or enhance environmental quality. Other issues turn on the structural and distributive consequences of technical change and productivity growth within the farm sector.

Finally, public concern is growing about the productivity-related effects of the mélange of public policies and programs that have evolved in the past half century. Agricultural policies and programs often appear to operate at cross purposes with respect to productivity, simultaneously stimulating and retarding growth. Agricultural, natural resource, and environmental policies sometimes reinforce and sometimes conflict in their effects on productivity; both qualitative and quantitative aspects of such policies often fail to be considered explicitly. Unanswered questions persist over long-term public agricultural R&D strategies and policies related to productivity; over the balance between applied and basic research; over tendencies toward technological determinism without sufficient regard for environmental, structural, and other distributional consequences; over the nature of public-private institutional arrangements; and over the organization and cost effectiveness of the R&D system.

With support from the Rockefeller Foundation, RFF undertook in 1983 a program of research and policy analysis to address such issues. The program had two principal objectives: measuring and explaining the principal sources of change in U.S. agricultural productivity since the end of World War II; and assessing the implications of alternative public policies bearing on agricultural productivity growth over the next several decades.

To initiate the program, a comprehensive review and assessment of the existing stock of data and research evidence was undertaken. In concert with work at RFF, papers were commissioned from leading economists in the field and presented and discussed at a two-day workshop held in Washington, D.C., in July 1984. Attended by analysts and research leaders from the United States and other countries, the workshop covered a broad range of subject matter from theory and research methodology to results of past and contemporary research and data and public policy considerations.

This book is based on the workshop but is not confined by it. That is, several chapters resulted from workshop presentations, with much original material written by editors Susan M. Capalbo and John M. Antle. The result is a comprehensive, integrated body of knowledge concerning agricultural productivity research, highlighting both its strengths and its limitations. This book should be of value to scholars and research leaders for the knowledge it conveys and in the design and conduct of future agricultural productivity research, at RFF and elsewhere.

Kenneth R. Farrell, Founding Director
National Center for Food and Agricultural Policy,
Resources for the Future; currently Vice President,
Agricultural and Natural Resources, University of California

ACKNOWLEDGMENTS

Many of the chapters in this volume have their origins in papers presented at a workshop funded by the Rockefeller Foundation and Resources for the Future (RFF), July 16–18, 1984. The workshop was organized by Susan M. Capalbo, coeditor of the volume, and Kenneth R. Farrell, then the director of the National Center for Food and Agricultural Policy at RFF. The focus of the workshop was twofold: to examine the strengths and weaknesses of previous research on agricultural productivity with respect to the theoretical foundations and the statistical data base; and to discuss methodologies for identifying and measuring the sources of productivity changes. Many of the chapters in this volume have benefited from the comments and criticisms of workshop participants.

Many of our colleagues at RFF and other research institutions whose names do not appear in this volume contributed substantially to it. In particular, we wish to acknowledge our appreciation and the collective thanks of the authors to Eldon Ball, Ernst Berndt, Michael Denny, Bruce Gardner, Zvi Griliches, Glenn Johnson, Stan Johnson, Yao-chi Lu, Chester McCorkle, John Miranowski, Henry Peskin, Tim Phipps, and Robin Sickles for their suggestions on earlier revisions to various chapters; and to the anonymous reviewers for their excellent critiques with respect to content and organization of the entire manuscript.

Throughout the publication process we have been greatly aided by the encouragement of John F. Ahearne, Kenneth R. Farrell, and Kenneth D. Frederick. Finally, our thanks go to James M. Banner, Jr., and the RFF Publications Committee; to Mary Ann Daly for her capable technical assistance in handling the numerous revisions; to Nancy Norton; to Sally Skillings, who coordinated the editing process; and to Barbara Karni, for an excellent job of editing this book.

Susan M. Capalbo
John M. Antle
Washington, D.C.

CONTRIBUTORS

John M. Antle is an associate professor of agricultural economics and economics, Montana State University. At the time of writing he was a visiting fellow at Resources for the Future on leave from the University of California, Davis.

Sandra O. Archibald is an assistant professor at the Food Research Institute, Stanford University.

Susan M. Capalbo is an assistant professor of agricultural economics and economics at Montana State University. She was a fellow at Resources for the Future at the time this book was prepared.

Robert E. Evenson is a professor at the Economic Growth Center, Yale University.

Yujiro Hayami is a professor in the Department of Economics, Tokyo Metropolitan University, Japan.

Michael Hazilla is a fellow in the Quality of the Environment Division, Resources for the Future.

Wallace E. Huffman is a professor of economics at Iowa State University.

Raymond J. Kopp is director and senior fellow in the Quality of the Environment Division, Resources for the Future.

Ramon E. Lopez is an associate professor in the Department of Agricultural and Resource Economics, University of Maryland.

Yair Mundlak is a professor of agricultural economics at Hebrew University, Rehovat, Israel, and professor of economics at the University of Chicago.

Vernon W. Ruttan is a professor in the Department of Agricultural and Applied Economics and in the Department of Economics, University of Minnesota.

C. Richard Shumway is a professor of agricultural economics at Texas A&M University.

Trang T. Vo is a graduate student in the Department of Agricultural and Resource Economics at the University of Maryland. She was formerly a research assistant at Resources for the Future.

1
INTRODUCTION AND OVERVIEW

JOHN M. ANTLE AND SUSAN M. CAPALBO

Productivity research encompasses a broad spectrum of theoretical and applied research aimed at better measuring, understanding, and predicting productive activity. Although productivity concepts have always been central to economic theory, until the middle of this century the explanation of economic activity remained largely a theoretical exercise. As economic theories began to receive increasing scrutiny by applied researchers, the nexus between productivity theory and measurement became increasingly clear. This confrontation of theory with reality has led to significant advances in both methodology and theory; as a result, productivity research has advanced rapidly during the past two decades. The application of duality theory to productivity measurement, the use of flexible functional forms in econometric research, and the development of theoretical linkages between index numbers and production technology have greatly augmented the tools economists have at their disposal to disentangle the complex factors involved in productivity change and have expanded the horizons of productivity research.

These advances in the conceptual foundations and empirical approaches to productivity measurement have also influenced the direction of agricultural economics research. For example, Hicks's induced innovation theory has provided a theoretical framework for the explanation of agricultural productivity differences across regions and over time. Agricultural productivity has also been explained in terms of investments in agricultural research, extension, and human capital. The rate and bias of technological change and the structural characteristics of the production technology for agriculture have been quantified using duality theory and advances in econometric methods.

The purpose of this book is to provide the reader with an understanding of agricultural productivity research in terms of its conceptual development and existing empirical evidence and to identify directions for future research. We accomplish this task by presenting a collection of essays that reflect the current research on measuring and explaining agricultural productivity. We believe the collection as a whole makes it possible to understand new developments in agricultural productivity research as extensions and generalizations of the existing body of theory and methodology.

The book is organized around three main topics. Part I contains three chapters of background to recent theoretical developments related to productivity measurement and the historical record on aggregate agricultural productivity research. The four chapters in part II report recent research on measuring productivity and technical change in the U.S. agricultural sector, with a focus on important problems that have emerged out of recent research efforts. Part III includes five chapters that identify the source of productivity changes. In seeking explanations for the observed growth rates, one is confronted by the induced-innovation hypothesis, agricultural research and extension effects, and the dynamics of the agricultural production process. As the chapters in parts II and III demonstrate, the measurement and explanation of productivity change are not unrelated. Improvements in the quantitative techniques for measuring productivity and technical change have aided researchers in investigating the sources of long-run productivity growth.

In this introduction we first provide an overview of each chapter. We conclude with summary observations on the current status and limitations of efforts to measure and explain agricultural productivity and offer suggestions for future research.

OVERVIEW OF THE BOOK

Because many of the chapters use recent production theory and econometric techniques, we provide in chapter 2 an introduction to these theoretical and methodological developments. We review the essential elements of modern production theory and approaches to the measurement of productivity and the explanation of technical change. In particular, we discuss and compare two distinct approaches to the measurement of total factor productivity—the growth accounting approach and the econometric approach—and we examine extensions of the static production models to their dynamic counterparts and the implications for productivity research. We hope through this chapter to make the contents of this volume accessible to readers who are not familiar with recent developments in productivity research.

In chapter 3, Susan Capalbo and Trang Vo review the agricultural productivity literature in terms of pre- and post-1970 contributions. This historical delineation coincides with some of the major advances in theory and methodology used for productivity measurement. The pre-1970 contributions emphasized the theoretical basis for the concept of total factor productivity and its relation to technical change. Empirically, the unexplained growth in U.S. agricultural productivity was attributed in equal proportions to three factors: input quality changes, economies of scale, and research and development. The post-1970 contributions focus to a large extent on improvements in measurement, primarily as a result of the advances in econometric methods. Capalbo and Vo conclude that, although differences in data and model specifications exist among the post-1970 contributions, a cross-study comparison of empirical evidence shows certain regularities. With regard to technical change biases, many studies find evidence of capital- and fertilizer-using biases, and labor-saving biases for the U.S. and Canadian agricultural sectors. Most agricultural inputs have an inelastic demand and are substitutes in the aggregate production process.

The data problems associated with agricultural productivity research are often cited as limitations to empirical analysis. These problems are considered in detail for the aggregate U.S. agricultural data base by C. Richard Shumway in chapter 4. That chapter provides an overview and evaluation of the U.S. Department of Agriculture (USDA) input and output series, and examines such fundamental issues as aggregation procedures, definition of the farm sector, input quality measurement, and the treatment of capital. Shumway discusses the changes that have been suggested recently to improve USDA data, and makes several recommendations. These include obtaining better information on the input allocation among commodities; assuring better documentation of procedures and assumptions used in constructing detailed data series; and facilitating access to data by researchers.

Part II presents a large body of theory and empirical evidence on the structure of aggregate agricultural technology and the rate and direction of technological change in U.S. agriculture. The use of dual cost and profit function models that has dominated the literature for the last decade undoubtably represents an important step forward in productivity measurement, especially in terms of the econometric models used for productivity measurement. In chapter 5, Susan Capalbo uses an aggregate data base, constructed along the lines suggested by Shumway, to estimate a series of primal and dual models of aggregate production for the U.S. agricultural sector. The price and quantity indexes for the inputs and outputs are Divisia aggregates, and thus reflect a much less restrictive set of assumptions on input substitution than either the Laspeyres or Paasche indexes currently used by the USDA. Not surprisingly, these data exhibit important differences in the relative rates of growth of aggregate inputs and outputs when compared with the USDA series. Capalbo compares the results from primal and dual production models for overall productivity measures and trends in rates of growth; for the direction and magnitudes of factor biases; for price and substitution elasticities; and for scale measures. She finds that many of the empirical results, including the consistency of the estimated technologies with economic theory, are sensitive to model specification.

In chapter 6, Ramon Lopez observes that although technological change influences market equilibrium and hence market prices, most econometric studies that measure biased technological change do not do so within an equilibrium framework. Lopez develops a model with which to integrate the measurement of technological change and the distribution of gains from technological change between output prices and land values in an equilibrium framework. The chapter thus represents an important step in the direction of generalizing the models surveyed by Capalbo and Vo in chapter 3. Lopez's empirical results suggest that taking the equilibrium perspective could result in significantly different measured biases in technological change. In particular, he finds a bias toward labor in U.S. agricultural technology after taking market equilibrium adjustments into account.

In chapter 7, Michael Hazilla and Raymond Kopp explore an approach for making interspatial and intertemporal agricultural productivity comparisons that combines index number and econometric methods. They illustrate these methods using corn and soybean data from the USDA's regional Firm Enterprise Data System (FEDS) and compare measures of intertemporal and interspatial productivity change based on both index number formulae and translog cost function models.

Hazilla and Kopp conclude that the difference between the productivity measures obtained from the index number and econometric approaches for this empirical application is small, thus lending support to the use of the index number approach for making these regional and spatial comparisons. Such an approach circumvents the problems and limitations associated with applying the more sophisticated econometric techniques to regional data sets. However, they caution researchers to avoid generalizing to other commodities based on this application.

In chapter 8, Wallace Huffman points out that use of a single aggregate output index imposes substantial structure on the technology. His objective is to provide a model that accounts for various economic structures of farmers' choice functions with respect to combinations of outputs and inputs and that can accommodate zero-value output levels. This is likely to be the case when disaggregated farm-level data are used in a multiproduct analysis. The chapters by Huffman and Hazilla and Kopp acknowledge the importance of measuring productivity growth at a disaggregate level and deal with the measurement problems associated with the limited nature of regional information.

The first two chapters of part III examine fundamental issues in the explanation of agricultural productivity growth. Chapter 9, by Vernon Ruttan and Yujiro Hayami, is an overview of one of the most influential theories of agricultural development—the theory of induced innovation. Readers not familiar with the induced innovation theory will find this a succinct introduction to the topic; other readers will find it a valuable review of the Hayami-Ruttan model. After outlining the principles of the induced innovation theory and its implications, Ruttan and Hayami discuss how the implications of the theory can be confronted by data. Using a cost function model, they reexamine their earlier comparison of U.S. and Japanese agricultural development from the late nineteenth century through the 1970s and conclude that both the United States and Japan developed agricultural technology to facilitate the substitution of relatively abundant factors for scarce factors in response to relative factor prices. The influence that Hayami and Ruttan's theory of induced technological and institutional innovation has had on the measurement and explanation of agricultural productivity is evident throughout part III.

The induced innovation theory provides an explanation of how economics influences the kind of research that a country or region undertakes. But how can research output be measured and used to help explain agricultural productivity? In chapter 10, Robert Evenson considers the concepts and issues that arise in attempting to measure research productivity and in relating research activity to agricultural productivity. In particular, he considers the temporal, spatial, and specialization dimensions of agricultural research and their relation to productivity, using a two-stage productivity decomposition method. In the first stage, the productivity residuals are calculated, utilizing primarily a growth accounting approach. In the second stage, these residuals are regressed on a set of decomposition variables, such as research and extension. Two case studies of U.S. agricultural productivity illustrate conceptual and practical issues in the measurement of research productivity, as well as its importance as a major factor in explaining the productivity residual. He concludes by urging more careful development of the R&D variables.

Although the induced innovation theory suggests that technological change and productivity growth are endogenous to the economic system, most econometric production models treat it as if it were exogenous, by representing it as a function of time. These time-drift models of technical change may be useful for measuring rates of growth, but they do not explain productivity change. In chapter 11 Yair Mundlak develops a conceptual and empirical framework for treating productivity growth as a process endogenous to the economic system. He begins by emphasizing that a sector such as agriculture has not one production function, but many. He then builds a theory of the aggregate production function in which the choice of technique made by individual firms is a function of exogenous state variables—technology, prices, and fixed inputs. Mundlak shows that this view of the aggregate production functions leads to a variable coefficient model in which the coefficients are functions of the state variables. This approach to aggregate productivity measurement and explanation provides an important alternative to the conventional econometric models and methods discussed in part I.

In chapter 12, John Antle explores the dynamic structure of agricultural production processes and its implications for the measurement and explanation of agricultural productivity. He finds that the recursive structure of farm-level production is preserved in aggregation. This result provides the basis for his investigation of the causal structure of aggregate production relations. An important implication of the dynamic structure of production concerns the testing of the induced innovation hypothesis. Antle finds that if production processes are dynamic, it is not possible to use models in final form to differentiate the implications of the induced innovation hypothesis from other theories of technological change; it is possible to test the hypothesis only if the structural parameters of the model can be identified. One important implication is that static models derived from duality theory, such as those commonly used in the literature to measure biased technological change, are essentially reduced form models and cannot be used to test the induced innovation hypothesis. Antle applies a dynamic structural model to U.S. time series data to test explicitly for the induced innovation hypothesis, under identifying assumptions. The results indicate some violations of the induced innovation theory. These findings raise questions about the use of results from static models to support the induced innovation theory.

In chapter 13, Sandra Archibald constructs a model with which she measures the effects of technology on the environment and the effects of policies such as environmental regulation on productivity. She proposes a dynamic, multiproduct model of production to account for the intertemporal and externality-producing characteristics of agricultural technology. This dynamic model is used to analyze the production and welfare implications of current environmental regulation and alternative schemes, such as externality taxes. A dynamic analogue to the conventional measure of total factor productivity is proposed that accounts for both the intertemporal nature of agricultural production processes and externality effects. Based on evidence from her study of cotton production in California, Archibald suggests that the conventional measure of total factor productivity (TFP) is likely to overstate productivity growth in this sector because it neglects to take account of pesticide resistance. Her findings also support the consideration of taxes as an alternative to standards, because of the latter's potential to induce a bias in technical change toward a less environmentally damaging technology.

SUMMARY OBSERVATIONS

At this juncture it is worthwhile to assess critically the current status of agricultural productivity research. Our observations seek to answer two questions: First, what have been the thrust and limitations of recent research? Second, what are some of the broader conceptual and methodological developments that are needed to put us in a better position to measure and explain productivity changes?

Contributions and Limitations of Recent Research

The research presented represents a substantial body of empirical evidence on the structure of aggregate technology and the rate and bias of technological change in U.S. agriculture. Considering that the empirical research reported in this volume is based on data that represent various time periods, regions, and methods of aggregation, among other factors, it is remarkable that certain empirical regularities emerge. In particular, aggregate factor demand and product supply elasticities generally do not exceed unity (in absolute terms); the average annual rate of total factor productivity growth in U.S. agriculture since 1945 is between 1.0 and 2.0 percent; and the direction of the technological change biases are generally consistent with the predictions of the induced innovation theory.

Nevertheless, in our view the existing econometric models and related index number formulae leave many important questions unanswered and suffer from at least five potentially serious shortcomings. First, the literature has been based largely on static models, in which production is treated as a static phenomenon and issues of expectations and dynamic adjustment are usually ignored.

Second, even though the models discussed in this volume are used to represent the agricultural sector, they are based largely on the microeconomic production theory of the profit maximizing firm, and thus treat prices as exogenous. The equilibrium of the sector is ignored in most cases.

Third, even though many of the flexible functional forms used in the static models are flexible, they nevertheless impose potentially important restrictions on the technology. One restriction, discussed in chapter 2, is the input-output separability required to construct a consistent aggregate output index. Other restrictions, such as homotheticity, neutral technical change, and separability, often are imposed largely for reasons of model tractability.

Fourth, most models treat technological change as an exogenous phenomenon that is a function of time, despite the fact that the major theoretical construct—the theory of induced technological change—posits technological change as endogenous to the economy's price system.

And finally, estimates of technology structure, total factor productivity, and technical change biases are clearly specific to the particular model, methodology, and data. Most contributions to the literature do not address the problem of reconciling the often large and significant differences in productivity measurements and estimates of the aggregate technology obtained from various studies.

The studies contained in this volume illustrate some of the limitations of existing research and suggest possible avenues for methodological advance. For example, Hayami and Ruttan use a homothetic production function to decompose the rate of change of the factor shares into the components resulting from relative price changes and those resulting from biased technical change. They find that the technical change biases were labor-saving and land-, mechanical power-, and fertilizer-using in U.S. agriculture in the period since the end of World War II. They cite the apparent relation between trends in relative factor prices and the biases for labor, power, and fertilizer as support of the induced innovation hypothesis. However, neither the Hayami-Ruttan study nor the other econometric production studies cited in chapter 3 provide a formal statistical test of predictions of the induced innovation hypothesis. In part, this is because of the model specifications. The econometric models measure biases using a time trend and thus cannot be used to test alternative explanations. Moreover, there is a certain inconsistency in the use of an exogenous time trend to measure technological change when the induced innovation theory states that technological change should be endogenous to prices.

The statistical decomposition models discussed by Evenson provide a means of decomposing and explaining the productivity residual, and thus represent a positive step in the direction of both measuring and explaining productivity. In his model, changes in the level of productivity are affected by the research process in three dimensions: research specialization, research location (country, state, or county), and time lags. Evenson's approach could be extended to use relative prices to explain changes in total factor productivity and thus to test for the influence of prices on innovation. This kind of approach has also been explored by Ben-Zion and Ruttan (1978) and Stevenson (1980). In chapter 11 Mundlak shows how the measurement and explanation of productivity can be integrated into a unified theoretical framework. In his model, the implemented technology is a function of the economy's state variables (prices, resource endowments, and stocks of scientific and technological knowledge), and therefore could provide the basis for more rigorous testing of the induced innovation theory in a model with endogenous technological change.

In the spirit of Mundlak's theory, in chapter 12 Antle proposes explicit statistical tests of the induced innovation theory in terms of the parameters of lagged relative price variables in a dynamic model. Using U.S. aggregate time series, he finds that some parameter estimates significantly violate the implications of the induced innovation theory, even though a time trend variable implies the usual pattern of residual biases. This finding raises serious questions about the interpretation of the evidence from static models in support of the induced innovation theory.

Another methodological problem is the current econometric practice of measuring productivity by modeling only the supply side of the economy or sector. As the chapter by Lopez suggests, making technology endogenous in production models raises the question of whether or not productivity can be modeled and measured without taking into account the equilibrium or disequilibrium properties of the sector and the overall economy. An important theoretical question raised by these observations is under what conditions, if any, endogenous productivity can be measured without employing a general equilibrium framework and without incorporating dynamic adjustments. When Lopez included the effects of technological change on market equilibrium prices in a static model, the measured biases in technological change were found to be different from those found by Ruttan and Hayami and by Capalbo. This finding suggests that researchers need to differentiate carefully between firm-level and aggregate concepts in both the measurement and the explanation of productivity.

An additional shortcoming of the existing models appears to be their reliance on a static theoretical paradigm to measure the aggregate effects of technological change, an inherently dynamic process. The restricted cost and profit function models discussed in chapter 5—referred to in the literature as "second-generation dynamic models"—provide no information on the time path of the adjustment from one period to the next.

Particularly regarding the explanation of technological change, the question of the intertemporal relations (or causality) among the various factors related to productivity change needs to be addressed. Incorporation of dynamics and expectations into tractable models is likely to remain a key issue in productivity research for the foreseeable future.

The existing econometric models estimated with time-series data are also subject to the limitations noted by Diamond and coauthors (1978).

> ...measurements of the bias of technical change and the elasticity of substitution...have either been based on cross-section data or have employed untested restrictions on the nature of production possibilities. This reflects a nonidentifiability of the elasticity and bias in the absence of *a priori* hypotheses on the nature of technical change (p. 125).

Furthermore, under the assumption of nonconstant returns to scale, the "rate of technical change and returns to scale can be identified under profit maximization...the elasticity of substitution, scale bias, and technical change bias are in general not identified," unless untested technological restrictions are imposed (p. 147). In most empirical research, these untested restrictions take the form of a parametric functional form and strong assumptions about the properties of technological change. Moreover, as Antle emphasizes in chapter 12, when dynamics and expectations are introduced, further untested maintained hypotheses must be imposed to be able to identify and estimate the rate or bias of technological change.

To overcome the various limitations of existing productivity research, we believe efforts must be made to improve the existing data bases. This can be done in at least two important ways. First, the aggregate time series data need to be revised, taking Shumway's suggestions into account. His suggestions include development of service prices for the flow inputs, such as land, equipment, and animal capital, and separation of the family and owner-operator labor from the hired labor components of the labor series.

Second, USDA and other responsible government agencies need to develop a pooled time series/cross-sectional national data base. An important question for such a data base is the level of aggregation for the cross-sectional dimension. USDA's current use of ten regions for some data series lacks a strong economic rationale, although it does have some regional validity. Likewise, using state divisions may not make sense: in many cases, states also need to be subdivided for meaningful analysis of productivity issues. The most important contribution a pooled time series/cross-sectional data base would make would be to increase the accuracy and policy relevance of productivity analysis. A pooled data base would also allow the static production models to be generalized in important ways. Key regional differences could be modeled and measured along the lines put forward by Hazilla and Kopp. A more disaggregated data set would also facilitate the measurement of productivity in a multiple-output framework, as suggested by Huffman. It would be possible to test for nonconstant returns to scale and nonhomotheticity—and to measure input substitutability—under weaker technological assumptions than those imposed in chapter 5. Moreover, with the increased degrees of freedom available from a pooled data set, further modeling refinements, such as extensions to dynamics, would be possible.

Finally, we believe that an important task facing productivity researchers is to test the robustness of their results with respect to their data procedures, modeling assumptions, and econometric techniques. Chapter 5 represents a first step in this direction; further research along these lines would be fruitful.

Broader Conceptual Developments

In addition to the need to generalize the existing models and to expand the data available for productivity analysis, agricultural economists need to address several issues that derive from the special characteristics of the agricultural sector. Agricultural production processes differ from manufacturing processes in many ways. The dependence on biological processes and the resulting dynamic dimension of agricultural production is one obvious difference. Historically, agriculture has played a unique role in economic development, and special institutional and economic relations have evolved. Some research on agricultural productivity has taken the special characteristics of agriculture into account. In particular, the institutions of publicly funded agricultural research and extension have been analyzed and their connection to agricultural productivity has been documented, beginning with Griliches (1964) and Schultz's earlier work (1964) and illustrated by Evenson's chapter in this volume. But other important characteristics of the agricultural sector have been ignored.

In particular, agriculture is a heavily regulated industry with extensive government intervention in both output and input markets. Price supports, below market loan rates, and acreage restrictions give rise to potential allocative distortions. These policies distort the allocation of inputs by driving a wedge between competitive prices and government induced prices.

These special features of agriculture have been largely ignored in recent attempts to measure and explain agricultural productivity. The studies reviewed by Capalbo and Vo in chapter 3, as well as most of the other chapters in this volume, continue to rely on the assumption that agriculture is an industry in competitive equilibrium. We know of virtually no research that has attempted to account for the effects of government intervention or regulation in agriculture on the measurement and explanation of agricultural productivity. Considering that government intervention is extensive in agriculture, especially in developed countries, we would expect that government policies may have substantial effects on agricultural productivity in the United States, Western Europe, Japan, and other developed areas.

One possible approach to modeling production processes with dynamics and expectations that could be fruitful in this regard is based on the rational expectations concept. Production models developed with the rational expectations theory can take a form similar to conventional models of acreage response or supply but differ in their interpretation. In principle, rational expectations theory shows how policy variables, such as changes in price support programs, affect the economic behavior of the firm or sector and thus could be related to productivity. The dynamic production models of Hansen and Sargent (1980) and Eckstein (1984) provide examples of this approach. Unfortunately, these models are very complex even when an extremely simple production technology is assumed. It is unclear whether or not more sophisticated production technologies can be embedded in such models. Furthermore, although there is no theoretical reason policy variables could not be embedded in these models, such a task has not yet been successfully accomplished.

Economists face a difficult task in attempting to quantify the effects of agricultural policy on productivity, because agricultural policies change frequently and the various policies are likely to have conflicting effects on productivity. For example, the productivity effects of specific acreage reduction policies become more difficult to isolate when price support programs exist. Economists need to explore ways to differentiate the productivity effects of these various policies. Regional analyses may provide one method of differentiating the effects of commodity policies and overcoming the limitations inherent in the use of aggregate time series data.

Recently, agricultural economists have become much more concerned with the relationships between the national and world economies and the performance of agriculture. The United States exports more agricultural products than any other nation in the world. And because exports account for a large proportion of the value of agricultural production, export policy and other influences on exports are important factors in explaining productivity growth in U.S. agriculture. Thus, researchers need to address more effectively the effects of macroeconomic policies on agricultural production.

Environmental regulations promise to become increasingly important (and binding) in agriculture. Although in some respects the issues are similar to those raised by environmental regulations facing manufacturing industries, there are important differences. Consider the case of pesticide regulation. Since pesticides are an "insurance" input, their contribution to productivity results partly from reduction of the risk farmers face. Therefore, to measure accurately pesticide productivity, and thus the costs of regulation, agricultural economists need to take the special properties of agricultural technology—such as production risk—into account.

Considering environmental issues raises the question of the validity of conventional TFP as an indicator of productivity change. The usual TFP indexes account for marketed outputs of goods and services but neglect nonmarketed goods and services, such as environmental quality. As yet, agricultural productivity research has largely ignored these potentially important adjustments to TFP. Thus, an important challenge facing productivity researchers is to determine how different measured productivity would be if these nonmarket goods and services were taken into account. The case study of cotton production in California's Imperial Valley reported in chapter 13 provides a framework for assessing the nonmarket effects of pesticide resistance on producers' welfare and should lead to further valuable research on this problem.

In concluding, we should like to return to the theme of this volume—the measurement and explanation of agricultural productivity. In researching and writing our own contributions to this volume, and in the editorial process, we have been impressed with the advances in theory and methodologies for productivity research in the past two decades. Much has been learned from the large body of applied productivity research performed since the early 1970s. It is clear that productivity research cannot stop with measurement if it is to be useful for assessing future changes in productivity growth. It is equally clear that the efforts toward better measurement should not be abandoned, because they facilitate the formulation of hypotheses with regard to major factors affecting the direction and magnitudes of growth patterns. Yet, we feel very strongly that little more can be gained by continuing to ply the existing aggregate time series data with the static, neoclassical models based on the theory of the firm. New, disaggregate data; new, more general models of the firm and sector; and appropriate measurement methods are needed to push back the frontiers of knowledge. We hope that this volume contributes to that process.

REFERENCES

Ben-Zion, U., and V. W. Ruttan. 1978. "Aggregate Demand and the Rate of Technical Change," in H. Binswanger and V. W. Ruttan, eds., *Induced Innovation* (Baltimore, Md., Johns Hopkins University Press).

Diamond, P., D. McFadden, and M. Rodriguez. 1978. "Measurement of the Elasticity of Factor Substitution and Bias of Technical Change," in M. Fuss and D. McFadden, eds., *Production Economics: A Dual Approach to Theory and Applications*, vol. 2 (Amsterdam, North-Holland).

Eckstein, Z. 1984. "A Rational Expectations Model of Agricultural Supply," *Journal of Political Economy* vol. 92, no. 1, pp. 1–19.

Griliches, Zvi. 1964. "Research Expenditures, Education, and the Aggregate Agricultural Production Function," *American Economic Review* vol. 54, no. 6, pp. 961–974.

Hansen, L. P., and T. J. Sargent. 1980. "Formulating and Estimating Dynamic Linear Rational Expectations Models," *Journal of Economic Dynamics and Control* vol. 2, no. 1, pp. 7–46.

Schultz, T. W. 1964. *Transforming Traditional Agriculture* (New Haven, Conn., Yale University Press).

Stevenson, R. E. 1980. "Measuring Technological Bias," *American Economic Review* vol. 70, no. 1, pp. 162–173.

PART I
BACKGROUND

2

AN INTRODUCTION TO RECENT DEVELOPMENTS IN PRODUCTION THEORY AND PRODUCTIVITY MEASUREMENT*

JOHN M. ANTLE AND SUSAN M. CAPALBO

Modern production theory is the basis for modern approaches to productivity measurement. This chapter seeks to introduce the reader to recent developments in production theory and productivity measurement used in later chapters of the book.[1] The range of material in this chapter reflects the rapid changes that have taken place in productivity research over the past two decades. It is hoped that this introduction will make the volume accessible to readers who have a basic knowledge of production theory, but who may not be familiar with the methods and applications used in this volume.

We begin with an introduction to some of the fundamental topics in modern production theory that relate to the productivity literature, including functional structure, functional form, duality theory, and measurement of technological change. The chapters in this volume show the contribution of these topics for measuring and analyzing productivity in the agricultural sector.

In the second section of this chapter we outline and compare the growth accounting approach and the econometric approach to productivity measurement. The basic concept in productivity measurement is total factor productivity (TFP), the ratio of an index of aggregate output to an index of aggregate input. Changes in TFP can be decomposed into

---

*The contributions and encouragement of Michael Denny, Michael Hazilla, and Raymond Kopp are gratefully acknowledged. Responsibility for the contents rests with the authors.

[1] It is beyond the scope of this chapter to provide the reader with a comprehensive survey of the literature, although we do provide references to the literature we discuss. For other introductions to and surveys of the literature, the reader may consult Walters (1963), Nadiri (1971), Kennedy and Thirlwall (1972), Kendrick (1973), Petersen and Hayami (1977), Fuss and McFadden (1978), and Jorgenson (1984).

In this chapter we use the terms "technological change" and "technical progress" interchangeably. Technical regress is also possible, of course.

components measuring changes in technical efficiency, scale, and the state of technology. The growth accounting approach uses index number measures of TFP to quantify the components of productivity change. Alternatively, econometric methods can be used to estimate the components of TFP using production, cost, and profit functions. Each approach requires certain assumptions, which must be considered in interpreting the findings of productivity studies.

The following three sections of this chapter provide background material for the specific extensions and generalizations of productivity analysis that are addressed elsewhere in this volume. One such generalization is the measurement of productivity across regions and over time. We review some recent developments in index number theory that are used by Kopp and Hazilla in chapter 7 to illustrate how intertemporal and interspatial agricultural productivity comparisons can be made using both the index number and econometric approaches.

Several chapters in this book deal with both the measurement and the explanation of productivity change. An important theoretical contribution to the explanation of productivity is Hicks' induced innovation theory, as elaborated by Ruttan and Hayami in chapter 9. Agricultural economists have made many other contributions to the literature on the explanation of agricultural productivity growth in terms of investments in human capital, agricultural research and extension, and physical infrastructure. This literature is briefly reviewed.

Another important generalization of productivity research is to account for the dynamic relations in production processes. Technological change in agriculture is a dynamic process, yet much productivity analysis abstracts from the dynamic dimension of agricultural production. The final two chapters of the book, by Antle and by Archibald, address some aspects of dynamic modeling of production processes for the analysis of productivity and technological change. The final section of this chapter introduces the reader to concepts that are used in the analysis of production as a dynamic phenomenon.

BASIC CONCEPTS IN MODERN PRODUCTION ECONOMICS

Functional Structure and Functional Form

Cobb and Douglas (1928) observed that the logarithms of output (Q) and inputs (X_i) in aggregate data appeared to be linearly related. This observation led them to hypothesize that the aggregate production function took the form:

$$\ln Q = \ln \alpha_0 + \alpha_1 \ln X_1 + \cdots + \alpha_n \ln X_n$$

or

$$Q = \alpha_0 X_1^{\alpha_1} \ldots X_n^{\alpha_n} \tag{2-1}$$

From the 1920s until the early 1950s the Cobb-Douglas production function was the function of choice for production analysis, both theoretical and empirical, due to its elegance, simplicity, and ease of interpretation and estimation (Douglas, 1976). But economists began to recognize some of the function's limitations and to explore alternatives. Heady (1952); Heady, Johnson, and Hardin (1956); and Heady and Dillon (1962) used the quadratic function, which is less restrictive in some respects than the Cobb-Douglas function. Halter, Carter, and Hocking (1957) introduced the transcendental production function, and Arrow, Chenery, Minhas, and Solow (1961) proposed the CES function, both of which are generalizations of the Cobb-Douglas function with a constant nonunitary elasticity of substitution. In the 1960s and early 1970s, various other generalizations of the Cobb-Douglas and other functional forms were introduced: the multifactor CES (Uzawa, 1962); generalized production functions (Zellner and Revankar, 1969); variable elasticity of substitution functions (Revankar, 1971); constant ratio elasticity of substitution-homothetic (Hanoch, 1971); and, in the late 1960s and early 1970s, the flexible functional forms, such as the translog (Christensen, Jorgenson, and Lau, 1973) and the generalized Leontief (Diewert, 1971). For an in-depth discussion of functional forms in production analysis, see Fuss, McFadden, and Mundlak (1978).

These developments in functional form reflect the growing understanding that the functional forms used in production analysis may impose restrictions on the economic relationships embedded in the corresponding behavioral relationships—that is, the firm's output supply and input demand functions. It was recognized that for empirical research it was desirable to impose as few restrictions on the functional form as possible while maintaining a function that is empirically tractable. The Cobb-Douglas has the virtue of simplicity, but this simplicity comes at the cost of imposing several restrictions, including unitary elasticities of substitution, constant production elasticities, and constant factor demand elasticities.

One important generalization of the Cobb-Douglas is the translog function that can be obtained by specifying the Cobb-Douglas production elasticities to be log-linear functions of the inputs. That is, in equation (2-1) add the following conditions:

$$\alpha_i = \alpha_{i0} + (.5) \sum_j \alpha_{ij} \ln X_j, \qquad i = 1, \ldots, n \tag{2-2}$$

Substitution of equation (2-2) into equation (2-1) gives the translog function:

$$\ln Q = \ln \alpha_0 + \sum_i \alpha_{i0} \ln X_i + (.5) \sum_i \sum_j \alpha_{ij}(\ln X_i)(\ln X_j) \tag{2-3}$$

Alternatively, the translog can be viewed as a second-degree (logarithmic) approximation to a general function about the point $\ln X_i = 0$, $i = 1, \ldots, n$, as discussed by Denny and Fuss (1977). The translog is attractive because it is, in the terminology of Fuss, McFadden, and Mundlak (1978), a "parsimonious flexible functional form," that is, it has the minimum number of parameters needed to represent economic behavior without imposing arbitrary restrictions on that behavior. This idea was originally that of Diewert (1971).

Theorists have developed various mathematical concepts to characterize the important dimensions of functional structure for economic analysis. An exhaustive survey of functional structure is beyond the scope of this introduction, and the reader is referred to Blackorby, Primont, and Russell (1978) for a comprehensive treatment of the subject. In this section we introduce the reader to homogeneity, homotheticity, separability, and substitutability for single-output technologies; for multiproduct technologies, we discuss the transformation function and the concepts of jointness and input-output separability.

We consider first the single-output production function $Q = F(X)$, where X is a vector of inputs and Q is output. Scale of production is conventionally defined in terms of the production function's *homogeneity*. The production function is *homogeneous of degree k* if and only if there are constants $k, \lambda > 0$ such that $Q\lambda^k = F(\lambda X)$. The number k determines the returns to scale of the production process, with $k = 1$ defining the case of linear homogeneity, or constant returns to scale. Note that many functions, such as the quadratic, are not homogeneous of any degree. A production process is *homothetic* if and only if there exists a homogeneous function $h(X)$ and a monotonically increasing function $g(\cdot)$ (that is, $g' > 0$) such that $F(X) = g(h(X))$. Homotheticity is a fundamentally important property, because it implies that the expansion path and the optimal (cost-minimizing or profit-maximizing) factor proportions and cost shares are independent of the rate of output. Homotheticity also plays an important role in the derivation of certain duality results. We shall see that such relationships are central to productivity measurement. Note that homogeneity implies homotheticity, but that the converse is not true.

Examples: The Cobb-Douglas function (2-1) is homogeneous of degree ($\Sigma_i \alpha_i$) and thus is also homothetic. To prove homotheticity, define

$g(h(X)) = h(X)$ and $h(X) = F(X)$, where F is the Cobb-Douglas function. A homothetic, nonhomogeneous function can be obtained from the Cobb-Douglas by letting $g(h(X)) = \ln h(X)$ and $h(X) = F(X)$ so $Q = \ln F(X)$. It is easily shown that both the Cobb-Douglas and the logarithm of the Cobb-Douglas have linear expansion paths, but that the labeling of the isoquants is different. From the definition of homogeneity given above, it can be shown that the translog function (2-3) is homogeneous of degree k if and only if $\Sigma_i \alpha_{i0} = k$ and $\Sigma_j \alpha_{ij} = 0$ for all i; otherwise it is neither homogeneous nor homothetic. To prove this, note that Euler's theorem states that F is homogeneous of degree k if and only if the sum of the production elasticities equals k. From equation (2-2), therefore, we must have $\Sigma_i \alpha_i = \Sigma_i \alpha_{i0} + 1/2 \Sigma_j (\Sigma_i \alpha_{ij}) \ln X_j = k$ for all $\ln X_j$.

A second fundamental structural property of production functions is *separability*. Consider the question of whether or not two inputs X_1 and X_2 can be combined into a single aggregate input X_A. This is the problem of *consistent aggregation*, first raised by Solow (1955). Upon reflection, one can observe that this is possible if the production function can be written in the form

$$F(X) = g(X_A, X_3, \ldots, X_n) \quad \text{where } X_A = f(X_1, X_2), \tag{2-4}$$

and the functions $g(\cdot)$ and $f(\cdot)$ satisfy the properties of production functions. When the production function F can be written in this way it is said to be *separable* in X_1 and X_2. Observe that if F is separable in X_1 and X_2, a change in another input, say X_3, shifts the isoquant between X_1 and X_2 homothetically (in a parallel fashion). Therefore, the marginal rate of technical substitution (MRTS) between X_1 and X_2 is independent of X_3. These observations suggest an alternative definition of separability. Let F_i denote the marginal product of X_i. Then

$$\partial \text{MRTS}_{1,2}/\partial X_3 = \partial(F_1/F_2)/\partial X_3 = 0, \tag{2-5}$$

if and only if the production function is separable in X_1 and X_2. Using equation (2-4), $F_i = g_1 f_i$, $i = 1, 2$, so $F_1/F_2 = f_1(X_1, X_2)/f_2(X_1, X_2)$, showing that equation (2-5) is satisfied.

More generally, separability can be represented in terms of a partition of X into groups $X^1, \ldots, X^s$, for $s > 2$. The production function is said to be *strongly separable* with respect to this partition if and only if the production function can be written

$$F(X) = f\left[\sum_{t=1}^{s} g^t(X^t)\right], \tag{2-6}$$

and it is *weakly separable* if and only if it can be written

$$F(X) = f[g^1(X^1), \ldots, g^s(X^s)] \tag{2-7}$$

Note that strong separability is a more restrictive structure, and that strong separability implies weak separability but the converse is not true. When the subfunctions $g^i(X^i)$ above are homothetic in their arguments, then the production function is said to be *homothetically separable*.

Several important distinctions arise in characterizing separable structures with flexible functional forms such as the translog. *Local* separability refers to the property at a point in the relevant variable space (such as at the expansion point $\ln X_i = 0$ of the translog production function); global separability refers to the property at all points in the relevant space. *Exact* separability refers to the conditions required when a function is interpreted as an exact representation of the true function; *approximate* separability conditions are based on the interpretation of functional form as an approximation to the true function. These concepts are discussed by Denny and Fuss (1977).

Examples: Letting $f[\cdot] = \exp(\cdot)$ and $g^t(X^t) = \alpha_t \ln X_t$ in equation (2-6), it follows that the Cobb-Douglas function is strongly separable in all inputs. Thus, the translog is strongly separable in all inputs if it satisfies the Cobb-Douglas restriction $\alpha_{ij} = 0$ for $i, j = 1, \ldots, n$. The translog is separable in an input group $\{X_1, \ldots, X_k\}$ if $\alpha_{ij} = 0$ for all $i = 1, \ldots, k$ and $j = k + 1, \ldots, n$. In this case, the subfunction of the separable groups has the translog form, but the aggregator function over the groups has the Cobb-Douglas form.

Input substitution is central to the economic behavior of firms and is critical in determining their ability to respond to changing economic conditions. Hicks (1963) introduced the *elasticity of substitution* to measure the degree to which firms can substitute inputs for each other. The modern mathematical definition of the elasticity of substitution in the two factor case was provided by Allen (1938), among others, and is generally referred to as the Allen elasticity of substitution (AES). In the two-factor case, the AES is defined as $d\ln(X_2/X_1)/d\ln \text{MRTS}_{1,2}$, that is, it is the proportionate change in factor proportions induced by a proportionate change in the MRTS. It can be shown that AES can be expressed in terms of the input quantities and the first and second derivatives of the production function (Ferguson, 1971). In the multifactor case, the *partial* AES is defined in terms of the bordered Hessian matrix of the production function:

$$\sigma_{ij} = \sigma_{ji} = (\Sigma_i X_i F_i) C_{ij}/X_i X_j H \tag{2-8}$$

where H is the bordered Hessian matrix of $F(X)$ and C_{ij} is the cofactor of F_{ij}. A useful interpretation of the partial AES is available in terms of compensated (output constant) factor demand elasticities: $\sigma_{ij} = E_{ij}/S_j$, where E_{ij} is the compensated elasticity of input i with respect to price j, and S_j is the cost minimizing share of factor X_j in total cost. Thus, the AES measures the degree of substitutability of inputs in an economically meaningful way, holding output constant and allowing inputs to adjust optimally to factor price changes.

The AES is also useful in characterizing separability. Intuitively, it can be observed that if the production function is separable in inputs X_1 and X_2, then a change in the quantity of another input does not change the optimal factor proportions between X_1 and X_2. Thus, separability must impose restrictions on input substitution that can be expressed in terms of the AES. It can be shown, for example, that if the production function is homothetic and separable in X_1 and X_2, then $\sigma_{1j} = \sigma_{2j}$ for $j = 3,\ldots,n$. Berndt and Christensen (1973a) show that this and several other important relations can be established between separability and the AES.

The AES is not the only substitution concept possible. As Mundlak (1968) has emphasized, various elasticities of substitution can be defined in the multifactor case, depending on what is allowed to vary and what is held constant. Moreover, from the perspective of economic analysis, one does not always want to use an input substitution concept that holds output constant. Uncompensated cross-price factor demand elasticities may be a more economically meaningful measure of input substitutability for many purposes, including policy analysis.

Examples: The Cobb-Douglas production function can be shown to imply $\sigma_{ij} = 1$ for all i and j; note that this result is consistent with the link between separability and the AES of a homothetic function mentioned above. In the translog case, applying the definition of the AES, it can be seen that the AES is a complicated function of the production function parameters and the inputs, and generally not a constant. The link between the AES and separability in the translog function is discussed by Berndt and Christensen (1973a).

In both agriculture and industry, many firms produce more than one output. Some firms produce multiple outputs to diversify their position in the market and reduce risk. Producing more than one product may allow capital and labor to be utilized more efficiently. And many products exhibit technological jointness, as in the joint production of grain and straw. The most general functional representation of multiple output production processes is based on the *transformation function* $F(Q,X) = 0$, introduced by Mundlak (1963). This function implicitly defines the firm's production possibilities in terms of its vector of outputs

$Q = (Q_1, \ldots, Q_m)$ and its vector of inputs $X = (X_1, \ldots, X_n)$ (here each X_i defines the total amount of input type i used in production of all outputs). Alternatively, one input or output can be chosen as numéraire and the transformation function can be written as $Q_1 = F(Q^2, X)$ with $Q^2 = (Q_2, \ldots, Q_n)$.

The production technology is said to be *nonjoint in inputs* if there exist individual production functions such that $Q_i = f^i(X_{1i}, \ldots, X_{ni})$ where X_{ji} denotes the quantity of input type X_j that is used only in production of output Q_i, so that $X_i = \Sigma_j X_{ji}$. Each production function can be defined in terms of the inputs specific to that output and independently of other inputs or outputs. The production technology is said to be *nonjoint in outputs* if there exist input requirement functions $X_j = g^j(Q_{1j}, \ldots, Q_{mj})$ such that $Q_i = \Sigma_j Q_{ij}$ (Lau, 1972).

One important question in the multiple-output case is whether or not it is possible to aggregate all outputs into a single-output index. Clearly, if there is *input-output separability*—that is, if the transformation function can be specialized to $F^1(Q) - F^2(X) = 0$—then an aggregate output index $F^1(Q)$ exists. However, the assumption of input-output separability imposes strong restrictions on the form of the technology: it implies that the technology cannot be nonjoint in inputs (Hall, 1973; Denny and Pinto, 1978). For a discussion of functional forms for multiple-output technologies, see Diewert (1974).

Technology Sets and Duality

The 1960s and early 1970s saw production theory move from the neoclassical approach, based on the production function and differential calculus, to the modern approach, based on the analysis of technology sets using the mathematical theory of convex sets. The modern approach permitted the duality relations between the production technology and cost, revenue, and profit functions to be derived rigorously and elegantly. As Diewert (1974) has emphasized, duality theory has two principal practical applications in economics. First, "it enables us to derive systems of demand [and supply] equations which are consistent with maximizing or minimizing behavior on the part of an economic agent... simply by differentiating a [cost, revenue, or profit] function" (p. 106). This result is due to the derivative properties of the cost, revenue, and profit functions and are known as Shephard's lemma and Hotelling's lemma. Second, duality theory "enables us to derive in an effortless way the 'comparative statics' theorems originally deduced from maximizing behavior" (p. 107).

The neoclassical definition of a production technology is based on the production function, which defines the maximum output obtainable from a specified set of inputs (Ferguson, 1971). More specifically, the single-output production function $F(X)$, where X is an input vector, is defined as a positive continuous twice differentiable function with certain monotonicity and concavity properties. In the economically relevant portion of the production surface defined by this function (where a profit maximizing firm would operate), the production function is strictly concave, with positive marginal products and a nonincreasing marginal rate of technical substitution. From these properties, and the assumption of profit maximization or cost minimization, the firm's economic behavior can be deduced in terms of its output supply and factor demand functions. The properties of the production function, in combination with the behavioral assumption of profit maximization or cost minimization, determine the properties of these functions.

Several difficulties are encountered in attempting to pursue the neoclassical, or primal, approach to the analysis of firm behavior. First, the closed-form solution of the first-order conditions of profit maximization or cost minimization for the firm's supply and demand functions is not generally possible. That is, given a system of first-order conditions $p\nabla_X F(X) - W = 0$, where p is output price and W is a vector of input prices, a closed-form solution for the demand functions exists if and only if there exist input demand functions $X^* = X^*(p, W)$ such that $p\nabla_X F(X^*) - W = 0$. This is true if the conditions of the implicit function theorem are satisfied (see Rudin, 1964, for these conditions). Second, even if the demand and supply functions can be solved, deriving comparative static results is difficult because the properties of the demand and supply functions must be derived from the properties of the production function. Third, the production function representation of a multiple-output technology is inconvenient for analytical and empirical purposes.

The modern approach to production theory is based on the mathematical properties of convex sets and the assumption of maximizing or minimizing behavior. The firm's production technology is defined in terms of a *production possibilities set*. The production possibilities set in turn defines the *producible output set*, which is bounded above by the production possibilities frontier, and the *input requirement set*, which is bounded below by the isoquant. The modern approach overcomes the difficulties inherent in the neoclassical approach by establishing correspondences between the properties of the production possibilities set and the firm's cost, revenue, and profit functions. Due to the derivative properties of the dual functions, closed-form expressions for input demand and output supply functions can be obtained for a variety of functional forms. Moreover, by virtue of the convexity and derivative properties of the dual functions, the comparative static properties of demand and supply functions are readily established. The multiple output versions of cost, revenue, and profit functions are straightforward generalizations of their single product counterparts.

Since the theory of convex sets is basic to the modern theory, it is useful to mention briefly some properties of convex sets. For a detailed discussion, see McFadden (1978a) or Green and Heller (1981). A nonempty subset S of n-dimensional Euclidean space $\mathbf{R}^n$ is a convex set if for any two elements $s_1, s_2 \in S$ and any number λ such that $0 \leq \lambda \leq 1$, the element $\lambda s_1 + (1 - \lambda) s_2 \in S$. The minimum convex set containing S, or the intersection of all sets containing S, is defined as the *convex hull* of S. For a nonzero element $y \in \mathbf{R}^n$ and a number λ, the *hyperplane* $H = \{s | y \cdot s = \lambda\}$ divides $\mathbf{R}^n$ into two *half-spaces* $S^+ = \{s | y \cdot s \geq \lambda\}$ and $S^- = \{s | y \cdot s \leq \lambda\}$, which are closed convex sets (here $y \cdot s$ denotes the inner product of the vectors y and s). H is called the *supporting hyperplane* of any convex set $T \in s$ whose boundaries intersect with H, and S is the *supporting half-space* of T. The mathematical basis for duality theory is a theorem by Minkowski that states that every closed convex set in $\mathbf{R}^n$ can be represented as the intersection of its supporting halfspaces. As we shall see below, there is a straightforward graphical interpretation of duality theory in terms of the supporting halfspaces of the production possibilities set.

The derivative properties of the neoclassical production function can be translated into restrictions on the input requirement set. Nonnegative marginal productivity is interpreted in terms of *monotonicity* or *free disposability*: if an input vector v can produce an output vector q, and if v^1 is as large as v for each element, then v^1 can produce q. The convexity of isoquants due to a nonincreasing marginal rate of technical substitution translates into the restriction that the input requirement set is convex from below.

Following McFadden (1978b), the general representation of production possibilities in the modern theory is built on the properties of the set T of possible production plans $x = (x_1, \ldots, x_n)$, where x is an N-vector of commodities that may be either *net outputs* (if positive) or *net inputs* (if negative). Prices are represented by an N-vector p conformable to x. This general representation of the firm's activities allows cost, revenue, and profit functions to be special cases of the same model. If x is a vector of outputs, then $p \cdot x$ represents the firm's revenue; if x is a vector of inputs, then $-p \cdot x$ represents cost of production; and if x is a vector of outputs and inputs, then $p \cdot x$ represents the firm's profits.

The firm's production possibilities are generally restricted or conditioned by fixed factors of production, managerial capacity, expectations, the state of technology, the natural environment, and so forth. Defining these conditioning factors as the M-vector $z \in Z$, where Z is a subset of R^m, the production possibilities set is written $T(z)$ to denote the dependence of the set of possible production plans on z. When the technology is thus conditioned on Z, restricted cost, revenue, and profit functions are obtained. In the following discussion, we direct our attention to the *restricted profit function*. As McFadden (1978b, p. 4) emphasized, "Cost, revenue, and profit functions can all be considered as special cases of a restricted profit function, defining maximum profits over a subset of inputs and outputs with competitive prices when quantities of the remaining inputs and outputs are fixed."

McFadden imposes several mathematical properties on the sets Z and $T(z)$: Z is a nonempty subset of M-dimensional Euclidean space; for each $z \in Z$, T is a closed, nonempty, convex subset of N-dimensional Euclidean space. As we proceed, it will be shown that these assumptions lead to several important behavioral relations.

The restricted profit function is defined as

$$\pi(p,z) \equiv \sup_x \{p \cdot x \mid x \in T(z), z \in Z, p \in P(z)\}$$

where $P(z)$ is the set of price vectors for which $p \cdot x$ is bounded above. In other words, the restricted profit function shows the least upper bound on $p \cdot x$ that is possible, given p and the vector of conditioning variables z. If $p \cdot x$ can attain its upper bound, then "sup" can be replaced by "max" in the above definition.

The geometry of the profit function is illustrated in figure 2-1 for the single-output, single-input, single-fixed factor case. Given $z = z^1$, the production possibilities set is constructed in figure 2-1 showing all possible outputs x_2 obtainable from input x_1. In this simple example, this curve is essentially the production function of neoclassical theory. Given price vector $p = p^1$, the *isoprofit line* is defined by $\pi = p^1 \cdot x =$

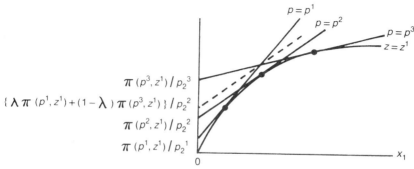

FIGURE 2-1. The properties of the restricted profit function

$p_1^1 x_1 + p_2^1 x_2$. Maximum profit obtainable, given the production possibility set with $z = z^1$, is attained at the tangency of the two curves and indicates the values of x_1 and x_2 at the maximum. The intercept of the isoprofit line on the x_2 axis measures maximum profit in terms of x_2.

The restricted profit function possesses a series of remarkable properties:

1. For each $z \in Z$, $\pi(p, z)$ is positively linear homogeneous, continuous, and convex in $p \in P(z)$.
2. $T(z) = T^*(z)$, where $T^*(z) = \{x | p \cdot x \leq \pi(p, z)$ for all $p \in P(z)\}$ is the closed convex hull of $T(z)$.
3. If the profit function is differentiable, the profit maximizing vector x^* satisfies $x^* = \nabla_p \pi(p, z)$.

Property 1 states that the profit function satisfies the obvious property of linear homogeneity in prices. Property 2 states a much stronger and less intuitive result: $\pi(p, z)$ is convex in prices. To see why this must be true, draw a series of isoprofit lines tangent to the production possibilities curve in figure 2-1 for several price vectors. By doing so, it can be seen graphically that a weighted average of profit $\lambda \pi(p^1, z^1) + (1 - \lambda) \pi(p^3, z^1)$ is necessarily greater than profit $\pi(p^2, z^1)$ at price $p^2 = \lambda p^1 + (1 - \lambda) p^3$. The geometry is such that the intersection of the $p = p^1$ and $p = p^3$ isoprofit lines gives the weighted average value of $\pi(p^1, z^1)$ and $\pi(p^3, z^1)$, which necessarily is greater than or equal to $\pi(p^2, z^1)$ at point x^2.

Property 2 states the duality relationship between the production possibilities set T and the profit function. To interpret this property, observe that the production possibilities set $T(z^1)$ in figure 2-1 is a convex set, and that each isoprofit line tangent to it defines a supporting halfspace. Property 2 says that the closed convex hull of $T(z^1)$—that is, the intersection $T^*(z^1)$ of all supporting halfspaces of $T(z^1)$—is equal to $T(z^1)$. This result is proved using the Minkowski theorem mentioned above. If we begin with the technology $T(z)$ and construct $\pi(p, z)$, and then use $\pi(p, z)$ to reconstruct the corresponding technology set $T^*(z)$, it must be true that $T(z) = T^*(z)$. Thus, the closed convex set $T(z)$ is equal to the intersection of its supporting halfspaces, $T^*(z)$, as stated in the Minkowski theorem.

It should be noted, however, that the duality results do not imply that there must be a unique correspondence between the technology set and the profit function. Suppose, for example, the technology were not globally convex, as assumed above. Then $T^*(z)$ would represent the convexification of $T(z)$; that is, all nonconvex segments would be replaced by a flat surface connecting the convex parts. Any technology with the same convexification would have the same profit function, so the firm's behavior would be the same regardless of the shape of the nonconvex segment. For example, a nonconvex segment of an isoquant has the effect of creating a discontinuity in the compensated factor demand functions, and the same discontinuous demand function would result regardless of the shape of the nonconvex portion of the isoquant. Thus, the information lost in convexifying the isoquant is not economically relevant, and all production possibility sets with the same convex regions are economically equivalent. Therefore, the duality of the production possibilities set and the profit function means that the profit function summarizes all of the economically relevant information contained in the technology.

Property 3 is the important derivative property of the profit function known as Hotelling's lemma (in the case of the cost function, it is known as Shephard's lemma). It tells us that the firm's system of output supply and input demand functions can be obtained by differentiating the profit function. These functions can in turn be used for comparative static analysis. For example, consider the commodity x_j, which is an output. Since an element of x that is an output is positive,

$$\partial \pi(p,z)/\partial p_j = x_j^*(p,z) > 0$$

is the supply function for x_j, and the own-price elasticity of supply is

$$\partial \ln x_j^* / \partial \ln p_j = (p_j \partial^2 \pi(p,z)/\partial p_j^2)/x_j^*(p,z).$$

Because the profit function is increasing and convex in p_j, it follows that this elasticity is positive as economic theory requires for a well-defined profit maximization problem.

Heuristically, property 3 can be derived by viewing the profit function as a function of p. Note that $\pi(p^0, z^0) = p^0 \cdot x^*$, where x^* solves the profit maximization problem at the given values p^0 and z^0. Viewing $\pi(p,z^0)$ as a function of p given z^0, it follows that $\pi(p,z^0) \geq p \cdot x^*$ for all p, with strict equality holding at p^0 only (this is because x^* is optimal only for $p = p^0$). Therefore at p^0 the function

$$\phi(p) = \pi(p,z^0) - p \cdot x^*$$

is globally minimized and the first order condition $\phi'(p) = \nabla_p \pi(p, z^0) - x^* = 0$ must be satisfied, showing that property 3 must be true.

Other properties can be derived for the profit function in relation to the conditioning variables z. If z is interpreted as a vector of commodities and free disposability of z is assumed, that is, if $z^1 \leq z^2$ and $x \epsilon T(z^1)$

implies $x \in T(z^2)$, then $\pi(p,z)$ is a nondecreasing function of z, given p. Thus, under free disposability of z, if we interpret z as the firm's capital stock, we obtain the commonsense result that profits should be increasing in the size of the firm's capital stock, all other things constant. Alternatively, if z is interpreted as a measure of the state of technology, free disposability of z implies that at fixed prices, technological progress increases the firm's profits.

Suppose Z is a convex set in the following sense: if $x^1 \in T(z^1)$ and $x^2 \in T(z^2)$, then $\lambda x^1 + (1-\lambda)x^2 \in T(\lambda z^1 + (1-\lambda)z^2)$. When the technology is convex in both X and Z in this sense, it can be shown that $\pi(p,z)$ is a *concave* function of z given p. Thus, if z represents the firm's capital stock, free disposability and convexity in x and z imply that profits are increasing in the firm's capital stock but at a decreasing rate. To prove this, we observe that if $\pi(p,z^1) = p \cdot x^1$ and $\pi(p,z^2) = p \cdot x^2$,

$$\begin{aligned}\pi(p,\lambda z^1 + (1-\lambda)z^2) &\geq p \cdot (\lambda x^1 + (1-\lambda)x^2) \\ &\geq \lambda p \cdot x^1 + (1-\lambda)p \cdot x^2 \\ &\geq \lambda \pi(p,z^1) + (1-\lambda)\pi(p,z^2)\end{aligned}$$

and therefore the profit function is concave in z.

Convexity of the technology in both x and z implies an even stronger restriction—that the technology does not exhibit increasing returns to scale. This is proved as follows. Let $T(z) = \{x | x_2 \leq F[x_1, z]\}$ so that $\lambda x^1 \in T(\lambda z^1)$ and $(1-\lambda)x^2 \in T((1-\lambda)z^2)$. We have $\lambda x^1 + (1-\lambda)x^2 \in T(\lambda z^1 + (1-\lambda)z^2)$ if and only if

$$\begin{aligned}F[\lambda x_1^1 + (1-\lambda)x_1^2, \lambda z^1 + (1-\lambda)z^2] &\geq \lambda x_2^1 + (1-\lambda)x_2^2 \\ &\geq \lambda F[x_1^1, z^1] + (1-\lambda)F[x_1^2, z^2]\end{aligned}$$

but this is the condition for F being a concave function in x_1 and z: increasing returns to scale are thus not possible. If the technology exhibits constant returns to scale—that is, if $x \in T(z)$ implies $kx \in T(kz)$ for $k \geq 0$—then $\pi[p, kz] = p \cdot kx^1 = kp \cdot x^1 = k\pi[p, z^1]$ and thus $\pi(p,z)$ is a linear homogeneous function of z given p. In this case $T(z)$ is said to be a *cone*.

Free disposability and convexity in z may or may not make economic sense, depending on how the z variables are defined. If z represents fixed factors of production, then both properties may be reasonable. But if a component of z represents the state of the technology, convexity may not be reasonable, because at fixed prices an advance in technology need not increase profits at a decreasing rate. Indeed, it may be the case that technological progress increases profit at an increasing rate.

Duality relations can be used to derive correspondences between the functional structure of the technology set and the cost or profit function, and thus can provide the basis for statistical tests of the technology structure. For example, it can be shown that the production function is homothetic in input quantities if and only if the normalized profit function is homothetic in normalized factor prices (Lau, 1978, theorem II-2). Similarly, it can be shown that the production function is homothetic if and only if the cost function is separable in output. In the multiproduct case, restrictions for input-output separability and nonjointness of the technology translate into restrictions on the profit and cost functions. Lau showed that the production function exhibits input-output separability if and only if the profit function exhibits separability in input and output prices, and that the production function is nonjoint in inputs if and only if its profit function is additive in output prices. Hall (1973) derived similar results for the implications of input-output separability and nonjointness in terms of the cost function.

Example: To see the correspondence between the firm's technology and the profit function, consider the Cobb-Douglas technology set

$$T(z) = \{x | x_0 \leq x_1^{\alpha_1} x_1^{\alpha_2} z^{\alpha_3}, x_1, x_2, z > 0\},$$

where x_0 is output, x_1 and x_2 are variable inputs, and z is a fixed factor. The dual restricted profit function can be derived by solving the primal profit maximization problem

$$\max_{x_0, x_1, x_2} \pi = p \cdot x = p_0 x_0 - p_1 x_1 - p_2 x_2 \text{ subject to } T(z)$$

It will be seen below that for a maximum profit to exist, the technology must satisfy decreasing returns to scale in x_1 and x_2, that is, $\alpha_1 + \alpha_2 < 1$. (Note, however, that the technology may nevertheless exhibit constant or increasing returns with respect to x_1, x_2 *and* z). Convexity of the technology in z requires in addition that $0 < \alpha_3 < 1$, as can be seen by applying the definition of convexity to $T(z)$.

The profit function can be derived from the production possibilities set in several ways. One method (the brute force method) is to obtain the first-order conditions for the above maximization problem and solve them for the optimal values $x_0^*(p, z)$, $x_1^*(p, z)$ and $x_2^*(p, z)$, substitute them into $p \cdot x$, and simplify to obtain $\pi(p, z)$. Another method is to solve for one of the factor demand functions, $x_i^*(p, z)$, $i = 1, 2$ from the first-order conditions, and then use Hotelling's lemma and the Fundamental Theorem of Calculus to obtain the profit function as

$$\pi(p, z) = -\int x_i^*(p, z) dp_i$$

Also this method could be applied using the output supply function $x_0^*(p,z)$. Following either of these approaches, one can obtain the Cobb-Douglas profit function

$$\pi(p,z) = b p_0^{\beta_0} p_1^{\beta_1} p_2^{\beta_2} z^{\beta_3}$$

where $b = (1-\mu)\alpha_1^{\alpha_1(1-\mu)^{-1}} \alpha_2^{\alpha_2(1-\mu)^{-1}}$, $\beta_0 = \mu/(1-\mu)$, $\beta_i = -\alpha_i/(1-\mu)$, $i = 1, 2$, $\beta_3 = \alpha_3/(1-\mu)$, and $\mu = \alpha_1 + \alpha_2$. The fact that the structure of the production technology determines the structure of the profit function is evident: the profit function parameters are composed of the production function parameters and the functional form of the profit function is determined by the functional form of the production function. Indeed, the Cobb-Douglas technology is an example of a *self-dual* technology, meaning that the dual has the same functional form as the primal.

It can readily be seen that the various properties of the Cobb-Douglas technology can be determined from the profit function. First, note that the profit function satisfies the conditions of property 1: it is linear homogeneous in prices (because $\beta_0 + \beta_1 + \beta_2 = 1$) and it is continuous and convex in prices. The fact that one could solve back from the profit function to the original technology set means that property 2 is satisfied; property 3 can be verified by showing that the supply and demand equations derived from the primal equal those obtained by differentiating the profit function.

The Cobb-Douglas profit function illustrates various duality results involving the structure of the primal and dual functions. For example, the function is homothetic in variable input quantities, and by duality it follows that the normalized profit function must be homothetic in variable input prices, as is readily verified in the above example.

It is readily verified that the Cobb-Douglas technology is monotonic in z and thus the profit function must be increasing in z. Moreover, if the restriction $0 < \alpha_3 < 1$ is imposed, then $T(z)$ is convex in z and the profit function is concave in z. If z is used to represent technological change at an exponential rate, that is, if $z = e^t$, where t represents time, then it is readily verified that the technology is not convex in t.

The usefulness of duality theory is even more apparent when we consider functional forms other than the Cobb-Douglas. Unlike the Cobb-Douglas and the quadratic functions, many forms of the production function are not self-dual, and many do not have closed-form solutions for the output supply and factor demand functions (an example is the translog). Conversely, many convenient functional representations of the cost, revenue, or profit function cannot be solved for a closed-form representation of the production function. Nevertheless, if the dual function satisfies the required conditions, there exists a well-defined production possibilities set corresponding to it, and its properties can be obtained from the dual function.

In concluding this discussion of duality theory, it should be noted that, although there are certain advantages to using the dual approach to production analysis, the approach has certain limitations. Pope (1982) has noted that if two distinct inputs or outputs have the same price, the dual approach will fail to yield distinct input demand or output supply functions, even though distinct functions may be obtainable from the primal approach. In addition, one can question the relevance of the dual approach, based as it is on the static, microeconomic theory of the firm, for the analysis of aggregate production relations and aggregate productivity. Should one expect microeconomic properties of dual functions to hold with aggregate data? These and other important questions as yet have not been answered satisfactorily.

Price collinearity may also be a problem from the point of view of econometric estimation. For example, with cross-sectional data representing individual firms, if all firms purchase inputs in the same markets, then they all face identical factor prices. Under these conditions it would be impossible to estimate a cost or profit function. A final observation is that, even though the properties of the technology can be recovered from each of the dual functions, they are not all equally relevant under all conditions. For example, if government policy essentially sets the output level of each farm, it would be erroneous to model the technology using a profit function based on the assumption that output levels are set to maximize profit. However, under these conditions it might be reasonable to assume that firms minimize costs for whatever output level is assigned to them by the government, so it would be possible to estimate a valid cost function. Researchers must thus carefully weigh the costs and benefits of the dual approach in light of the research question at hand.

Technological Change

Technological change refers to the changes in a production process that come about from the application of scientific knowledge. These changes in the production process can be realized in various ways at the firm level: through improved methods of utilizing existing resources such that a higher output rate per unit of input is obtained, often referred to as *disembodied* technological change; through changes in input quality, referred to as *embodied* technological change; or through the introduction of new processes and new inputs.

If disembodied technological change occurs in an existing production process, then it can be modeled in terms of a shift in the production surface. Even in this simple case, difficult measurement problems are encountered. For instance, the measured rate of disembodied technological change generally depends on the point in input space (the factor levels and proportions) at which it is measured. Typically a new technology is efficient at different factor levels and intensities from the old technology, and vice versa. Embodied technological change introduces other measurement problems, such as the measurement of input and output quality changes. In the economics literature, this problem has been often discussed in terms of the measurement of physical capital (Christensen and Jorgenson, 1969; Diewert, 1980a). It is clearly a much more general problem, however, as is illustrated by the quality changes taking place in chemical inputs, seeds, and human labor in agricultural production.

If technological change occurs through the introduction of new processes and inputs, then the production possibilities set is both multiproduct and multifactor. In some respects it is simpler to use the dual cost or profit function to represent the multiproduct technology and to define technological change. The effects of technological change are expressed in terms of a reduction in the cost of production (given outputs and factor prices) or an increase in profits (given output and input prices). However, dual measures of technological change also involve certain difficulties. When technological change involves the adoption of new inputs or production of new outputs, for example, firms may be observed at corner solutions, where some inputs are not used or some outputs are not produced. Conventional duality theory assumes interior solutions and thus is not appropriate for analysis of such situations without appropriate modifications (as discussed in chapter 8 and in Lee and Pitt, 1986). Since the dual cost and profit functions are based on the assumption of maximizing behavior, they provide measures of technological change at the firm's optimal input and output levels. As we explain below, dual measures may differ from primal measures due to a scale effect introduced as firms adjust optimally in response to technological change. These scale effects must be taken into account if primal and dual measures of the rate and bias of technological change are to be comparable. Moreover, if firms make systematic maximization errors, those errors are introduced into the measured rate and bias of technological change.

The Single-Product Case. Much of the theoretical literature on technological change, as well as most empirical research, is conducted at the aggregate level of the industry or sector. In the aggregate approach, an

DEVELOPMENTS IN PRODUCTION THEORY & PRODUCTIVITY MEASUREMENT

aggregate production function is often postulated to be of the form $Q = F(X, t)$, where Q and X are aggregate output and an aggregate input vector and t denotes the state of the technology. In agriculture, as in most industries, there are many products; the use of an industry production function thus implicitly requires the assumption that the transformation function is separable in inputs and outputs. We pursue the multiple-output case in detail below. In this section, we consider the case in which the technology is nonjoint in inputs, so that each output can be represented in terms of its own production function. When the input categories are broadly defined (such as land and structures, machinery, labor, chemicals) input quality may change with technological change while the categories themselves remain unchanged. Thus, in the aggregate it is reasonable to interpret aggregate disembodied technological change as the upward shift in the production surface. In terms of the aggregate production function, if production is efficient, we can interpret $\partial \ln F(X, t)/\partial t$ as the *primal rate of technological change*, which can be expressed as

$$\partial \ln F/\partial t = d\ln Q/dt - (1/F) \sum_{i=1}^{n} F_i dX_i/dt$$

where F_i is the marginal product of X_i. If the industry is in competitive equilibrium, so that price equals marginal cost, and inputs are paid the value of their marginal products, the above equation can be written as

$$\partial \ln F/\partial t = d\ln Q/dt - (\partial \ln C/\partial \ln Q)^{-1} \sum_{i=1}^{n} S_i d\ln X_i/dt \tag{2-9}$$

where $S_i \equiv W_i X_i / \sum_i W_i X_i$ is the factor cost share. Note that the term $\partial \ln C/\partial \ln Q$ is the elasticity of cost with respect to output, and can be used to classify returns to scale. Equation (2-9) says that the primal rate can be expressed as the rate of change in output minus a scale-adjusted index of the rate of change in inputs. We shall see that this fact allows the primal rate to be related to the concept of total factor productivity, as discussed in the following section.

The dual cost or profit function can also be used to measure the rate of technological change. Define the cost function as $C = C(Q, W, t)$. Differentiating with respect to time gives

$$\frac{d\ln C}{dt} = \sum_{i=1}^{n} \frac{\partial \ln C}{\partial \ln W_i} \frac{d\ln W_i}{dt} + \frac{\partial \ln C}{\partial \ln Q} \frac{d\ln Q}{dt} + \frac{\partial \ln C}{\partial t}$$

Using Shephard's lemma, $X_i = \partial C(Q, W, t)/\partial W_i$, and noting that $S_i = \partial \ln C/\partial \ln W_i$,

$$-\frac{\partial \ln C}{\partial t} = \sum_{i=1}^{n} S_i \frac{d \ln W_i}{dt} + \frac{\partial \ln C}{\partial \ln Q}\frac{d \ln Q}{dt} - \frac{d \ln C}{dt} \tag{2-10}$$

where $-\partial \ln C/\partial t$ is defined as the *dual rate of technological change*. Equation (2-10) shows that the dual rate equals an index of the rate of change in factor prices plus a scale effect minus the rate of change of total cost. Totally differentiating $C = \Sigma W_i X_i$ with respect to time gives

$$d\ln C/dt = \sum_{i=1}^{n} S_i d\ln W_i/dt + \sum_{i=1}^{n} S_i d\ln X_i/dt \tag{2-11}$$

Combining equations (2-9), (2-10), and (2-11), it follows that the primal and dual rates of technological change are related as follows:

$$-\partial \ln C/\partial t = (\partial \ln C/\partial \ln Q)\partial \ln F/\partial t \tag{2-12}$$

Thus it has been demonstrated that the primal and dual rates of technological change are equal if and only if $\partial \ln C/\partial \ln Q = 1$. Note that the cost function takes the form $C(Q, W, t) = Qg(W, t)$ and $\partial \ln C/\partial \ln Q = 1$ if and only if the technology exhibits constant returns to scale. It follows that the primal and dual rates of technological change must be equal if and only if the technology is constant returns to scale.

In a multiple-input production process, technological change can affect input productivities and factor utilization differentially. This observation has led to the distinction between *neutral* and *biased* technological change. The original definitions of neutral and biased technological change are due to Hicks (1963), who defined them in terms of the marginal rate of technical substitution (MRTS):

> If we concentrate on two groups of factors, "labour" and "capital," and suppose them to exhaust the list, then we can classify inventions according as their initial effects are to increase, leave unchanged, or diminish the ratio of the marginal product of capital to that of labour. We may call these inventions "labour-saving," "neutral," and "capital-saving," respectively (Hicks, 1963, p. 121).

Figure 2-2 illustrates how Hicks' concept of biased technological change may be interpreted in the case of a homothetic technology. Starting at the firm's initial equilibrium point *A* with the old technol-

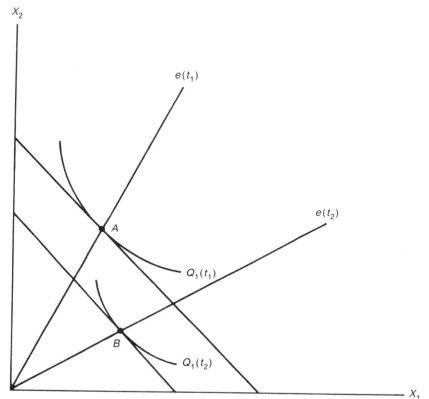

FIGURE 2-2. Biased technological change

ogy, a shift to a new technology allows the firm to move to point B, where it can produce the same output (Q_1) at the same factor prices at a lower total cost. Since the cost-minimizing firm producing Q_1 moves from A to B and lowers its input of X_2 relative to X_1, Hicks calls the technological change factor X_1-using and X_2-saving. Technological change is neutral if the MRTS and optimal factor proportions are unaffected.

While Hicks' definition seems straightforward, careful scrutiny reveals that it is open to several different interpretations, as evidenced by the various definitions of "Hicks-neutral" and biased technological change in the literature (Blackorby, Lovell, and Thursby, 1976). In particular, the above definition does not specify whether the effect of technological change is to be measured at some specified point in input space, or at the optimal (cost minimizing or profit maximizing) factor proportions, or along the firm's expansion path. If the technology is homothetic, so that the firm's expansion path is linear, then neutrality can be defined equivalently in terms of either the MRTS at a point, optimal factor proportions, or the expansion path. But if the technology is nonhomothetic, then the expansion path is nonlinear, and optimal factor proportions change along the expansion path. Thus, a technological change that is neutral with respect to the expansion path (that is, the isoquants are simply renumbered) is also neutral in terms of the MRTS measured at a point on the expansion path, but is not neutral in terms of optimal factor proportions due to changes in the optimal scale of production.

Blackorby, Lovell, and Thursby argue that Hicks intended neutrality to be defined as the invariance of the expansion path to technological change. This concept of neutrality is useful because it is not dependent upon assumptions about the homotheticity of the technology. Henceforth we adopt this definition of Hicks-neutral technological change as expansion-path preserving. Blackorby, Lovell, and Thursby observe that the technology is Hicks-neutral in this sense if and only if the production function is weakly separable in X, that is, $Q = F[f(X), t]$. In contrast to this definition of Hicks neutrality, they define *extended Hicks-neutral technological change* as the case in which the production function can be written in the strongly separable form $Q = A(t)f(X)$. Note that neither of these neutrality concepts implies or is implied by input homotheticity. If the technology is homothetic, then Hicks neutrality implies extended Hicks neutrality, but the converse is not true.

Consider the following measure of the bias in technological change, which we refer to as the *primal measure*:

$$B_{ij}(X_1, t) \equiv \partial \ln(F_i/F_j)/\partial t = \partial \ln F_i(X, t)/\partial t \\ - \partial \ln F_j(X, t)/\partial t, \qquad i \neq j \qquad (2\text{-}13)$$

Hicks neutrality implies $B_{ij} = 0$ for all i and j. Note that this measure of the bias is defined at a given point in input space. Essentially, it measures the rotation of the isoquant at this point in response to technological change, as illustrated in figure 2-3. The initial expansion path is $e(t_1)$ and the firm is producing at point A. Technological change leads to a new expansion path $e(t_2)$. The isoquant Q_2 on the new expansion path passes through point A. B_{ij} measures the change in the slope of the isocost line C_1 tangent to Q_1 to the slope of the isocost line C_2 tangent to Q_2. Thus, $B_{ij} = 0$ if technical change is Hicks neutral so that the expansion path is unchanged by technological change.

One difficulty with Hicks' definition of bias is that it requires pairwise comparisons of all factors' marginal products; it is thus not clear whether

DEVELOPMENTS IN PRODUCTION THEORY & PRODUCTIVITY MEASUREMENT 39

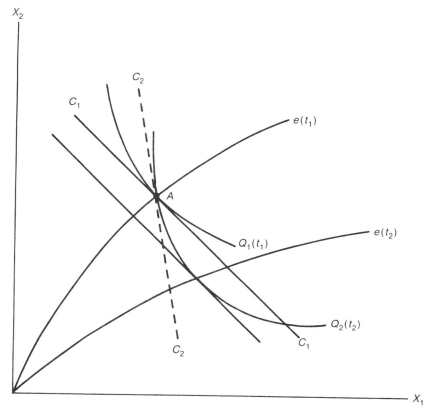

FIGURE 2-3. The primal measure of the bias in technological change

technological change is using or saving in each input. To overcome this difficulty, an overall bias measure for each factor is needed. Define

$$B_i(X, t) \equiv \sum_{j \neq i}^{n} S_j B_{ij}$$

where S_j is a factor cost share. In order to interpret B_i, note that if factor prices changed such that the point X was on the firm's expansion path with the new technology—that is, letting $dW_i/dt = dF_i/dt$—then

$$\sum_{j \neq i}^{n} S_j B_{ij} = \partial \ln F_i(X, t)/\partial t - \sum_{j=1}^{n} \partial \ln F_j(X, t)/\partial t$$

$$= d\ln W_i/dt - \sum_{j=1}^{n} S_j d\ln W_j \, dt$$

$$= \partial \ln S_i/\partial t \big|_x \tag{2-14}$$

Since the pairwise biases B_{ij} measure the rotation of the isoquant at the given point in input space, the overall bias $B_i(X, t)$ can be interpreted as the change in the ith factor cost share that would occur if factor prices changed so that the firm's original cost-minimizing input X remained on the expansion path.

The overall bias measure B_i is useful because it tells us that if on average the marginal product of factor i is increasing relative to all others, then $B_i > 0$ and technological change is factor-i using overall. It follows that if technological change is neutral—that is, if all $B_{ij} = 0$—then all $B_i = 0$. Therefore, $B_i = 0$ for all i is a necessary condition for Hicks neutrality and can be used to formulate statistical tests for neutrality. Moreover, since the S_i sum to unity, $\ln(\Sigma_i S_i) = 0$, and differentiation with respect to t shows that $\Sigma_i S_i B_i = 0$. Thus, if technological change is biased, at least one B_i must be positive and one must be negative. This result reflects the fact that the B_i are relative measures of bias, and do not necessarily provide information about the absolute changes in marginal productivities.

Bias measures can also be defined in terms of the optimal factor proportions. Consider the following dual measure of bias:

$$B_{ij}^c(Q, W, t) = \partial \ln X_i(Q, W, t)/\partial t - \partial \ln X_j(Q, W, t)/\partial t \tag{2-15}$$

where the $X_i(Q, W, t)$ is the compensated (output constant) factor demand function obtained by applying Shephard's lemma to the cost function $C(Q, W, t)$. Lau (1978) refers to the condition $B_{ij}^c = 0$ as "indirect Hicks neutrality." Using a derivation similar to that of equation (2-14), it can be shown that

$$B_i^c(Q, W, t) \equiv \sum_{j \neq i}^n S_j B_{ij} = \partial \ln S_i^c(Q, W, t)/\partial t \tag{2-16}$$

where $S_i(Q, W, t)$ is the optimal (cost-minimizing) factor cost share. B_i^c is defined as a *dual cost* measure of the bias in technological change. This measure was proposed by Binswanger (1974, 1978) for use with a homothetic cost function.

Observe that if the cost function is homothetic, $C(Q, W, t) = g^1(Q, t)g^2(W, t)$, and it follows that the cost shares are independent of the rate of output, so $B_i^c(W, t) = \partial \ln S_i(W, t)/\partial t$ is independent of output. However, if the technology is nonhomothetic, optimal factor proportions and thus factor shares depend on the output level, and a change in scale of production induces a change in optimal factor proportions. Thus, neutral technical change generally affects optimal factor proportions, in the nonhomothetic case, because it represents a renumbering of the isoquant map. Consequently, in the case of nonneutral technological change, the bias measure B_i^c represents two distinct effects, a *scale effect*, due to the movement along the expansion path, and a *bias effect*, due to the shift in the expansion path. If the technology is homothetic, the scale effect is zero.

In order to measure the shift in the expansion path, it is necessary to adjust B_i^c for the scale effect in the nonhomothetic case. Observe that the change in the factor proportions due to technological change is exactly equal to the change that would occur if output were reduced by a corresponding percentage $d \ln Q^*$. Thus the scale effect of technological change due to the movement along the expansion path can be defined as $[\partial \ln S_i(Q, W, t)/\partial \ln Q] d\ln Q^*/dt$. A measure of the shift in the expansion path can therefore be defined along the isocost line as

$$B_i^{ce} \equiv \partial \ln S_i(Q, W, t)/\partial t |_{dC=0} = B_i^c - (\partial \ln S_i/\partial \ln Q)(d \ln Q^*/dt) \qquad (2\text{-}17)$$

To measure B_i^{ce}, we need to express $d\ln Q^*$ in terms of the cost function. The cost change due to an output change, holding constant input prices and technology—that is, a cost change due to a movement along the expansion path—is

$$d\ln C|_{W,t} = (\partial \ln C/\partial \ln Q) d\ln Q \qquad (2\text{-}18)$$

The corresponding output change is thus

$$d\ln Q = (\partial \ln C/\partial \ln Q)^{-1} d\ln C|_{W,t}$$

The cost change due to technological change is

$$d\ln C|_{Q,W} = (\partial \ln C/\partial t) dt \qquad (2\text{-}19)$$

Thus, assuming the two cost changes (2-18) and (2-19) are equal, we obtain

$$d\ln Q^* = (\partial \ln C/\partial \ln Q)^{-1} (\partial \ln C/\partial t) dt$$

The dual Hicks measure of factor bias is thus

$$B_i^{ce} = B_i^c - (\partial \ln S_i/\partial \ln Q)(\partial \ln C/\partial \ln Q)^{-1} \partial \ln C/\partial t \qquad (2\text{-}20)$$

Note that the scale effect of technological change depends on the elasticity of the cost share with respect to output, the elasticity of total cost with respect to output, and the dual rate of technological change. In the homothetic case, $\partial \ln S_i/\partial \ln Q = 0$ so $B_i^{ce} = B_i^c$.

A graphical interpretation of the scale and bias effects is illustrated in figure 2-4. The scale effect, due to the movement along the expansion path, is represented by the movement from A to B or from C to D. The bias effect, B_i^{ce}, can be interpreted as the movement from points A to C or from B to D since it measures the change in factor share holding total cost constant. If the primal measure of Hicks bias, B_i, also is measured at A, then both B_i and B_i^{ce} are zero for all i if and only if technological change is Hicks neutral; and if technological change is biased towards (against) factor i, then both bias measures are positive (negative). Note that the numerical values of the two bias measures generally are not equal.

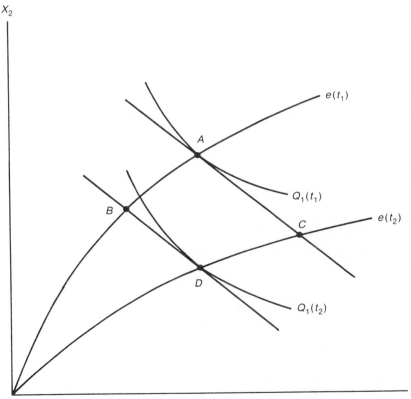

FIGURE 2-4. Scale and bias effects of technological change

A related dual measure of technological change bias can be defined in terms of the profit-maximizing factor proportions. Let

$$B_{ij}^p(p, W, t) = \partial \ln X_i(p, W, t)/\partial t - \partial \ln X_j(p, W, t)/\partial t, \quad i \neq j \qquad (2\text{-}21)$$

Thus, $B_{ij}^p = 0$ is the condition for indirect Hicks neutrality based on the profit-maximizing factor proportions. Taking the cost-share weighted sum,

$$B_i^p(p, W, t) \equiv \sum_{j \neq i} S_j(p, W, t) B_{ij}^p = \partial \ln S_i(p, W, t)/\partial t \qquad (2\text{-}22)$$

where $S_i(p, W, t) = X_i(p, W, t) W_i \Big/ \sum_{j=1}^{n} X_j(p, W, t) W_j$ is the profit maximizing cost share derived from the profit function. This bias measure was introduced by Antle (1984a).

Like the cost-share measure B_i^c, B_i^p can be interpreted as a measure of the shift in the expansion path if the technology is homothetic. However, if the technology is nonhomothetic B_i^p is composed of both a scale effect and a bias effect. A measure of the bias effect due to a shift in the expansion path can be defined as

$$B_i^{pe} \equiv \partial \ln S_i(p, W, t)/\partial t \big|_{dC=0} \qquad (2\text{-}23)$$

In a manner analogous to the cost share case, it can be shown that B_i^{pe} equals the profit maximizing cost share change B_i^p minus the scale effect,

$$B_i^{pe} = B_i^p - (\partial \ln S_i/\partial \ln p)(\partial \ln \pi/\partial \ln p)^{-1}(\partial \ln \pi/\partial t) \qquad (2\text{-}24)$$

Note that when the technology is homothetic, $\partial \ln S_i/\partial \ln p = 0$ so $B_i^{pe} = B_i^p$.

The interpretation of B_i^{pe} is similar to B_i^{ce} except that B_i^{pe} measures the shift in the expansion path assuming a constant level of profit rather than a constant level of cost. The two measures differ quantitatively but necessarily provide the same qualitative results: they are both zero if technological change is Hicks neutral and both are positive (negative) if technological change is on average factor-i using (saving).

The following conclusions can be drawn about the measurement of the rate and bias of technological change in the single-product case. First, the primal and dual measures of the rate of technological change are necessarily numerically equal only if the industry is in competitive equilibrium and the aggregate technology satisfies constant returns to scale. If the technology is not characterized by constant returns to scale, the difference between the primal and dual measures can be estimated as a scale effect. Second, the primal and dual measures of the bias in technological change based on cost shares are not numerically equal. With homotheticity, the conventional primal and dual measures based on cost shares give the same qualitative indication of Hicksian neutrality or bias. In the nonhomothetic case, measures of Hicksian biases can be constructed by adjusting the dual cost-share measures for the scale effect of technological change.

The Multiple-Output Case. Recall that the multiproduct technology can be represented in terms of the transformation function. Let the transformation function be $Q_1 = F(Q^2, X, t)$, where $Q^2 = (Q_2, \ldots, Q_m)$ and $X = (X_1, \ldots, X_n)$. Following Hulten (1978), we can define the *primal multiproduct rate of technological change* as

$$R_1 \frac{\partial \ln F}{\partial t} \tag{2-25}$$

where R_1 is defined as the revenue share of Q_1 in total revenue. Note that the rate of technological change is measured in terms of Q_1.

Define the multiproduct cost function dual to the transformation function as $C(Q, W, t)$, where $Q = (Q_1, \ldots, Q_m)$ is the vector of outputs and W is the vector of factor prices. The dual rate of multiproduct technological change can be defined as $(-\partial \ln C/\partial t)$. Following the derivations in equations (2-9) through (2-12), it can be shown that a condition analogous to equation (2-12) holds in the multiproduct case. Totally differentiating the transformation function, we derive

$$\frac{d\ln Q_1}{dt} = \sum_{i=2}^{m} \frac{\partial \ln F}{\partial \ln Q_i} \frac{d\ln Q_i}{dt} + \sum_{i=1}^{n} \frac{\partial \ln F}{\partial \ln X_i} \frac{d\ln X_i}{dt} + \frac{\partial \ln F}{\partial t} \tag{2-26}$$

From the profit maximization problem form, the Lagrangean is

$$L = \sum_{i=1}^{m} p_i Q_i - \sum_{i=1}^{n} W_i X_i - \lambda(Q_1 - F[Q^2, X])$$

The first-order conditions for multiproduct profit maximization (for the choice of the Q_i and X_i) are thus

$$p_1 = \lambda, \; p_i = -\lambda \frac{\partial F}{\partial Q_i}, \quad i = 2, \ldots, m,$$

$$W_i = \lambda \frac{\partial F}{\partial X_i}, \quad i = 1, \ldots, n$$

Defining the numéraire price $p_1 = 1$, and using the equilibrium condition $p_i = \partial C/\partial Q_i$ (price equals marginal cost), equation (2-26) can be rewritten

$$\sum_{i=1}^{m} R_i \frac{d\ln Q_i}{dt} = \left(\sum_{i=1}^{m} \frac{\partial \ln C}{\partial \ln Q_i}\right)^{-1} \sum_{i=1}^{n} S_i \frac{d\ln X_i}{dt} + R_1 \qquad (2\text{-}27)$$

where $R_i \equiv P_i Q_i / \sum_{i=1}^{m} P_i Q_i$

Now combining equation (2-27) with the multiproduct generalizations of equations (2-10) and (2-11), we obtain

$$-\partial \ln C/\partial t = \left(\sum_{i=1}^{m} \partial \ln C/\partial \ln Q_i\right) R_1 \partial \ln F/\partial t \qquad (2\text{-}28)$$

Thus, the primal and dual measures of the rate of technological change are equal if $\Sigma_i \partial \ln C/\partial \ln Q_i = 1$, that is, if the technology exhibits constant returns to scale.

In the multiproduct case, technological change can be Hicks neutral in two senses: First, it can leave the expansion path unchanged in the input space, as discussed above in the single-product case; second, it can leave the expansion path unchanged in output space, that is, it can result in a neutral shift in the production possibilities frontier.

Consider first the multiproduct generalization of the cost-share biases. As in the single-product case, we can define both primal and dual input bias measures. Define the primal input bias measure as

$$MB_{ij}(Q^2, X, t) = \partial \ln F_i(Q^2, X, t)/\partial t - \partial \ln F_j(Q^2, X, t)/\partial t$$

where F_i refers to the derivative with respect to X_i. Taking the cost-share weighted average we obtain

$$MB_i(Q^2, X, t) = \sum_{j \neq i}^{n} S_j MB_{ij}$$

The primal measure of bias is thus defined at given output and input levels. Note that if the technology satisfies input-output separability, then F_i depends only on X and t and hence the primal measure depends only on inputs and not on outputs.

A dual measure of bias can be defined in terms of changes in optimal (cost minimizing) factor proportions, in a manner analogous to the single product case. Taking a cost-share weighted sum of changes in factor proportions gives

$$MB_i^c \equiv \partial \ln S_i(Q, W, t)/\partial t = \sum_{j \neq i} S_j \partial \ln (X_i/X_j)/\partial t$$

As in the single-product case, the change in optimal cost shares due to technological change can be decomposed into a scale effect (a movement along the expansion path) and a bias effect (a shift of the expansion path). In the multiproduct case, the scale effect is eliminated if the technology is separable in inputs and outputs. It can be shown (Hall, 1973) that the cost function is written $C = g^1(Q, t)g^2(X, t)$ if and only if input-output separability holds; the cost shares are thus functions $S_i(W, t)$, which do not depend on Q. If the technology is nonseparable, then the scale effect must be netted out of MB_i^c to obtain the bias effect. Define the measure of Hicksian bias representing a shift in the expansion path along the isocost line as

$$MB_i^{ce} \equiv \partial \ln S_i(Q, W, t)/\partial t \big|_{dC=0}$$

Following the same steps as in the single-product case (equations 2-17 through 2-20),

$$MB_i^{ce} = MB_i^c - \left[\sum_{j=1}^{m} (\partial \ln S_i(Q, W, t)/\partial \ln Q_j)(\partial \ln C/\partial \ln Q_j)^{-1} \right] \partial \ln C/\partial t$$

MB_i^{ce} is the generalization of the dual bias measure B_i^{ce} to the multiple output case. Since input-output separability implies $\partial \ln S_i/\partial \ln Q_j = 0$ for all i and j, it also implies $MB_i^{ce} = MB_i^c$.

An analogous result can be obtained for the bias measures derived from the multiproduct profit function. Define

$$MB_i^p \equiv \partial \ln S_i(p, W, t)/\partial t$$
$$MB_i^{pe} \equiv \partial \ln S_i(p, W, t)/\partial t \big|_{dC=0}$$

hence

$$MB_i^{pe} = MB_i^p + \left[\sum_{j=1}^{m} (\partial \ln S_i(p, W, t)/\partial \ln p_j)(\partial \ln \pi/\partial \ln p_j)^{-1} \right] \partial \ln \pi/\partial t$$

where p is the vector of output prices corresponding to Q; W is the input price vector; and $\pi = \pi(p, W, t)$ is the multiproduct profit function dual to the transformation function. Lau (1978) has shown that when the technology is input-output separable, the profit function can be written $\pi = G(g^1(p, t), g^2(W, t))$. It immediately follows that the profit-maximizing cost shares do not depend on p. Therefore, in the case of input-output separability, $\partial \ln S_i/\partial \ln p_j = 0$ and $MB_i^{pe} = MB_i^p$.

A similar analysis can be conducted for the measurement of the bias in the shift of the production possibilities frontier due to technological change. Letting the transformation function be $F(Q, X, t) = 0$, define the *primal output bias* as

$$RB_{ij} \equiv \partial \ln F_{Q_i}(Q, X, t)/\partial t - \partial \ln F_{Q_j}(Q, X, t)/\partial t, \quad i, j = 2, \ldots, m, \quad i \neq j$$

where F_{Q_i} refers to the derivative of F with respect to Q_i. Assuming that the optimal output mix is produced, so that $p_i/p_j = F_{Q_i}/F_{Q_j}$,

$$RB_i = \sum_{j \neq i}^{m} R_j RB_{ij}$$

RB_i measures the rotation of the production possibilities frontier due to technological change at the point (Q, X).

Dual measures of output bias also can be constructed using the multiproduct cost, revenue, or profit function. In the case of the cost function, an output bias measure can be constructed to detect a movement of the expansion path in output space. Recall that the profit maximizing firm produces such that $(p_i/p_j) = (\partial C/\partial Q_i \div \partial C/\partial Q_j)$. Thus, define $CB_{ij}(Q, W, t) \equiv \partial \ln (\partial C/\partial Q_i \div \partial C/\partial Q_j)/\partial t$. CB_{ij} measures the change in the slope of the isocost curve due to technological change, at a given point in output space. This dual output bias measure differs from the dual input bias measure based on the cost function in an important way: because the measure is defined at a point in output space, it does not embody a scale effect. Thus, defining the overall bias measure

$$CB_i \equiv \sum_{j \neq i} R_j CB_{ij}$$

it follows that technological change is output Hicks neutral if $CB_i = 0$ for all i, and is biased otherwise.

Output biases can be derived from the revenue or profit function as well, but since they allow output to adjust in response to technological change, they involve both a pure bias effect and a scale effect. For example, define the *restricted revenue function* $R(p, X, t)$, where p is the output price vector and X is the input vector. The ith revenue share is

$$R_i(p, X, t) = p_i Q_i / R = \partial \ln R(p, X, t)/\partial \ln p_i$$

Define the output bias in technological change as

$$RB_i^r = \partial \ln R_i(p, X, t)/\partial t$$

If the technology is input-output separable, the revenue shares are independent of X and given by $R_i(p, t)$. Thus, in a manner exactly analogous to the factor biases, it can be seen that unless the technology is input-output separable, RB_i will contain both a bias effect (measured as a movement along the isorevenue line in output space) and a scale effect (measured as a movement along the expansion path in output space). Therefore, the pure Hicksian bias effect can be defined as

$$RB_i^{re} \equiv \partial \ln R_i(p, X, t)/\partial t|_{dR=0}$$
$$= RB_i^r - \left[\sum_{i=1}^{n} (\partial \ln R_i/\partial \ln X_j)(\partial \ln R/\partial \ln X_j)^{-1} \right] \partial \ln R/\partial t$$

When the technology is input-output separable, $\partial \ln R_i(p, X, t)/\partial \ln X_j = 0$ for all i and j and hence $RB_i^{re} = RB_i^r$.

The general conclusion to be reached is that the primal and dual measures are equivalent if and only if the technology is linearly homogeneous. The primal and dual measures of factor biases defined in terms of cost, revenue, and profit shares are not numerically equal but provide the same bias classification if there is input-output separability. For output biases, the primal and dual cost measures yield the same qualitative results in general. If the technology is not input-output separable, then the scale effect of technological change must be subtracted from the profit and revenue share changes to obtain a valid estimate of the output Hicksian bias in technological change.

APPROACHES TO TOTAL FACTOR PRODUCTIVITY MEASUREMENT

In elementary economic theory, productivity is defined in terms of the rate of output produced per unit of input utilized in the production process. Beyond the single-output, single-input case, however, the definition and measurement of productivity become less straightforward. One approach is to generalize the average productivity concept by comparing an aggregate output index to an aggregate input index. Early productivity measures often were stated in terms of the value of aggregate output per man-hour of labor input. More recently, productivity research has been concerned with *total factor productivity* (TFP) measures, based on comprehensive aggregates of outputs and inputs.

Production theory provides a basis for analyzing the factors that explain output level changes. Generally, the rate of output depends on three factors: the state of technology, or kind of production process utilized; the quantities and types of resources put into the production

DEVELOPMENTS IN PRODUCTION THEORY & PRODUCTIVITY MEASUREMENT 49

process; and the efficiency with which those resources are utilized. A simple example illustrates how each of these factors influences measured productivity. Figure 2-5 shows single-output neoclassical production functions $F^1(X)$ and $F^2(X)$, which represent the technically efficient combinations of input X and output Q for two different production processes. Let Q_1 and Q_2 be the outputs observed in periods 1 and 2, and assume production process F^1 was used in period 1 and F^2 was used in period 2. Since these two observations lie on different rays from the origin, TFP, measured as the average product of factor X, is greater in period 2 than in period 1. This measured productivity change can be attributed to three distinct phenomena. First, Q_1 is below $F^1(X)$, indicating technical inefficiency; efficient production would have resulted in output Q_1'. Second, output Q_2 was produced with a greater input than was Q_1, so there is a difference in scale of production, which explains the difference between, say, Q_1' and Q_2'. Third, production function F^2

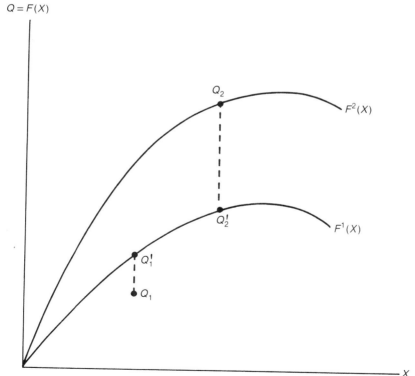

FIGURE 2-5. Productivity differences in the neoclassical model

exhibits a higher total productivity than F^1, which explains the gap between Q_2' and Q_2.

The above example shows that differences in productive efficiency, the scale of production, and the state of technology all may explain part of the observed differences in TFP over time (or across production units at a point in time). Thus, it is clear that however productivity is defined, its measurement must be based on a valid representation of the production technology. Two approaches to TFP measurement have been identified in the literature: the growth accounting (index number) approach and the econometric approach. We briefly outline and compare each of these approaches.

The Growth Accounting Approach and Index Number Theory

Growth accounting was a natural extension of the early research by Kuznets and others to develop consistent national accounts data. The starting point of this approach was the view that, in the absence of technological advance, the growth in total output can be explained in terms of the growth in total factor input. This view was buttressed by the neoclassical theory of production and distribution: competitive equilibrium and constant returns to scale imply that payments to factors exhaust total product. Given technological advance, however, payments to factors would not exhaust total product, and there would remain a residual output not explained by total factor input. This famous "residual," as Domar (1961) termed it, was associated with productivity growth in the early growth accounting literature and remains a fundamental concept in the measurement and explanation of productivity growth. Research by numerous economists has been devoted to measuring and explaining the residual (for example, Kendrick, 1961, 1973; Denison, 1967, 1979; Jorgenson and Griliches, 1967; for literature surveys, see Nadiri, 1970 and Dogramaci, 1981).

The growth accounting approach involves compiling detailed accounts of inputs and outputs, aggregating them into input and output indexes, and using these indexes to calculate a TFP index. In determining aggregate output and aggregate input measures, the method by which the raw data are combined into a manageable number of subaggregates, and in turn reaggregated, is important. The theory of index numbers addresses this issue. Recent advances have made it possible to identify the economic assumptions about the underlying aggregation functions that are implicit in the choice of an indexing procedure (Diewert, 1976, 1981).

For example, the Laspeyres indexing procedure used in much of the early productivity studies by the USDA has been shown to be exact for, or imply, either a linear production function in which all inputs are perfect substitutes, or a Leontief production function in which all inputs are used in fixed proportions. Similarly, the geometric index is exact for a Cobb-Douglas production function, and the Tornqvist-Theil index, which is also an approximation to the Divisia index, is exact for a homogeneous translog production function. Thus, the importance of index number theory lies in the linkage of the growth accounting approach to production theory.

Our discussion of the economic theory of index numbers is taken from the work by Diewert (1976, 1980b, 1981). We briefly define the concept of an index number, relate various functional forms to index number formulas, and discuss the use of these indexes for productivity measurement.[2] The readers are referred to the Diewert citations for formal proofs.

Define the *aggregator function* $F(X)$, which specifies the way in which X_i combine to produce the aggregate output. The producer's problem is to first minimize the cost of achieving a given output level; the solution to this problem defines the cost function C:

$$C(Q, W) \equiv \min_X \{W \cdot X : F(X) \geq Q, X \geq 0\} \tag{2-29}$$

where W and X denote input price and quantity vectors. If the aggregator function F is continuous, increasing, and quasiconcave, C satisfies the conditions outlined in appendix A. Furthermore, if F too is linearly homogeneous, then the cost function can be expressed as

$$C(Q, W) \equiv Q \cdot c(W) \tag{2-30}$$

where $c(W)$ is the unit cost function corresponding to F. Following Diewert, we refer to F and c in this more restricted case as a neoclassical aggregator function and a neoclassical unit cost function respectively.

The Konüs (1924) cost of production index, PI_k is defined as

$$PI_k(W^0, W^1, X) \equiv C[F(X), W^1]/C[F(X), W^0] \tag{2-31}$$

PI_k depends on W^0 and W^1, the vectors of input prices in period 0 and period 1, and X, which is a reference vector of quantities. PI_k is the

---

[2] Our discussions pertain to the producer theory application of index numbers; the consumer theory applications are identical, as noted by Diewert (1981).

minimum cost of producing $F(X)$ given prices W^1 relative to the minimum cost of producing the same quantity when prices are W^0, that is, on the same isoquant. The properties of the cost of production index are determined by the properties of the aggregator and cost function.

Some special cases of this index are noted. First, if all prices in period 1 are proportional to prices in period 0, the index is equal to that factor of proportionality. Second, a theorem due to Malmquist (1953), Pollack (1971), and Samuelson and Swamy (1974) states that PI_k is independent of X if and only if $F(X)$ is homothetic. The reasoning is as follows: if $F(X)$ is homothetic, then there exists a monotonically increasing function G such that $G(F(X))$ is a neoclassical aggregator function (let $F(X) = H(h(X))$, $h(X)$ linear homogeneous, and let $G = H^{-1}$). Then the associated cost function is $C(F(X), W) = G(F(X))c(W)$. After substituting this cost function into equation (2-31) we are left with an index that is equal to the ratio of unit cost functions, $c(W^1)/c(W^0)$.[3]

Fisher's (1922) weak factor reversal property states that given either a price index or a quantity index, the other function can be implicitly defined using the expenditure ratio:

$$PI \cdot QI = W^1 \cdot X^1 / W^0 \cdot X^0 \tag{2-32}$$

(To simplify notation, we use " $\cdot$ " to indicate vector multiplication; it is understood that all vectors are conformable.) The Fisher property can be used to derive the *Laspeyres-Konüs implicit quantity index* as

$$QI_k(W^0, W^1, X^0, X^1, X^0) = (C[F(X^1), W^1]/C[F(X^0), W^0])/PI_k(W^0, W^1, X^0)$$
$$= W^1 \cdot X^1 / W^1 \cdot X^0$$

More generally, the Konüs implicit quantity index $\tilde{QI}_k$ is defined as the cost ratio deflated by the Konüs cost of production index,

$$\tilde{QI}_k(W^0, W^1, X^0, X^1, X)$$
$$\equiv [C(F(X^1), W^1)/C(F(X^0), W^0)]/[C(F(X), W^1)/C(F(X), W^0)] \tag{2-33}$$

Note that the Konüs implicit quantity index is defined relative to a base period price index $PI_k(W^0, W^1, X) = (C[F(X), W^1]/C[F(X), W^0])$ and thus is a function of the base period or reference input vector X. If the reference input vector is chosen to be either X^0 or X^1, then $\tilde{QI}_k$ will be well-behaved in the sense that $\tilde{QI}_k \gtreqless 1$ if $F(X^1) \gtreqless F(X^0)$.

---

[3] For a formal proof, see Diewert (1981).

Following the above arguments for the price index, it follows that if $F(X)$ is homothetic, such that $G(F(X))$ is a neoclassical aggregator function, and producers are cost minimizers, then

$$\tilde{QI}_k(W^0, W^1, X^0, X^1, X) = G(F(X^1))/G(F(X^0)) \tag{2-34}$$

That is, the implicit quantity index is independent of input prices.

A quantity index, QI, is defined to be *exact* for a neoclassical aggregator function, $f(X) = G(F(X))$, if for every $W^0 > 0$, $W^1 > 0$, $X^r > 0$, which is a solution to the problem $\max_X\{f(X): W^r \cdot X \leq W^r \cdot X^r\} = f(X^r)$, for $r = 0, 1$, we have

$$QI(W^0, W^1, X^0, X^1) = f(X^1)/f(X^0) \tag{2-35}$$

A price index is defined to be exact for a neoclassical aggregator function $f(X)$ that has a dual unit cost function, $c(W)$, if

$$PI(W^0, W^1, X^0, X^1) = c(W^1)/c(W^0) \tag{2-36}$$

If QI is exact for a neoclassical aggregator function, then it can be interpreted as an implicit Konüs quantity index, and equation (2-36) can be interpreted as the Konüs price index.

As an example of an exact index number formula, consider the following Laspeyres price and quantity indexes.

$$PI_k(W^0, W^1, X^0) = W^1 \cdot X^0/W^0 \cdot X^0 \tag{2-37}$$

$$QI_k(W^0, W^1, X^0, X^1, X^0) = W^1 \cdot X^1/W^1 \cdot X^0 \tag{2-38}$$

It can be shown that these indexes are exact for a Leontief (fixed coefficient) aggregator function, $f(x) = \min_i[X_i/b_i]$ for all $i = 1, \ldots, N$, where b is a vector of positive constants. Note that $f(X)$ being a fixed coefficient function implies the unit cost function is $c(W) \equiv W \cdot b$, and $W^r \cdot X^r = f(X^r)c(W^r) = C(f(X^r), W^r)$ for $r = 0, 1$. Furthermore, since we are assuming cost minimizing behavior, we can employ Shephard's lemma:

$$X^r/W^r \cdot X^r = \nabla_W C[f(X^r), W^r]/C[f(X^r), W^r] = \nabla c(W^r)/c(W^r) \tag{2-39}$$

Using equation (2-39), we can rewrite equation (2-37) as

$$\begin{aligned} PI_L &\equiv W^1 \cdot X^0/W^0 \cdot X^0 \\ &= W^1 \cdot [\nabla c(W^0)/c(W^0)] \\ &= W^1 \cdot b/c(W^0) \\ &= c(W^1)/c(W^0) \end{aligned} \tag{2-40}$$

Thus, since PI_L is exact for the fixed-coefficient aggregator function, QI_L also is exact for this aggregator function. A similar argument can be used to show that the Paasche indexes also are exact for the Leontief function, and that both the Laspeyres and Paasche indexes also are exact for the linear aggregator function.

Other important examples of indexes that are exact for well-known production functions include the geometric index, which is exact for a Cobb-Douglas aggregator function, and the Tornqvist-Divisia index (discussed below), which is exact for a linear homogeneous translog production function. Generally, one would not want to select an indexing procedure (such as the Paasche or Laspeyres) that implies unnecessarily restrictive assumptions about the substitutability of inputs. A better alternative would be to use an index that is exact for a linear homogeneous flexible functional form for the aggregator function. Indexes with this latter property have been termed *superlative* by Diewert (1976).

One of the most popular of the superlative quantity indexes is the Tornqvist-Theil quantity index, expressed in logarithmic form as

$$\ln QI_T \equiv \ln [f(X^1)/f(X^0)] = \frac{1}{2}\sum_i (S_i^1 + S_i^0) \ln (X_i^1/X_i^0) \qquad (2\text{-}41)$$

where S_i^j is the share of the ith input in total payments for period j. This index is a discrete approximation to a Divisia index (discussed below), which is frequently used in the measurement of TFP. The flexible functional form for which equation (2-41) is exact is the linear homogeneous translog function. Similarly the Tornqvist-Theil price index,

$$\ln PI_T \equiv \ln [C(W^1)/C(W^0)] = \frac{1}{2}\sum_i (S_i^1 + S_i^0) \ln (W_i^1/W_i^0) \qquad (2\text{-}42)$$

is exact for a translog unit cost function. Further discussion of these indexes can be found in appendix C.

Using the factor reversal rule (2-32), one also can determine a Tornqvist-Theil implicit price or quantity index using the observable prices and quantities each period and the index defined in equations (2-41) or (2-42). The implicit Tornqvist-Theil index will differ from the corresponding explicit index because the translog function is not self-dual.[4] Diewert showed, however, that these differences are small

---

[4] If the homogeneous translog function is restricted to the Cobb-Douglas functional form, then it is self-dual.

provided that the one-period variation in the prices and quantities is not too great.

If the aggregator functions are nonhomothetic, the Tornqvist-Theil index is still attractive, since the translog function can provide a second-order differential approximation to an arbitrary twice-differentiable function (theorems 26 and 27 in Diewert, 1981). Since nonhomothetic production functions are not characterized by a constant distance between any given pair of isoquants along a ray from the origin, input bundles are not directly comparable without reference to the output levels in each period. Diewert's results show that when the reference point is the isoquant for the geometric mean of output in the two periods, the Tornqvist-Theil index is exact for a nonhomothetic translog function.

Use of Index Numbers for TFP Measurement.[5] The Tornqvist-Theil indexing procedure is often used to obtain an index of TFP because it can be viewed as a discrete approximation to the continuous Divisia index (Hulten, 1973; Diewert, 1976). The Divisia indexes of aggregate output (Q) and aggregate input (X) are defined in terms of proportional rates of growth,

$$\dot{Q} = \sum_j \left(\frac{p_j Q_j}{\sum_i p_i Q_i} \right) \dot{Q}_j \tag{2-43}$$

$$\dot{X} = \sum_j \left(\frac{W_j X_j}{\sum_i W_i X_i} \right) \dot{X}_i \tag{2-44}$$

Since TFP = Q/X, the proportionate rate of growth of TFP is

$$\dot{\text{TFP}} = \dot{Q} - \dot{X} \tag{2-45}$$

The Tornqvist-Theil quantity index given in equation (2-41) can be used to approximate equations (2-43) and (2-44) as:

$$\ln(Q_t/Q_{t-1}) = \frac{1}{2} \sum_j (S_{jt} + S_{jt-1}) \ln\left(\frac{Q_{jt}}{Q_{jt-1}}\right) \tag{2-43'}$$

$$\ln(X_t/X_{t-1}) = \frac{1}{2} \sum_i (S_{it} + S_{it-1}) \ln\left(\frac{X_{it}}{X_{it-1}}\right) \tag{2-44'}$$

---

[5] The material in this section draws on Diewert (1980b, 1981).

and the discrete approximation to equation (2-45) is

$$\ln(\text{TFP}_t/\text{TFP}_{t-1}) = \ln(Q_t/Q_{t-1}) - \ln(X_t/X_{t-1}) \qquad (2\text{-}45')$$

Choosing the index to equal 100.00 in a particular year and accumulating the measure according to equations (2-45') provides an index of TFP. Diewert (1981) indicated that the Divisia approach does not have a unique estimate of TFP because the Tornqvist-Theil approximation is only one way to approximate equations (2-43) and (2-44).

TFP defined by equation (2-45') can be used to approximate technological progress if producers behave competitively and if the production technology is input-output separable; exhibits extended Hicks-neutral technological change; and is linearly homogeneous. Using the equation developed earlier for the primal rate of technological change (2-9) and equations (2-43) and (2-44), we have

$$\frac{\partial \ln F}{\partial t} = \dot{Q} - \left(\frac{\partial \ln C}{\partial \ln Q}\right) - 1_{\dot{x}} \qquad (2\text{-}46)$$

The second term on the right-hand side of equation (2-46) is the Divisia index of aggregate input growth defined in equation (2-44) times the scale elasticity; the first term is the growth rate of aggregate output. Therefore, equation (2-46) gives the proportional rate of output growth unexplained by the growth in inputs. If the production function is characterized by input-output separability and extended Hicks-neutral technological change so that the aggregate technology can be written $Q(t) = A(t) \cdot F(X)$ (or alternatively $A(t) = Q(t)/F(X)$), and constant returns to scale holds so that $\partial \ln C/\partial \ln Q = 1$, equation (2-46) becomes

$$\dot{A} = \dot{Q} - \dot{X} \qquad (2\text{-}47)$$

Thus, TFP defined by equation (2-45') measures technological change or a shift in the production function due to technical progress under the following assumptions: competitive behavior, constant returns to scale, extended Hicks-neutral technological change, and input-output separability.

An alternative approach to measuring changes in TFP is based on a direct application of the Tornqvist-Theil index number theory, rather than indirectly as an approximation to continuous-time derivatives. This alternative approach leads to an exact formula for TFP that is suitable for discrete data, but the formula is contingent on the cost function being of the translog form (see Diewert, 1981 and appendix C to this chapter).

Thus, the exact index number approach to TFP measurement also involves an approximation, as it is unlikely that the technology can be precisely represented by a translog cost function over the entire range of prices and quantities.

The exact index number formula provides a measure of the percentage change in variable costs at time t that cannot be explained by changes in inputs or outputs or changes in variable input prices. Since the Tornqvist-Theil indexes are based on cost and revenue shares, and we use Shephard's lemma, the exact index number approach implicitly assumes competitive behavior.

More recently, Caves, Christensen, and Diewert (1982a) have shown that the translog input and output indexes are exact for the geometric mean of the Malmquist input and output indexes, respectively, for periods $t = 0$ and $t = 1$, "when the underlying aggregator functions are both translog (not necessarily homogeneous), but with different parameters" (p. 1411). This result implies that the Tornqvist-Theil productivity index is superlative in a considerably more general sense than shown by Diewert (1976).

The Econometric Approach

The econometric approach to productivity measurement is based on econometric estimation of the production technology. As the example in figure 2-5 showed, technological change can be inferred from shifts in the production function, and under the assumption that production is efficient (or that the degree of inefficiency is constant), productivity change and technological change are synonymous. As discussed above, duality relations provide a means by which to infer production function shifts from direct estimation of the production function, or from estimates of the dual cost or profit function. To grasp this, consider the conventional production function approach in which a production function is specified in terms of inputs X_{it} and time t is used as a proxy for technological change. Let the production function be specified in the Cobb-Douglas form with a constant exponential rate of extended Hicks-neutral technological change:

$Q_t = A_t X_{1t}^{\alpha_1} \ldots X_{nt}^{\alpha_n}$ where $A_t = e^{\alpha_0 t}$

Thus, the measured rate of productivity growth under constant returns to scale is, following equation (2-47)

$$\dot{\text{TFP}} = \frac{\partial \ln A_t}{\partial t} = \alpha_0$$

As shown earlier, a corresponding profit function exists of the form:

$$\pi_t = B_t P_t^\gamma W_{1t}^{\beta_1} \ldots W_m^{\beta_m} X_{m+1}^{\beta_{m+1}} \ldots X_n^{\beta_n}$$

where $B_t = e^{\beta_0 t}$, P_t is the price of Q_t, W_{it} is the price of X_{it}, $\beta_0 = -\alpha_0/(1-\mu)$ where $\mu = \sum_{i=1}^m \alpha_i$, and $X_{m+1}, \ldots, X_n$ are fixed factors of production. Therefore, it follows that the rate of TFP growth can be inferred from econometric estimates of either the primal (production) function or the dual (cost or profit) function.

The translog model also can be used to measure changes in total factor productivity in a manner analogous to the Cobb-Douglas model. For example, a time trend can be added to equation (2-3):

$$\ln Q = \alpha_0 + \sum_i \alpha_i \ln X_i + \frac{1}{2}\sum_i \sum_j \alpha_{ij} \ln X_i \ln X_j + \beta_0 t + \frac{1}{2}\beta_1 t^2 + t \sum_i \gamma_i \ln X_i$$

Recall from the previous section (equation (2-45)) that TFP is defined as Q/X, so that extended Hicks-neutral technical change must be assumed. As specified in the translog model above, technological change is not extended Hicks neutral, so the production function cannot be used to measure TFP. The parametric restrictions for extended Hicks neutrality are $\gamma_i = 0$ for all i. If the restrictions are not imposed, one can calculate an alternative measure of TFP growth, $\text{TFP}^* = \dot{Q} - \dot{X}$, where $\dot{Q}$ and $\dot{X}$ are calculated without assuming neutral technical change. Using the above function,

$$\dot{Q} = \frac{d\ln Q}{dt} = \sum_i \alpha_i \frac{d\ln X_i}{dt} + \sum_j \alpha_{ij} \ln X_j \frac{d\ln X_i}{dt} + t\sum_i \gamma_i \frac{d\ln X_i}{\partial t}$$
$$+ \beta_0 + \beta_1 t + \sum_i \gamma_i \ln X_i \qquad (2\text{-}48)$$

and

$$\frac{\partial \ln Q}{\partial \ln X_i} = \frac{W_i X_i}{pQ} = \frac{C}{pQ} S_i = \alpha_i + \sum_j \alpha_{ij} \ln X_j + \gamma_i t \qquad (2\text{-}49)$$

where S_i is the factor cost share and the equilibrium condition $p = \partial C/\partial Q$ was used. From equations (2-44), (2-48), and (2-49),

$$\dot{Q} = \sum_i \left(\alpha_i + \sum_j \alpha_{ij} \ln X_j + \gamma_i t\right) \frac{d\ln X_i}{dt} + \beta_0 + \beta_1 t + \Sigma \gamma_i \ln X_i$$
$$= \frac{C}{pQ}\dot{X} + \beta_0 + \beta_1 t + \sum_i \gamma_i \ln X_i$$

Thus the translog production function model implies

$$\text{T}\dot{\text{F}}\text{P}^* = \dot{Q} - \dot{X} = \left[\left(\frac{\partial \ln C}{\partial \ln Q}\right)^{-1} - 1\right]\dot{X} + \beta_0 + \beta_1 t + \Sigma \gamma_i \ln X_i$$

With constant returns to scale, $\partial \ln C/\partial \ln Q = 1$ and

$$\text{T}\dot{\text{F}}\text{P}^* = \beta_0 + \beta_1 t + \Sigma \gamma_i \ln X_i$$

If extended Hicks neutrality is assumed, then $\gamma_i = 0$ for all i and $\text{T}\dot{\text{F}}\text{P} = \beta_0 + \beta_1 t$.

As pointed out above, duality theory ensures that productivity change also can be measured in terms of the cost or profit function. Let $C(Q, W, t)$ be the cost function. Using equations (2-9), (2-10), and (2-11), we have

$$-\frac{\partial \ln C}{\partial t} = \left(\frac{\partial \ln C}{\partial \ln Q}\right)\dot{Q} - \dot{X}$$

and hence combined with the definition of TFP*,

$$\text{T}\dot{\text{F}}\text{P}^* = -\frac{\partial \ln C}{\partial t} + \left(1 - \frac{\partial \ln C}{\partial \ln Q}\right)\dot{Q} \qquad (2\text{-}50)$$

thus showing that $\text{T}\dot{\text{F}}\text{P}^* = -\partial \ln C/\partial t$ under constant returns to scale.

If extended Hicks neutrality is assumed, then the cost function is of the form $C = g(Q, W)B(t)$ and

$$\text{T}\dot{\text{F}}\text{P} = -\dot{B} + \left(1 - \frac{\partial \ln C}{\partial \ln Q}\right)\dot{Q}$$

Thus, both T$\dot{\text{F}}$P* and T$\dot{\text{F}}$P can be measured as the dual rate of technological change plus a scale effect equal to the second term on the right-hand side of equation (2-50). These results show that, by specifying and econometrically estimating either a production or a cost function, the rate of TFP growth can be estimated. For example, using the translog cost function specified in appendix A, parametric expressions for $-\partial \ln C/\partial t$ and $\partial \ln C/\partial \ln Q$ can be obtained, enabling T$\dot{\text{F}}$P* or T$\dot{\text{F}}$P to be computed.

A significant advance in the econometric approach was achieved by combining developments in duality theory and flexible functional forms with econometric theory to improve estimation efficiency. The methodology used by most studies employing flexible functional forms was introduced in Berndt and Christensen's (1973b) seminal paper on the translog production function. This methodology involves specifying a function representing the technology (a production, cost, or profit function) and econometrically estimating it or its derivatives. Let the representation of the technology be

$$y = f(p, z, t; \beta) + \epsilon \qquad (2\text{-}51)$$

where y is the dependent variable (output, cost, or profit); p is a vector of variable arguments (variable input quantities, factor prices, or output and factor prices); z is a vector of fixed factors or outputs; ϵ is a random error for econometric estimation; and β is a vector of unknown parameters to be estimated. The first-order conditions for profit maximization or cost minimization imply the system of equations

$$S_i = g_i(p, z, t; \beta_i) + \epsilon_i, \qquad i = 1, \ldots, n \qquad (2\text{-}52)$$

where β_i is a subset of the parameters in β. S_i may be a cost share, a function of quantities, or a profit or revenue share, depending on the model. In its most general case, the model can be viewed as a subsystem of a larger structural system. The econometric task is to obtain estimates of the β vector.

The econometric issues that arise in estimating equations (2-51) and (2-52) involve (a) the determination of the identification status of the system; (b) the establishment of statistical assumptions for the error terms; (c) the choice of functional form for equation (2-51); and (d) the choice of an appropriate estimator. The identification question is usually resolved by assuming that the model is either a self-contained system and that p, z, and t are exogenous or a predetermined system, in which case the model can be interpreted as a multivariate (but not simultaneous) regression problem (either linear or nonlinear in the parameters, but additive in the errors). Alternatively, a set of exogenous and predetermined variables are specified, identification is assumed, and estimation methods such as three-stage least squares are used that account for the possible endogeneity of some of the right-hand-side variables. In both cases, the system given by equations (2-51) and (2-52) is overidentified, because the parameter vector β, or a subset of it, occurs in each of the $n + 1$ equations. In both the multivariate regression model and the simultaneous model, the conventional assumption for the error terms is that they satisfy the seemingly unrelated regression (SUR) model, that is, that the errors of each equation are independently and identically distributed, but that there are constant across-equation correlations. Although some studies have used models that are nonlinear in the parameters, most studies have used flexible functional forms that have the convenient property that the function (2-51) and its first-order conditions (2-52) are linear in the parameters.

Given the SUR structure for the covariance matrix of the system, several alternative estimation approaches are possible. One is to make the further assumption of normality of the errors and to apply the maximum likelihood (MLE) technique to estimate the unknown parameters (β and the elements of the covariance matrix) of the system. This generally involves solution of a nonlinear optimization problem and can be costly. In the linear-in-parameters case, MLE is asymptotically equivalent to generalized least squares (GLS), with a consistently and efficiently estimated covariance matrix. To obtain such a feasible GLS estimator, one can apply what has become known as the "iterated SUR" procedure (for example, see Magnus, 1978). The first step is to use least squares to estimate the system, and construct a consistent estimate of the covariance matrix from the least squares residuals. Next the SUR estimator is computed, using the estimated covariance matrix from the first step, and a new covariance matrix estimate is constructed. This procedure continues iterating from estimates of β to estimates of the covariance matrix, until it converges, that is, until the parameter estimates change less than a specified tolerance.

If the S_i are share equations, such as cost shares, then they satisfy an adding-up restriction, such as $\Sigma_i S_i = 1$, with probability one. This means that the errors sum to zero, $\Sigma_i \epsilon_i = 0$, and that the system error covariance matrix of the system of share equations is singular. Thus, the SUR estimator, which requires the inverse of the error covariance matrix, does not exist for the share equation subsystem. Similarly, if the technology equation can be derived from the first-order conditions (for example, if the S_i are cost shares and y is total cost, $y\Sigma_i S_i = y$), an adding-up constraint holds for the entire system and the covariance matrix is singular. The solution to this singularity problem is to observe that one of the share equations is redundant, since the nth equation's parameters can be solved from the other $n - 1$ equations. But this raises the problem of knowing which equation to drop for estimation. Because the iterated SUR estimates converge to MLEs and the MLEs are unique, it follows that the iterated SUR estimates are invariant with respect to which share equation is dropped for estimation (the skeptical reader can readily verify this computationally using standard econometric software programs).

The statistically most efficient estimator is obtained when all relevant data and *a priori* information are used. In the SUR model, efficiency requires estimating the full system while imposing the across-equation restrictions implied by profit maximization or cost minimization. However, in some cases it may not be possible to estimate the full system. For example, when equation (2-50) is written as a flexible functional form, it may contain many parameters and there may be insufficient degrees of freedom for its estimation. Yet in such a case the first-order conditions, represented by the subsystem (2-52), contain fewer parameters and may be estimable. If the parameters needed for the analysis are contained in the subvectors β_i, then estimation can proceed using the subsystem, although the resulting parameter estimates will be less efficient than those obtained from the full system.

The adoption of flexible functional forms to represent dual cost or profit functions also means that the theoretical properties of linear homogeneity in prices, monotonicity, and curvature, can be tested or imposed on the model. The various parametric restrictions that are involved are discussed in appendixes A and B to this chapter for the translog cost and profit functions.

Comparing the Growth Accounting and Econometric Approaches

Both the growth accounting and the econometric approaches have strengths and limitations. As in all applied research, the choice between the two approaches should be based on the research objectives, the data requirements, the availability of data, and the appropriateness of assumptions.

In deriving the Divisia index numbers and their relation to production theory, several strong assumptions were made about the technology and the industry, namely, Hicks-neutral technological change, constant returns to scale, and long-run competitive industry equilibrium. Clearly, these assumptions are not likely to be appropriate in many cases, and thus the interpretation of the index numbers can be questioned. However, as Caves, Christensen, and Diewert (1982a,b) indicate, the Tornqvist-Theil index numbers are also superlative for some very general production structures, that is, nonhomogeneous and nonconstant returns to scale. Future research may be able to show that conventional index numbers are more generally valid than we now believe.

Another disadvantage of the index number approach is that the calculations are not based on statistical theory, so statistical methods cannot be used to evaluate their reliability. However, index number calculations can be used when econometric methods are infeasible. For example, very detailed data with many inputs and output categories can be used regardless of the number of observations over time; there are no degrees-of-freedom problems or statistical reliability problems in working with small samples.

The econometric approach allows the researcher to relax some of the assumptions required for index numbers, but only at the cost of necessitating other assumptions. An econometric production function, such as the translog, can be estimated without making any assumptions about neutrality of technological change, returns to scale, or industry equilibrium. Moreover, because the estimated model has known statistical properties, confidence intervals can be constructed around the estimates. However, estimation of a production function such as the translog with aggregate data requires that the outputs be aggregated into a single index, so input-output separability must be assumed. For sufficient degrees of freedom, and to mitigate multicollinearity problems, it is necessary to aggregate input data into a small number of categories, which can be done only under input separability assumptions. Moreover, a strong assumption is typically made about the nature of technological change, namely, that it can be represented as a function of time, as in the translog example above. In order to estimate cost or profit function models, the additional assumptions of competitive pricing and efficient input utilization must be made. Finally, assumptions about the statistical properties of the data must be made.

Both approaches can be useful and should be considered appropriate. The fact that both index numbers and econometric models are linked through production theory means that each approach can be used to evaluate the other. In the next section, we discuss a methodology for making interspatial and intertemporal productivity comparisons that exploits this link between the two approaches.

MEASURING INTERSPATIAL AND INTERTEMPORAL TOTAL FACTOR PRODUCTIVITY

In this section we present a methodology developed by Denny and Fuss (1980; 1983a,b) to analyze sources of intertemporal and/or interspatial differences in productivity. Assuming efficient production, intertemporal productivity differences can be attributed to technological change. Interspatial productivity differences arise from differences in the (efficient) production functions in each region.

Let two different sets of observations on input use be denoted by the vectors X^r and X^s. The production function, $Q = f(X_1, \ldots, X_n)$ can be approximated by a second-order Taylor series expansion around X^r,

$$Q = Q^r + \sum_i f_i^r [X_i - X_i^r] + \frac{1}{2}\Sigma\Sigma f_{ij}^r [X_i - X_i^r][X_j - X_j^r] + R_3^r \qquad (2\text{-}53)$$

where $R_3{}^r$ is the third-order remainder term. By replacing X_i with X_i^s in equation (2-53), an approximation to Q^s is obtained. In a similar fashion, an approximation to Q^r can be obtained by initially expanding the production function around X^s. The difference between the two approximations is

$$Q^s - Q^r = \sum_i \frac{1}{2}(f_i^r + f_i^s)[X_i^s - X_i^r] + \sum_i \sum_j \frac{1}{4}(f_{ij}^s - f_{ij}^r)[X_i^s - X_i^r]$$
$$\cdot [X_j^s - X_j^r] + \frac{1}{2}[R_3^r - R_3^s] \tag{2-54}$$

If $f(X)$ is either linear or quadratic, equation (2-54) simplifies to

$$Q^s - Q^r = \sum_i \frac{1}{2}(f_i^r + f_i^s)(X_i^s - X_i^r) \tag{2-55}$$

Equation (2-55) corresponds to Diewert's (1976) "quadratic lemma" and states that, if the functions are quadratic, the difference in output levels is a function of the first derivatives and the input levels.

Equation (2-55) can be evaluated on an index number basis only by replacing the f_i with an expression containing only prices and quantities. Expressing Q and X in logarithms, it follows that in competitive equilibrium $f_i^r = S_i^r$, where S_i^r is the factor cost share of X_i^r. Thus equation (2-55) can be expressed in logarithmic form as

$$\ln Q^s - \ln Q^r = \frac{1}{2}\sum_i (S_i^r + S_i^s)(\ln X_i^s - \ln X_i^r) \tag{2-56}$$

Equation (2-56) is in the form of the Tornqvist approximation of the Divisia index discussed earlier in this chapter.

For analyzing productivity differences across different regions and time periods, suppose the production function can be written $\ln Q = f(\ln X_1, \ldots, \ln X_n, T, D)$, where T is a technological change variable and D denotes a vector of regional variables, or "efficiency difference indicators." In order to use the growth accounting method for productivity measurement, we assume the function f to be a quadratic function for which only the zero and first order parameters vary by regions. The logarithmic version of equation (2-55) is then

$$\Delta \ln Q = \frac{1}{2}(f_i^r + f_j^s)(\ln X_i^s - \ln X_i^r) + \rho_{sr} + \tau_{sr} \tag{2-57}$$

where

$$\rho_{sr} = \frac{1}{2}(f_D^r + f_D^s)(D^s - D^r) \tag{2-58}$$

is the interspatial effect and

$$\tau_{sr} = \frac{1}{2}(f_T^r + f_T^s)(T^s - T^r) \tag{2-59}$$

is the intertemporal effect; f_i, f_D, and f_T denote $\partial f/\partial \ln X_i$, $\partial f/\partial \ln D$, and $\partial f/\partial \ln T$, respectively. Equation (2-57) is a general second-order expression that can be used to account for productivity changes. Differences in output levels can be attributed to differences in input levels, regional effects, and intertemporal (technological change) effects. Estimation of equation (2-57) requires econometric estimation of the first derivatives of the function f.

Assuming competitive markets and a constant returns to scale technology, $f_i = S_i$, and equation (2-57) can be expressed as

$$\Delta \ln Q = \frac{1}{2}(S_i^r + S_j^s)(\ln X_i^s - \ln X_i^r) + \rho_{sr} + \tau_{sr} \tag{2-60}$$

Solving for τ_{sr} within the same region ($D^w = D^r$) yields the discrete Divisia index of intertemporal productivity differences; solving for ρ_{sr} within the same time period yields the discrete Divisia index of interspatial productivity differences.

Denny and Fuss showed that if the second-order parameters in a Taylor series approximation (f_{ij}^r in equation (2-53)) vary across regions and time periods, the intertemporal and interspatial productivity measures, τ_{sr} and ρ_{sr}, from the second-order (or econometric) method will differ from the discrete Divisia index approximations. The bias that results from using the discrete Divisia intertemporal index for a given region is

$$B_\tau = \frac{1}{4}\left[\sum_i \sum_j (f_{ij}^r - f_{ij}^s)(\ln X_i^s - \ln X_i^r)(\ln X_j^s - \ln X_j^r)\right]$$

for s, r denoting two time periods; and the interspatial bias is

$$B_p = \frac{1}{4}\left[\sum_i \sum_j (f_{ij}^r - f_{ij}^s)(\ln X_i^s - \ln X_i^r)(\ln X_j^s - \ln X_j^r)\right]$$

for s, r denoting two different regions.

EXPLAINING TECHNOLOGICAL CHANGE

The evolution of on-farm technology is widely believed to be the product of both the supply and demand for technological innovations that reduce production costs. Although some technological change is clearly the result of independent scientific discovery, most of the theoretical and empirical research has been devoted to the explanation of technological change in response to relative price changes. In *The Theory of Wages*, Hicks suggested that technological innovations "are the result of a change in the relative prices of the factors" (Hicks, 1963, p. 125), and referred to these as *induced innovations*. Since Hicks' time, a considerable theoretical literature on induced innovation has evolved (for a survey of this literature, see Binswanger, 1978).

In the twentieth century in the United States and elsewhere, the supply of agricultural technology has come from both public institutions, such as the land grant universities, and from private industry. Schultz (1964) argued that the economies of scale in the production of new agricultural technology, especially in terms of basic research, and the positive social externalities associated with such research, justify public subsidization of investment in agricultural research. This reasoning is supported by studies that find that the rate of return to investments in agricultural research has been above market rates of return (Schuh and Tollini, 1979) although the validity of such findings can be questioned, as Antle notes in chapter 12.

The importance of the public sector in agricultural research and development raises problems for the Hicksian theory of induced innovation based on the innovative responses of private firms to relative prices. Hayami and Ruttan (1971) argued that Hicks' induced innovation theory can be generalized to both the public and private sectors. The measurement of the productivity of agricultural research, and the allocation of resources to research, is therefore a fundamental issue in the explanation of agricultural productivity. Griliches (1964), Evenson (1968), Hayami and Ruttan (1971), and Evenson and Kislev (1975) used measures of agricultural research activity to explain agricultural productivity differences within and across countries.

Once invented, a technological innovation must be adopted by producers, and it is here that the demand side of the technology market plays an important role in agricultural productivity. Beginning with Griliches' (1957) seminal research on the adoption of hybrid corn varieties, many studies have been devoted to explaining technology adoption and technological change in agriculture. In addition to Griliches' explanation of adoption as a function of profitability, numerous other variables have been introduced: human capital and farmer education

(Griliches, 1964; Welch, 1970; Huffman, 1974; Petzel, 1978; Schultz, 1975; Lockheed, Jamison, and Lau, 1980); agricultural extension (Griliches, 1964; Huffman, 1974; Evenson and Kislev, 1975); transport and communications infrastructure (Easter, Abel, and Norton, 1977; von Oppen, Rao, and Rao, 1985; Antle, 1984b); and risk (Roumasset, Boussard, and Singh, 1979; Feder, 1982).

DYNAMICS

Economic theory suggests several reasons why static production models may not provide an adequate basis for the measurement and explanation of agricultural productivity. Hayami and Ruttan (1971) described induced innovation in agriculture as a process of "dynamic factor substitution." Indeed, the process of technological change, in agriculture as well as in other sectors of the economy, involves investment in new kinds of capital. Thus, technological change must involve capital investment (in both physical and human capital) and multiperiod decision making. In addition, agricultural production itself is inherently dynamic due to its dependence on biological processes and the long time periods between input utilization and output realizations. In this respect, agricultural production processes are fundamentally different from many industrial processes. The time dimension in agricultural production generally means that aggregate agricultural productivity analysis requires aggregation over time as well as across producing units.

Another reason dynamic models are needed for agricultural productivity analysis is that production activities are likely to affect not only current output but also future output and social welfare due to the externalities associated with agricultural production. Obvious examples of external costs are pesticide pollution, pest resistance, and soil erosion. But equally important may be positive externalities associated with learning-by-doing and technological change. These concerns motivate Archibald's dynamic analysis in chapter 13.

A useful classification of dynamic models into three "generations" has been made by Berndt, Morrison, and Watkins (1981). First generation models are partial adjustment models, as originated by Nerlove and applied extensively in the agricultural economics and economics literature. The partial adjustment model is based on the assumption that the value of a choice variable in period t, say x_t, differs from its value in $t-1$ in proportion to the difference between x_{t-1} and the optimal value x_t^*. Thus

$$x_t - x_{t-1} = \alpha(x_t^* - x_{t-1}), \quad 0 < \alpha < 1 \qquad (2\text{-}61)$$

Solving this difference equation gives

$$x_t = \alpha \sum_{i=1}^{\infty} (1-\alpha)^{i-1} x_{t-i}^* \qquad (2\text{-}62)$$

Although the behavior suggested by this model has intuitive appeal, it leaves unanswered the fundamental question of whether or not this behavior is consistent with the firm's economic choice problem. In general, it is found to be inconsistent. However, if the firm chooses x_t to minimize the quadratic loss function

$$L = \frac{1}{2}(x_t - x_t^*)^2 + \frac{1}{2}(1-\alpha)(x_t - x_{t-1})^2 \qquad (2\text{-}63)$$

it can be shown that behavior will satisfy equation (2-61). (Note that the first term in equation (2-63) can be interpreted as the cost of a suboptimal choice of x_t and the second term represents the cost of adjustment from period $(t-1)$ to t.) But introducing equation (2-63) to rationalize equation (2-61) does not explain how the firm determines x_t^*. If x_t^* is derived from the solution to the firm's optimization problem, then x_t should also be. This is a serious shortcoming of the partial adjustment model.

Second generation models are the dual cost and profit function models with fixed factors similar to the models estimated in chapter 5. Although this generation of models is not explicitly dynamic, these models can be used to estimate both short-run and long-run input demand and output supply relationships and to approximate the long-run equilibrium levels of the fixed factors (see Brown and Christensen, 1981; Hazilla and Kopp, 1986b). Although both first and second generation models provide some insight into dynamic behavior, they are not based on solutions to dynamic optimization problems.

Third generation dynamic models are those based on dynamic optimization paradigms. These models consist of an objective function and a specification of the technology. If perfect foresight is not assumed, these models also require specification of the expectations formation process.

One of the most important third generation models is known as the cost-of-adjustment model. It originated with Lucas' (1967) study of investment behavior, in which he formulated the production function as

$$Q(t) = f[L(t), K(t), I(t), t]$$

where Q is output, L is labor input, K is capital input, and I is the rate of investment in capital, that is, $I(t) = dK(t)/dt + \delta K(t)$, where δ is the capital depreciation rate. Lucas hypothesized that output would be decreasing in I due to disruptions to the production process, hence there was a cost of adjusting the capital stock in the form of forgone output, in addition to the explicit cost of the new capital itself. Lucas assumed the firm chose the sequence of labor and investment inputs over time that maximized the present discounted value of its profits.

The quadratic cost-of-adjustment model is the one most frequently used in applied research because it yields closed-form solutions to the dynamic optimization problem (for example, see Hansen and Sargent, 1980; Epstein and Denny, 1982; Eckstein, 1984; Epstein and Yatchew, 1985). More generally, dynamic production models have been specified as quadratic functions of inputs because the quadratic model is one of the few dynamic models with a closed-form solution (see Chow, 1975; Bertsekas, 1976). To illustrate concretely the issues that arise in formulating and analyzing dynamic production models, we consider here a simple two-period linear-quadratic model. This model is developed to illustrate the key components of dynamic models and their solution.

The production function in the first period is a linear function:

$$Q_1 = \alpha_0 + \alpha_1 x_1 + \epsilon_1 \tag{2-64}$$

where Q_1 is first-period output, x_1 is input, and ϵ_1 is a random disturbance with mean zero. The production function in period 2 is the quadratic function:

$$Q_2 = \beta_0 + \beta_1 x_2 + \frac{1}{2}\beta_2 x_2^2 + \beta_3 Q_1 + \beta_4 x_2 Q_1 + \frac{1}{2}\beta_5 Q_1^2 + \epsilon_2 \tag{2-65}$$

where x_2 is period 2 input and ϵ_2 is the random shock to production in period 2. Note that concavity of Q_2 in x_2 and Q_1 requires $\beta_2 < 0$, $\beta_5 < 0$, and $\beta_2 \beta_5 > \beta_4^2$.

In Antle's terminology (see chapter 12), equations (2-64) and (2-65) represent a model of *output dymamics*, because output in period 2 depends on output in period 1. Several interpretations may be given to this simple dynamic structure. The two production functions can represent two sequential stages in one production process, with Q_1 being interpreted as an intermediate product that is combined with other inputs (x_2) to produce the final product Q_2. An alternative interpretation is that Q_1 and Q_2 are outputs from two distinct production processes, and productivity in period 2 depends on output in period 1 because of crop rotation, learning-by-doing, or some other dynamic technological externality.

A special case of this model can be interpreted as a cost-of-adjustment model. Set $\alpha_0 = 0$, $\alpha_1 = 1$, and $\epsilon_1 = 0$, so that $Q_1 = x_1$, and write

$$Q_2 = \gamma_0 + \gamma_1 x_2 + \gamma_2 \Delta x_2 + \gamma_3 (\Delta x_2)^2 \tag{2-66}$$

where $\Delta x_2 = x_2 - x_1$. Substituting for Δx_2 and simplifying shows that equation (2-66) is a special case of equations (2-64) and (2-65) with certain parameter restrictions. This shows that quadratic cost-of-adjustment models can be formulated as special cases of more general dynamic models. The parameter restrictions implied by the cost-of-adjustment interpretation can be used as a basis for a test of the cost-of-adjustment hypothesis (for example, see Eckstein, 1984).

For this illustrative example, let us interpret the production process given by equations (2-64) and (2-65) as a two-stage production process with Q_1 as an intermediate output and Q_2 as a final output that is sold. Economic returns to the firm are

$$\pi = pQ_2 - w_1 x_1 - w_2 x_2 \tag{2-67}$$

where p is the value of output and w_t, $t = 1, 2$ are the input prices. Without loss of generality, assume prices are independently distributed. The risk-neutral manager chooses inputs x_1 and x_2 to maximize the present value of expected profit. Assume the decision maker solves the input choice problem sequentially using all information available at the time each decision is made. At the beginning of stage 1, w_1 is known but p, w_2, Q_1, and Q_2 are unknown. At the beginning of stage 2, w_2 and Q_1 are known, but p and Q_2 are not yet known. Therefore, the firm's price and output expectations must be part of the dynamic production model. Generally, the firm manager can be assumed to formulate production plans based on his or her subjective distributions of unknown variables, such as future prices and outputs. Over time, firm managers can be expected to update their subjective distributions, based on economic and technological information they acquire. One of the fundamental problems in economics is to know how firms formulate and update their expectations, and how their expectations relate to observable economic data. We return to this issue below in our discussion of rational expectations.

With the production problem thus formulated, the input choice problem is an optimal stochastic control problem and can be solved by applying the dynamic programming algorithm (Bertsekas, 1976). In stage 2 the optimal value for x_2 is found by maximizing expected profit subject to the information available at the beginning of stage 2. Output Q_1 and input price w_2 are known when x_2 is chosen so the problem is

$$\max_{x_2} E_2[\pi | Q_1, w_2] \qquad (2\text{-}68)$$

where E_2 denotes the expectation operator based on the decision maker's subjective probability distribution functions of p and Q_2 at the beginning of stage 2, given Q_1 and w_2. Assuming p and Q_2 are independently distributed, the solution x_2^0 is linear in $w_2/E_2[p]$ and Q_1:

$$x_2^0 = \frac{1}{\beta_2}\left(\frac{w_2}{E_2[p]} - \beta_4 Q_1 - \beta_1\right) \qquad (2\text{-}69)$$

The optimal value of x_1 is chosen to maximize the present value of expected profit, given w_1 and the knowledge that x_2^0 is a function of x_1 through Q_1. However, at the beginning of stage 1, when x_1 is chosen, only the distributions of w_2 and Q_1 are known, not their values. The optimal value of x_1 is found by solving

$$\max_{x_1} E_1[\pi_2 | w_1, x_2 = x_2^0] = E_1[p] E_1[Q_2 | x_2 = x_2^0] - E_1[w_2 x_2^0] - w_1 x_1 \qquad (2\text{-}70)$$

where E_1 is the expectations operator based on the joint probability distribution of w_2, p, Q_1, and Q_2 at the beginning of stage 1. Using equations (2-64), (2-65), and (2-69) it can be shown that

$$E_1[w_2 x_2^0] = \frac{1}{\beta_2}\{E_1[w_2^2]/E_2[p]\} - E_1[w_2]\frac{\beta_1}{\beta_2} - E_1[w_2] E_1[Q_1]\frac{\beta_4}{\beta_2} \qquad (2\text{-}71)$$

and

$$E_1[Q_2 | x_2 = x_2^0] = \left(\beta_0 - \frac{1}{2}\frac{\beta_1^2}{\beta_2^2}\right) + \left(\beta_3 - \frac{\beta_1 \beta_4}{\beta_2}\right)E_1[Q_1] - \frac{1}{2}\frac{\beta_4}{\beta_2}E_1\left[\frac{w_2}{p}\right]E_1[Q_1]$$

$$+ \frac{1}{2\beta_2}E_1\left[\left(\frac{w_2}{p}\right)^2\right] + \left(\beta_5 - \frac{\beta_4^2}{\beta_2}\right)E_1[Q_1^2] \qquad (2\text{-}72)$$

By virtue of equation (2-64), $E_1[Q_1]$ is linear in x_1 and $E_1[Q_1^2]$ is quadratic in x_1. Thus, as long as Var $[Q_1]$ is at most a quadratic function of x_1, expected profit in period 1, given by equation (2-70), is quadratic in x_1. Therefore, the value of x_1 that solves equation (2-70) is a linear function of w_1 and expectations of future prices

$$x_1^0 = \delta_0 + \delta_1 \frac{w_1}{E_1[p]} + \delta_2 E_1\left[\frac{w_2}{p}\right] + \delta_3 \frac{E_1[w_2]}{E_1[p]} \qquad (2\text{-}73)$$

where the δ_i are composed of the production function parameters α_i and β_i.

Several lessons can be learned from this simple example. First, the structure of the technology, the stochastic properties of prices and production, and the way in which the firm uses information all determine the firm's behavior. Second, *ex ante* expected output and thus expected productivity differ from *ex post* output and productivity. Firms base decisions on the former, whereas economists generally measure the latter. This observation suggests that the problem of decomposing productivity change into technical efficiency, scale, and technological change components is even more complex in the dynamic case than in the static case. For example, what economists measure as technical inefficiencies and scale changes in static production models may be due to misspecifications of firm behavior and expectations formation.

It is clear from the above example that an inherent problem in dynamic analysis is the modeling of expectations. Indeed, the Nerlovian partial-adjustment model can be interpreted as a particular model of expectations formation (adaptive expectations). More recently, the concept of rational expectations has been utilized in the economics profession. Simply put, a rational expectations model under uncertainty is analogous to a neoclassical model under certainty. In the neoclassical model, the firm is assumed to know prices and technology; under rational expectations, the firm's subjective distribution of prices and technology is assumed to be the actual distribution of prices and outputs that generates observed data. In the words of Lucas and Sargent (1981, p. xvi),

> In postulating the objective function... for our exemplary decision maker we used the expectation operator $E\{\cdot\}$, adding that it is "taken with respect to *the* distribution of $z_1, z_2, \ldots$" From the point of view of normative decision theory, "*the* distribution of $z_1, z_2, \ldots$" means just whatever distribution (i.e., environment f) the decision maker *thinks* is appropriate. From the point of view of an outside observer who wishes to predict the agent's response to changes in f... we are using "f" to denote the *actual* environment, as observed in the data and as altered by the hypothetical policy changes we wish to assess.... The hypothesis of rational expectations amounts to equating the subjective z-distribution to the objective distribution f.

Thus, in the above example, the rational expectations hypothesis would mean that the expectation operators were based on the distributions of the actual stochastic processes generating the price and output data. However, it should be emphasized that rational expectations is not equivalent to perfect foresight. Agents with rational expectations do not know what realizations of random variables will be; rather, they know the distribution from which the realizations will be drawn.

The rational expectations hypothesis is important to productivity research for several reasons. One reason is that the modeling of expectations is a formidable problem in economics. If a firm's subjective expectations are unrelated to observable economic data, it will be extremely difficult to model producer behavior. On the other hand, if subjective expectations are in some way related to observable data, then economic data can be used to draw inferences about firms' expectations formation. The rational expectations hypothesis is one means of linking subjective expectations to observed (objective) distributions of stochastic phenomena.

Rational expectations is also important because of its implications for policy analysis. Lucas' criticism of econometric policy analysis based on the rational expectations view, is that if economic agents' (consumers', firms') expectations are based on the actual processes generating the data, then their behavior will depend on (be endogenous to) changes in economic policy. Hence, an econometric model estimated under one policy regime will not be the correct model for the analysis of a different policy regime. Since many economic policies affect productivity, especially in agriculture, Lucas' critique means that production models estimated under one set of agricultural policies will not necessarily represent producer behavior under alternative policies.

For example, in the above model, the distribution of output prices under a price support program would be different from the output price distribution generated by a free market. Under rational expectations, farmers would realize the difference the two policy regimes would have on the output price distribution, and thus their decision rules for input demand and output supply would change. Therefore, a production model estimated under one policy regime would not be an appropriate model for productivity analysis or policy analysis under another policy regime.

A second major challenge in the formulation of dynamic models is the problem of tractability. It was noted above that the quadratic cost-of-adjustment model has been used in applied research because it is known to have a closed-form solution to the factor demand equations. However, the quadratic model has serious limitations as a flexible representation of a production technology (see chapter 12 in this volume) and has been found to be inconsistent with aggregate data (see Epstein and Yatchew, 1985). Currently, besides the quadratic model, the only production model with a closed-form solution to the input demand equations appears to be the Cobb-Douglas model (see Hatchett, 1984). In addition to searching for primal production-function models that have closed-form solutions, researchers are exploring other avenues. One approach is to utilize the first-order conditions of the dynamic optimization problem as the basis for model estimation, without attempting to solve for the factor demand equations (see Hansen and Singleton, 1982; Pindyck and Rotemberg, 1983). Another approach is to develop a dynamic version of duality theory (Epstein, 1981). This latter approach seems promising, but as yet has not yielded empirically tractable approaches to production analysis and productivity measurement. One difficulty with the dual approach is that it is not well defined under uncertainty. The search for flexible, tractable dynamic production models thus remains a crucial topic.

APPENDIX 2-A.
THE TRANSLOG COST MODEL

This appendix formally presents the translog cost function and the restrictions required for theoretical consistency and other structural properties. Development of the translog function is due to Christensen, Jorgenson, and Lau (1971, 1973). In order to justify using the duality relations between the primal and dual representations of the technology, a dual model's consistency with theoretical properties must be verified. Due to the flexible nature of the translog function, the translog cost function does not globally satisfy the theoretical properties of monotonicity or concavity in factor prices. Therefore, we must check for these properties locally. In this appendix, we discuss the evaluation of theoretical properties of the translog function at the point of approximation, that is, with all variables set equal to unity so that the translog can be interpreted as a second-order Taylor series expansion in the logarithms about the point zero.

A cost function $C(Q, W)$ is continuous in factor prices (W) and output (Q); monotonic, nondecreasing in Q and W; and linearly homogenous and concave in W. Moreover, decreasing returns to scale imply that the cost function is convex in Q. Differentiation of $C(Q, W)$ with respect to factor prices produces a set of cost-minimizing factor demands (Shephard's lemma); differentiation with respect to each output produces a set of marginal cost functions. From an econometric point of view, Shephard's lemma is useful because a set of cost minimizing factor demands can be obtained by first specifying a functional form for the cost function that satisfies the theoretical properties and then differentiating $C(Q, W)$ with respect to the factor prices.

Translog Cost Model

Let the price of the factors be denoted by W_i, $i = 1, \ldots, N$, and their respective quantities by X_i; denote the level of each output by Q_k, $k = 1, \ldots, M$, total cost as $C = \Sigma_i W_i X_i$ and the effects of disembodied technological progress as t. The translog cost function is written

$$\ln C = \alpha_0 + \sum_i \alpha_i \ln W_i + \frac{1}{2} \sum_i \sum_j \gamma_{ij} \ln W_i \ln W_j + \sum_k \beta_k \ln Q_k$$
$$+ \frac{1}{2} \sum_k \sum_l \beta_{kl} \ln Q_k \ln Q_l + \sum_i \sum_k \rho_{ik} \ln W_i \ln Q_k + \alpha_t t$$
$$+ \frac{1}{2} \alpha_{tt} t^2 + \sum_i \alpha_{it} \ln W_i t + \sum_k \beta_{kt} \ln Q_k t \qquad (2A-1)$$

where i, j denote factors; k, l denote outputs; and ln is the natural logarithm. Imposing the symmetry conditions on the cross-price effects and on the cross-output effects implies the following restrictions:

$$\gamma_{ij} = \gamma_{ji} \quad \text{for all } i,j = 1, \ldots, N$$
$$\beta_{kl} = \beta_{lk} \quad \text{for all } l, k = 1, \ldots, M \qquad (2A-2)$$

Homogeneity of degree one in factor prices, given Q and t, implies the following linear restrictions:

$$\sum_i \alpha_i = 1$$

$$\sum_i \gamma_{ij} = 0 \quad \text{for all } j = 1, \ldots, N$$

$$\sum_i \rho_{ik} = 0 \quad \text{for all } k = 1, \ldots, M$$

$$\sum_i \alpha_{it} = 0 \qquad (2A-3)$$

The factor share equations are derived using Shephard's lemma:

$$\frac{\partial \ln C}{\partial \ln W_i} = S_i = \alpha_i + \sum_j \gamma_{ij} \ln W_j + \sum_k \rho_{ik} \ln Q_k + \alpha_{it} t \qquad i = 1, \ldots, N \quad (2A-4)$$

It is noted that if all the γ_{ij} and ρ_{ik} are zero the translog cost function becomes a Cobb-Douglas function that is globally concave in prices. However, if these parameters are not all zero, the concavity property may only be satisfied over a range of prices. This is discussed in greater detail below.

Monotonicity and Curvature Conditions

Unlike the restrictions for linear homogeneity of the cost function in factor prices, the monotonicity and curvature restrictions are not easily handled within the econometric framework. This is because both involve inequality restrictions on the parameter set or share equations. As a result, the conventional approach has been to check the estimated model for these properties rather than imposing the restrictions in estimation. However, because of the widespread failure of the curvature conditions in many empirical applications involving the translog as well as other flexible functional forms, researchers are developing algorithms for imposing these inequality restrictions (for example, see Hazilla and Kopp, 1986b, and the references cited therein). In the econometric models reported in chapter 5, the curvature conditions are not imposed, but are checked locally.

One should note that checking for compliance of the curvature conditions is not a statistical test of these properties. If the curvature conditions are satisfied, then a statistical test of these properties would fail to reject the null hypothesis that the production technology was concave. However, the converse is not true. If the curvature conditions are not satisfied they will not necessarily be rejected by a statistical test. The failure of the model to be concave in W or convex in Q can be interpreted as a violation of the cost minimization postulate underlying the development of the cost function model. But researchers must be aware that violations of curvature or other theoretical properties by an econometrically estimated model may simply be due to sampling error or to other problems, such as data errors, model misspecification, or simultaneous equation bias.

The cost function is said to be monotonically increasing in prices if $C(Q, W^i) > C(Q, \overline{W}^i)$ where $W^i > \overline{W}^i$. This is equivalent to $\partial C/\partial W_j > 0$ for all $j = 1, \ldots, N$. The monotonicity conditions for the translog are

$$\frac{\partial C}{\partial W_j} = \frac{\partial \ln C}{\partial \ln W_j} \cdot \frac{C}{W_j} > 0 \qquad \text{for all } j = 1, \ldots, N \tag{2A-5}$$

Equation (2A-5) is equivalent to

$$\frac{\partial C}{\partial W_j} = \frac{S_j C}{W_j} > 0$$

Note that $C > 0$ and $W_j > 0$ by definition. Therefore, a necessary and sufficient condition for monotonicity (in prices) is that the cost shares are greater than zero

$$S_i = \alpha_i + \sum_j \gamma_{ij} \ln W_j + \sum_k \rho_{ik} \ln Q_k + \alpha_{it} t > 0 \tag{2A-6}$$

In a similar manner, monotonicity in output requires that the partial derivative of the cost function with respect to the kth unit of output is nonnegative. This translates into the following condition on the translog cost function:

$$\frac{\partial \ln C}{\partial \ln Q_k} = \beta_k + \sum_l \beta_{kl} \ln Q_l + \sum_i \rho_{ik} \ln W_i + \beta_{kt} t > 0 \tag{2A-7}$$

The inequality conditions (2A-6) and (2A-7) are functions of the observations. However, at the point of approximation—that is, where all W_i and Q_k are indexed to 1.0 and t is indexed to zero—condition (2A-6) becomes $S_i = \alpha_i > 0$. Similarly, the condition that the cost function is nondecreasing in output at the point of approximation translates into the restriction on $\beta_k > 0$, $k = 1, \ldots, M$.

The curvature conditions also involve the arguments of the cost function. The concavity conditions of the cost function with respect to the factor prices are based on the property that the Hessian matrix of second partials of C with respect to factor prices, H_{WW}, is negative semidefinite. The convexity with respect to the output quantities is established if the Hessian matrix of second partials of C with respect to output quantities, H_{QQ}, is positive semidefinite.

Looking first at the concavity conditions, quasiconcavity of the cost function in W is established through the necessary condition that the direct Allen elasticity of substitution (AES) is negative and the necessary and sufficient condition that the determinants of the principal minors of the estimated Hessian alternate in signs.[1]

The symmetric Hessian matrix, H_{WW}, has $\partial^2 C/\partial W_i^2$ as the diagonal elements and $\partial^2 C/\partial W_i \partial W_j$ ($i \neq j$) as the off-diagonal elements. Since the concavity requirement is with respect to the factor prices only, we need only be concerned with the terms in the cost function that involve the factor prices, that is, the terms involving the α_i and γ_{ij} parameters. Because the translog cost function involves the logarithms of the input prices, some additional manipulations are required. First note that

---

[1] A third condition relating to the Cholesky decomposition of the estimated Hessian is discussed in Hazilla and Kopp (1984).

$\partial^2 \ln C/(\partial \ln W_i)^2 = \gamma_{ii}$, and $\partial S_i/\partial W_i = \gamma_{ii}/W_i$. Therefore, $\gamma_{ii} = W_i \cdot \partial S_i/\partial W_i$. Using $S_i \equiv (\partial C/\partial W_i) \cdot (W_i/C)$, we have

$$\gamma_{ii} = W_i[\partial(\partial C/\partial W_i \cdot W_i/C)/\partial W_i] \tag{2A-8}$$

Differentiating the term in brackets yields

$$\gamma_{ii} = W_i\left[\frac{W_i}{C}\frac{\partial^2 C}{\partial W_i^2} + \frac{\partial C}{\partial W_i}\left(\frac{1}{C} - \frac{W}{C^2}\frac{\partial C}{\partial W_j}\right)\right] \tag{2A-9}$$

Substituting $\partial C/\partial W_i = X_i$ into equation (2A-9) yields

$$\gamma_{ii} = \left(\frac{W_i^2}{C}\frac{\partial^2 C}{\partial W_i^2}\right) + \frac{W_i X_i}{C} - \frac{W_i^2}{C^2}X_i^2 \tag{2A-10}$$

At the point of approximation ($W_i X_i/C = \alpha_i$),

$$\gamma_{ii} = \frac{W_i^2}{C} \cdot \frac{\partial^2 C}{\partial W_i^2} + \alpha_i - \alpha_i^2 \tag{2A-11}$$

Equation (2A-11) can be rewritten to isolate the diagonal elements of the required Hessian matrix:

$$\frac{\partial^2 C}{\partial W_i^2} = \frac{C}{W_i^2}(\gamma_{ii} + \alpha_i^2 - \alpha_i) \tag{2A-12}$$

Derivation of the off-diagonal elements follows a similar procedure (see Hazilla and Kopp, 1984):

$$\frac{\partial^2 C}{\partial W_i \partial W_j} = \frac{C}{W_i W_j}(\gamma_{ij} + \alpha_i \alpha_j) \quad \text{for all } i \neq j \tag{2A-13}$$

The sign of each element in the Hessian is determined by the terms in parentheses in equations (2A-12) and (2A-13); thus we can evaluate a modified Hessian,

$$H^*_{ww} = \begin{pmatrix} \gamma_{11} + \alpha_1^2 - \alpha_1 & \gamma_{12} + \alpha_1\alpha_2 & \cdots & \gamma_{1n} + \alpha_1\alpha_n \\ \gamma_{12} + \alpha_1\alpha_2 & & & \\ \vdots & & \ddots & \\ \gamma_{1n} + \alpha_1\alpha_n & & & \gamma_{nn} + \alpha_n^2 - \alpha_n \end{pmatrix} \tag{2A-14}$$

As $C/W_i W_j = C/W_j W_i > 0$ for all i,j, the Hessian matrix H_{ww} is negative semidefinite if and only if H^*_{ww} is negative semidefinite. A necessary and sufficient condition for quasi-concavity of the translog cost function translates into a condition that the principal minors of equation (2A-14) must alternate between nonpositive and nonnegative values, starting with $h_{11} \leq 0$, $(h_{11} h_{22} - h_{12}^2) \geq 0$, where h_{ij} is the ijth element of equation (2A-14).

The necessary condition for curvature relating to the own AES can also be determined from the diagonal elements of equation (2A-14). Uzawa (1962) showed that the elasticities of substitution can be computed from the partial derivatives of the cost function:

$$\sigma_{ij} = \frac{C \cdot \partial^2 C/\partial W_i \partial W_j}{(\partial C/\partial W_i)(\partial C/\partial W_j)} \quad \text{for } i,j$$

For the translog cost model, the AES are given by

$$\sigma_{ii} = (\gamma_{ii} + S_i^2 - S_i)/S_i^2$$
$$\sigma_{ij} = (\gamma_{ij} + S_i S_j)/S_i S_j \quad (2A\text{-}15)$$

Consistent with our earlier discussion, the AES evaluated at the point of expansion are

$$\sigma_{ii} = (\gamma_{ii} + \alpha_i^2 - \alpha_i)/\alpha_i^2$$
$$\sigma_{ij} = (\gamma_{ij} + \alpha_i \alpha_j)/\alpha_i \alpha_j \quad (2A\text{-}16)$$

The estimated AES are related to the elements of the modified Hessian,

$$\sigma_{ii} = h_{ii}/\alpha_i^2$$
$$\sigma_{ij} = h_{ij}/\alpha_i \alpha_j \quad (2A\text{-}17)$$

The necessary condition for negative semidefiniteness of the Hessian translates into a negativity restriction on each σ_{ii}. This is ascertained by noting that the first principal minor of the Hessian is $h_{11} = \sigma_{11} \alpha_1^2$, which must be negative. Furthermore, any i can be positioned as the first column of this matrix.

One can also note that the nonsymmetric, constant-output price elasticities of demand are related to the AES, $\epsilon_{ij} = S_j \sigma_{ij}$. For the translog cost function these can be expressed at the point of approximation as

$$\epsilon_{ii} = (\gamma_{ii} + \alpha_i^2 - \alpha_i)/\alpha_i$$
$$\epsilon_{ij} = (\gamma_{ij} + \alpha_i \alpha_j)/\alpha_i \quad \text{for all } i \neq j \quad (2A\text{-}18)$$

The condition for convexity of the translog cost function with respect to output quantities is developed in a similar manner. The symmetric Hessian matrix, H_{QQ}, has $\partial^2 C/\partial Q_k^2$ as the diagonal elements and $\partial^2 C/\partial Q_k \partial Q_l$ ($k \neq l$) as the off-diagonal elements. Noting that $\partial C/\partial Q_k$ is the marginal cost function pertaining to Q_k, and assuming perfect competition (output price is equal to marginal cost), we find that the H_{QQ} at the point of expansion has the following diagonal and off-diagonal elements:

$$\frac{\partial^2 C}{\partial Q_k^2} = (\beta_{kk} - \beta_k + \beta_k^2) \cdot \frac{C}{Q_k^2}$$

$$\frac{\partial^2 C}{\partial Q_k \partial Q_l} = (\beta_{kl} + \beta_k \cdot \beta_l) \cdot \frac{C}{Q_k^2} \qquad \text{(2A-19)}$$

We can simplify the analysis by noting that H_{QQ} is proportional to H_{QQ}^*, where H_{QQ}^* is a matrix composed of the terms in parentheses in equation (2A-19). H_{QQ} is positive semidefinite if and only if H_{QQ}^* is positive semidefinite, because $C/Q_k Q_l > 0$ for all k, l, which translates into a condition that the principal minors of H_{QQ}^* must have nonnegative values.

The elements of the H_{QQ}^* matrix are related to Diewert's (1974) inverse elasticities of transformation between outputs, τ_{kl}, the change in the marginal cost of the kth unit of output with respect to a change in the supply of output l. For all observations,

$$\tau_{kk} = (\beta_{kk} + S_k^2 - S_k)/S_k^2$$

$$\tau_{kl} = (\beta_{kl} + S_k S_l)/S_k S_l \qquad k \neq l$$

where S_k is the cost share of the kth unit of output. At the point of approximation, the estimated elasticities τ_{kk} and τ_{kl} are determined by replacing S_k with β_k. Finally, the necessary conditions for positive semidefiniteness of H_{QQ} at the point of approximation translates into a positive restriction on τ_{kk} for each k.

Additional Restrictions

In this section we consider the parameter restrictions required for nonjointness and Hicks-neutral technical change of the translog cost function, equation (2A-1). If a technology is nonjoint in inputs, the marginal cost of each unit of output k is independent of the level of any other output (Hall, 1973). The cost function can be expressed as $C = \Sigma_k G^k(Q_k, W, t)$, where G^k is the cost function for the kth unit of output. Before deriving the parameter restrictions for nonjointness in inputs, we derive the more restrictive conditions required for nonjointness and constant returns to scale. If the technology has constant returns to scale, the nonjoint cost function can be expressed as

$$C = \sum_k Q_k G^k(W, t) \qquad (2A\text{-}20)$$

A necessary and sufficient condition for equation (2A-20) is that the sub-Hessian, H_{QQ}, is a null matrix (Kohli, 1981). If H_{QQ} is a null matrix, then each term in equation (2A-19) must be zero. This suggests that in addition to the following restrictions for linear homogeneity of the translog cost function in output:

$$\sum_k \beta_k = 1$$

$$\sum_k \beta_{kl} = 0 \qquad \text{for all } l = 1, \ldots, M \qquad (2A\text{-}21)$$

$$\sum_k \beta_{kt} = 0$$

we need the restrictions for nonjointness of the translog

$$\beta_{kl} = -\beta_k \cdot \beta_l \qquad \text{for all } k \neq l \qquad (2A\text{-}22)$$

When we do not impose constant returns to scale, the restrictions for nonjointness are given by equation (2A-22). The restrictions on the translog model given by equations (2A-1) through (2A-4) result in H_{QQ} being a diagonal matrix.

The final set of restrictions relate to modeling technical change. As discussed in chapter 2, Hicks-neutral technical change is defined as "expansion-path-preserving" and is equivalent to the total effect of technical change on the equilibrium cost shares, B_i^c, minus the effects due to scale changes:

$$B_i^{ce} = B_i^c - \left[\sum_k \left(\frac{\partial \ln S_i}{\partial \ln Q_k} \right) \left(\frac{\partial \ln C}{\partial \ln Q_k} \right)^{-1} \right] \frac{\partial \ln C}{\partial t}$$

At the point of approximation, we have

$$B_i^{ce} = \alpha_{it} - \left(\sum_k \rho_{ik}/\beta_k \right) \alpha_t \qquad (2A\text{-}23)$$

for the translog cost model. Under input-output separability, the cost function is known to be strongly separable in outputs. Therefore $\rho_{ik} = 0$, the second term on the right-hand side of equation (2A-23) is zero, and the conditions for Hicks neutrality are that $\alpha_{it} = 0$ for all i. Under nonseparability, Hicks neutrality implies equation (2A-23) must be zero for all i, or

$$\alpha_{it} = \sum_k (\rho_{ik}/\beta_k)\alpha_t \quad \text{for all } i = 1, \ldots, N \tag{2A-24}$$

The preceding discussion is for the translog long-run (unrestricted) cost function; similar restrictions hold for the translog restricted (short-run) cost function. The theoretical properties of the restricted cost function are summarized in Diewert (1985). The translog restricted cost model is used by Brown and Christensen (1981) and discussed in Berndt (1984).

APPENDIX 2-B.
THE TRANSLOG PROFIT MODEL

In this appendix the theoretical properties of the profit function are translated into restrictions on a translog profit function. These restrictions are either imposed *a priori* on the model or tested for *ex post*. The development parallels the exposition in appendix 2A for the translog cost function.

Translog Profit Model

The translog profit function is written

$$\ln \pi = \alpha_0 + \sum_i \alpha_i \ln P_i + \frac{1}{2}\sum_i \sum_j \gamma_{ij} \ln P_i \ln P_j + \alpha_t t + \frac{1}{2}\alpha_{tt} t^2 + \sum_i \phi_{it} \ln P_i t \tag{2B-1}$$

where P_i denotes the price of the ith unit of output or input, $i = 1, \ldots, M + N$. Denote the levels of the corresponding output and inputs by X_i, where $X_i > 0$, $i = 1, \ldots, M$, and $X_i < 0$ for $i = M + 1, \ldots, N$, respectively. Total profits, π, are equal to $\Sigma_i P_i X_i$, and the effects of technical progress are denoted by t.

Symmetry and linear homogeneity of π in prices imply the following sets of restrictions

$$\gamma_{ij} = \gamma_{ji} \quad \text{for all } i,j \tag{2B-2}$$

$$\Sigma \alpha_i = 1$$

$$\sum_i \gamma_{ij} = 0 \quad \text{for all } j \tag{2B-3}$$

$$\sum_i \phi_{it} = 0$$

Differentiating equation (2B-1) with respect to $\ln P_i$ yields the following system of equations

$$S_i = \frac{P_i X_i}{\sum_i P_i X_i} = \alpha_i + \sum_j \gamma_{ij} \ln P_j + \phi_{it} t \quad i = 1, \ldots, N + M \tag{2B-4}$$

Monotonicity and Curvature Conditions

The monotonicity conditions on the profit function require that it be nondecreasing (nonincreasing) in P_i if X_i is an output (input). This can be verified by examining the predicted profit shares using equation (2B-4). At the point of expansion (or approximation) these shares are simply equal to the corresponding α_i. Thus, if X_i is an output, $\alpha_i \geq 0$; if X_i is an input, $\alpha_i \leq 0$.

As with the translog cost function, the translog profit model will not globally satisfy the convexity property. The condition for local compliance of the convexity property involves the Hessian matrix of second-order partial derivatives with respect to prices. For convexity, the Hessian needs to be positive semidefinite: all principal minors must have nonnegative determinants. The positive semidefiniteness of the Hessian matrix translates into a positive semidefinite condition on the modified Hessian, H^*_{PP}, with the ith diagonal element, $\gamma_{ii} + \alpha_i^2 - \alpha_i$, and the ijth off-diagonal element, $\gamma_{ij} + \alpha_i \alpha_j$. The convexity and monotonicity conditions can either be imposed or checked *ex post*. The former involves estimation of the profit model subject to inequality constraints.

The Marshallian price elasticities of the output supply and factor demands can be expressed in terms of the parameters of the translog profit function. Diewert (1974) showed that the partial elasticity of X_i with respect to P_j is

$$\epsilon_{ij} = S_j + \gamma_{ij}/S_i$$
$$\epsilon_{ii} = S_i + \gamma_{ii}/S_i - 1 \qquad (2\text{B-}5)$$

At the point of approximation, $S_i = \alpha_i$ and the elasticities are functions only of the estimated parameters γ_{ij}, γ_{ii}, α_i, α_j. As with the translog cost function, these elasticities are related to the elements of H^*_{PP}.

$$\alpha_i \epsilon_{ii} = h_{ii}$$
$$\alpha_i \epsilon_{ij} = \alpha_j \epsilon_{ji} = h_{ij} \qquad (2\text{B-}6)$$

A second (necessary) condition that can be used to test for price convexity relates to the own- (uncompensated) price elasticities, $\epsilon_{ii} > 0$ for an output, and $\epsilon_{ii} < 0$ for an input. This can be tested using the predicted profit shares for all observations or at the point of approximation.

Additional Restrictions

Lau (1978) showed that a necessary and sufficient condition for a technology to be nonjoint in inputs is

$$\pi(P, t) = \sum_{i=1}^{M} G_i(P_i, P_{M+1}, \ldots, P_{M+N}, t) \qquad (2\text{B-}7)$$

Equation (2B-7) implies that the terms in the Hessian H^*_{PP} corresponding to the cross-output price derivatives are zero, that is,

$$\frac{\partial \pi}{\partial P_i \partial P_j} = \frac{\partial^2 G_i}{\partial P_i \partial P_j} = 0, \qquad i,j = 1, \ldots, M; i \neq j$$

Therefore, the restrictions for nonjointness in inputs on the translog profit function at the point of approximation are

$$\gamma_{ij} = -\alpha_i \alpha_j \qquad \text{for all } i,j = 1, \ldots, M \qquad (2\text{B-}8)$$

It is important to note that the translog profit function cannot be globally nonjoint, because it is additive in the logarithms and not in the original variables.

Input-output separability can also be imposed on the translog profit model. In general, the profit function for a technology that is input-output separable is homothetically separable in input and output prices

$$\pi(P, t) = \pi[G_1(P_1, \ldots, P_M, t), G_2(P_{M+1}, \ldots, P_{M+N}, t), t] \tag{2B-9}$$

Under input-output separability optimal output proportions (and shares) do not depend on input prices; likewise, optimal input proportions (and shares) are independent of output prices. For the translog profit function, input-output separability requires

$$\gamma_{ij} = 0 \quad \text{for all } i = 1, \ldots, M, \\ j = M+1, \ldots, N \tag{2B-10}$$

Finally, we can restrict the translog profit model for Hicks-neutral technical change. Consider first the case of Hicks neutrality with respect to the inputs. If we have nonjointness in inputs, equation (2B-8), the restrictions for Hicks neutrality at the point of approximation are $\phi_{it} = 0$ for all $i = M+1, \ldots, N$; if the technology is joint, equation (2-24) implies that the needed restrictions are

$$\phi_{it} - \left(\sum_{j=1}^{M} \gamma_{ij}/\alpha_j \right) \alpha_t = 0 \quad \text{for all } i = M+1, \ldots, N \tag{2B-11}$$

For Hicks neutrality with respect to the outputs, an analogous set of restrictions would hold with the indexing over $i = 1, \ldots, M$. For Hicks neutrality with respect to both inputs and outputs, the indexing is over all $i = 1, \ldots, N$.

APPENDIX 2-C.
TFP MEASUREMENT USING EXACT INDEX NUMBERS

The purpose of this appendix is (a) to show that the Tornqvist-Theil quantity and price indexes are exact for homogeneous translog aggregator functions and translog unit cost functions, and (b) to use an exact index number formula to measure TFP. The basis for this appendix is the work by Diewert (1976, 1980b, 1981).

Tornqvist-Theil Indexes

Diewert defines a quantity (price) index to be superlative if it is exact for an aggregator (unit cost) function that is capable of providing a second-order approximation to an arbitrary twice continuously differentiable linearly homogeneous aggregator (unit cost) function. If PI is a superlative price index and $\tilde{QI}$ is the corresponding implicit quantity index derived by the weak factor reversal property, equation (2-32), the pair of index number formulas (PI, $\tilde{QI}$) is superlative.

Before developing the Tornqvist-Theil superlative indexes, we make note of the following property: if $f(Z)$ is a quadratic function, then

$$f(Z^1) - f(Z^0) = \frac{1}{2}[\nabla f(Z^1) + \nabla f(Z^0)](Z^1 - Z^0) \tag{2C-1}$$

where $\nabla f(Z^r)$ is the gradient vector of f evaluated at Z^r, $r = 0, 1$.

The Tornqvist-Theil price and quantity indexes, PI_T and QI_T, are defined in equations (2-41) and (2-42).

Theorem 1 (Diewert, 1976, p. 119). QI_T is exact for the homogeneous translog aggregator function

$$\ln f(X) = \alpha_0 + \sum_i \alpha_i \ln X_i + \frac{1}{2} \sum_i \sum_j \alpha_{ij} \ln X_i \ln X_j \tag{2C-2}$$

where $\Sigma \alpha_i = 1$, $\alpha_{ij} = \alpha_{ji}$ for all i, j, and $\Sigma_j \alpha_{ij} = 0$ for all i. The proof is as follows: assume that the observed prices and quantities are the profit-maximizing prices and quantities during each period. Now define $Z_i \equiv \ln X_i$ for $r = 1, 0$, and note that equation (2C-2) is a quadratic function in Z. Noting that $\partial f(Z)/\partial(Z_j) \equiv \partial \ln f(X)/\partial \ln X_j$, we translate equation (2C-1) into the following identity involving the partial derivatives of the aggregator function defined in equation (2C-2)

$$\ln f(X^1) - \ln f(X^0) = \frac{1}{2} \sum_i \left[\frac{\partial \ln f(X^1)}{\partial \ln X_j^1} + \frac{\partial \ln f(X^0)}{\partial \ln X_j^0} \right] (\ln X_i^1 - \ln X_i^0)$$

$$= \frac{1}{2} \sum_i \left[\frac{\partial f(X^1)}{\partial X_i^1} \cdot \frac{X_i^1}{f(X^1)} + \frac{\partial f(X^0)}{\partial X_i^0} \cdot \frac{X_i^0}{f(X^0)} \right]$$

$$(\ln X_i^1 - \ln X_i^0) \tag{2C-3}$$

We can rewrite equation (2C-3) as

$$\ln [f(X^1)/f(X^0)] = \frac{1}{2} \Sigma \left(\frac{X_i^1 W_i^1}{W^1 \cdot X^1} + \frac{X_i^0 W_i^0}{W^0 \cdot X^0} \right) \ln (X_i^1/X_i^0) \tag{2C-4}$$

or

$$f(X^1)/f(X^2) = \prod_i (X_i^1/X_i^0)^{1/2(S_i^1 + S_i^0)} \equiv QI_T$$

The implicit Tornqvist-Theil price index, $\tilde{PI}_T$, can be defined using equation (2-32). Since QI_T is exact for the flexible functional form

DEVELOPMENTS IN PRODUCTION THEORY & PRODUCTIVITY MEASUREMENT 87

(2C-2), the pair of index numbers ($\tilde{\text{PI}}_T$, QI_T) is a superlative pair of index number formulas.

Theorem 2 (Diewert, 1976, p. 121). PI_T is exact for the translog unit cost function

$$\ln C(W) = \beta_0 + \sum_i \beta_i \ln W_i + \frac{1}{2} \sum_i \sum_j \beta_{ij} \ln W_i \ln W_j \qquad (2\text{C-5})$$

where $\Sigma_i \beta_i = 1$, $\beta_{ij} = \beta_{ji}$ for all i and j, and $\Sigma_j \beta_{ij} = 0$ for all i. (The proof is similar to the proof for theorem 1.) Given that PI_T is exact, the implicit quantity index $\tilde{\text{QI}}_T$ can be defined using equation (2-32) and (PI_T, $\tilde{\text{QI}}_T$) is a superlative pair of index number formulas.

Exact Index Number Approach to TFP Measurement

The exact number approach to TFP measurement involves deriving an exact index number formula for TFP that is consistent with the specific functional form of the production function or cost function, namely the translog.

Assume that the firm's cost function is the translog model described by equations (2A-1) through (2A-3). Noting that $\ln C$ is quadratic in $\ln W_i$, $\ln Q_k$, and time t, we can apply equation (2C-1)

$$\ln C^1(\cdot) - \ln C^0(\cdot) = \frac{1}{2} \sum_i [\partial \ln C^1(\cdot)/\partial \ln W_i^1 + \partial \ln C^0(\cdot)/\partial \ln W_i^0] \ln \left(\frac{W_i^1}{W_i^0}\right)$$

$$+ \frac{1}{2} \sum_k [\partial \ln C^1(\cdot)/\partial \ln Q_k^1 + \partial \ln C^0(\cdot)/\partial \ln Q_k^0] \ln \left(\frac{Q_k^1}{Q_k^0}\right)$$

$$+ \frac{1}{2} [\partial \ln C^1(\cdot)/\partial t + \partial \ln C^0(\cdot)/\partial t](1-0) \qquad (2\text{C-6})$$

where $\ln C^r(\cdot)$ denotes the rth period cost function. If we assume profit maximizing behavior, then $X^r = \nabla_W C^r(\cdot)$, $r = 0, 1$ (Shephard's lemma) and $p^r = \nabla_Q C^r(\cdot)$, $r = 0, 1$ and $W^r \cdot X^r = C^r(\cdot)$. Equation (2C-6) can be rewritten as

$$\frac{1}{2}(\tau^1 + \tau^0) = \ln\left(\frac{W^1 \cdot X^1}{W^0 \cdot X^0}\right) - \frac{1}{2} \sum_i \left(\frac{W_i^1 X_i^1}{W^1 \cdot X^1} + \frac{W_i^0 X_i^0}{W^0 \cdot X^0}\right) \ln \frac{W_i^1}{W_i^0}$$

$$- \frac{1}{2} \Sigma \left(\frac{P_k^1 Q_k^1}{W^1 \cdot X^1} + \frac{P_k^0 Q_k^0}{W^0 \cdot X^0}\right) \ln \frac{Q_k^1}{Q_k^0} \qquad (2\text{C-7})$$

where $\tau^r \equiv \partial \ln C^r(\cdot)/\partial t$ for $r = 0, 1$.

Applying equation (2-41), we can express equation (2C-7) in terms of the Tornqvist-Theil input quantity index, QI_T

$$\frac{1}{2}(\tau^1 + \tau^0) = \ln\left(\frac{W^1 X^1}{W^0 \cdot X^0}\right) - QI_T - \frac{1}{2}\sum_k \left(\frac{P_k^1 Q_k^1}{W^1 \cdot X^1} + \frac{P_k^0 Q_k^0}{W^0 \cdot X^0}\right) \ln\frac{Q_k^1}{Q_k^0} \quad (2C\text{-}8)$$

Furthermore, if the technology is characterized by constant returns to scale, $P^k Q^k = W^r X^r$, and the last term in equation (2C-8) is the Tornqvist-Theil output quantity index. Note, however, that equation (2C-8) can provide a measure of the shift in technology in both the constant and nonconstant returns to scale cases, provided that there is competitive behavior.

REFERENCES

Allen, R. G. D. 1938. *Mathematical Analysis for Economists* (London, Macmillan).

Antle, J. M. 1984a. "The Structure of U.S. Agricultural Technology, 1910–78," *American Journal of Agricultural Economics* vol. 66, no. 4, pp. 414–421.

———. 1984b. "Human Capital, Infrastructure, and the Productivity of Indian Rice Farmers," *Journal of Development Economics* vol. 14, no. 2, pp. 163–181.

Arrow, K. J., B. H. Chenery, B. S. Minhas, and R. M. Solow. 1961. "Capital-Labor Substitution and Economic Efficiency," *Review of Economics and Statistics* vol. 43, no. 3, pp. 225–250.

Berndt, E. R. 1984. "A Memorandum on the Estimation and Interpretation of the Translog Short-Run Variable Cost Function," Center for Economic Study, U.S. Bureau of the Census.

———, and L. R. Christensen. 1973a. "The Internal Structure of Functional Relationships: Separability, Substitution, and Aggregation," *Review of Economic Studies* vol. 40, no. 123, pp. 403–410.

———, and L. R. Christensen. 1973b. "The Translog Function and the Substitution of Equipment, Structures, and Labor in U.S. Manufacturing, 1929–68," *Journal of Econometrics* vol. 1, no. 1, pp. 81–114.

———, C. J. Morrison, and G. C. Watkins. 1981. "Dynamic Models of Energy Demand: An Assessment and Comparison," in E. R. Berndt and B. C. Field, eds., *Modeling and Measuring Natural Resource Substitution* (Cambridge, Mass., MIT Press).

Bertsekas, D. P. 1976. *Dynamic Programming and Stochastic Control* (New York, Academic Press).

Binswanger, H. P. 1974. "The Measurement of Technological Change Biases with Many Factors of Production," *American Economic Review* vol. 64, no. 6, pp. 964–976.

———. 1978. "Measured Biases of Technological Change: The United States," in H. P. Binswanger, V. W. Ruttan, and co-editors, eds., *Induced Innovation: Technology, Institutions, and Development* (Baltimore, Md., Johns Hopkins University Press).

Blackorby, C., C. A. K. Lovell, and M. C. Thursby. 1976. "Extended Hicks Neutral Technological Change," *Economic Journal* vol. 86, no. 344, pp. 845–852.

———, D. Primont, and R. R. Russell. 1977. "On Testing Separability Restrictions with Flexible Functional Forms," *Journal of Econometrics* vol. 5, no. 3, pp. 195–209.

———, D. Primont, and R. R. Russell. 1978. *Duality Separability and Functional Structure: Theory and Economic Applications* (New York, N.Y., North-Holland).

Brown, R. S., and L. R. Christensen. 1981. "Estimating Elasticities of Substitution in a Model of Partial Static Equilibrium: An Application to U.S. Agriculture, 1947 to 1974," in E. R. Berndt and B. C. Fields, eds., *Modeling and Measuring Natural Resource Substitution* (Cambridge, Mass., MIT Press).

Capalbo, S. M., and M. Denny. 1984. "Testing Productivity Models for the Canadian and U.S. Agricultural Sectors." Paper presented at the annual meeting of the American Association of Agricultural Economists, Dallas, Texas, December.

———, and M. Denny. 1985. "Testing Long-Run Productivity Models," Discussion Paper No. RR-8503 (Washington, D.C., Resources for the Future).

Caves, D. W., L. R. Christensen, and W. E. Diewert. 1982a. "The Economic Theory of Index Numbers and the Measurement of Input, Output, and Productivity," *Econometrica* vol. 50, no. 6, pp. 1393–1414.

———, L. R. Christensen, and W. E. Diewert. 1982b. "Multilateral Comparisons of Output, Input and Productivity Using Superlative Index Numbers," *Economic Journal* vol. 92, no. 365, pp. 73–86.

———, L. R. Christensen, and J. A. Swanson. 1981. "Productivity Growth, Scale Economies, and Capacity Utilization in U.S. Railroads, 1955–1974," *American Economic Review* vol. 71, no. 5, pp. 994–1002.

Chow, G. C. 1975. *Analysis and Control of Dynamic Economic Systems* (New York, N.Y., Wiley).

Christensen, L. R., and D. W. Jorgenson. 1969. "The Measurement of U.S. Real Capital Input 1929–1967," *Review of Income and Wealth* vol. 15, no. 4, pp. 293–320.

———, D. W. Jorgenson, and L. J. Lau. 1971. "Conjugate Duality and the Transcendental Logarithmic Production Function," *Econometrica* vol. 39, no. 4, pp. 255–256.

———, D. W. Jorgenson, and L. J. Lau. 1973. "Transcendental Logarithmic Production Frontiers," *Review of Economics and Statistics* vol. 55, no. 1, pp. 28–45.

Cobb, C. W. and P. H. Douglas. 1928. "A Theory of Production," *American Economic Review* vol. 18, no. 1 supplement, pp. 139–165.

Denison, E. F. 1967. *Why Growth Rates Differ* (Washington, D.C., The Brookings Institution).

———. 1979. *Accounting for Slower Economic Growth* (Washington, D.C., The Brookings Institution).

Denny, M., and M. Fuss. 1977. "The Use of Approximation Analysis to Test for Separability and the Existence of Consistent Aggregates," *American Economic Review* vol. 67, no. 3, pp. 404–418.

———, and M. Fuss. 1980. "Intertemporal and Interspatial Comparisons of Cost Efficiency and Productivity," Working Paper No. 8018 (Toronto, Ontario, Institute for Policy Analysis, University of Toronto).

———, and M. Fuss. 1983a. "A General Approach to Intertemporal and Interspatial Productivity Comparisons," *Journal of Econometrics* vol. 23, no. 3, pp. 315–330.

———, and M. Fuss. 1983b. "Intertemporal Changes in the Level of Regional labor Productivity in Canadian Manufacturing," in A. Dogramaci, ed., *Developments in Econometric Analysis of Productivity* (Boston, Mass., Kluwer Nijhoff).

———, and M. Fuss. 1983c. "The Use of Discrete Variables in Superlative Index Number Comparisons," *International Economic Review* vol. 24, no. 2, pp. 419–421.

———, and C. Pinto. 1978. "An Aggregate Model with Multiproduct Technologies," in M. Fuss and D. McFadden, eds., *Production Economics: A Dual Approach to Theory and Application* (Amsterdam, North-Holland).

———, M. Fuss, and L. Waverman. 1983. "The Measurement and Interpretation of Total Factor Productivity in Regulated Industries, with an Application to Canadian Telecommunications," in T. Cowing and R. Stevenson, eds., *Productivity Measurement in Regulated Industries* (New York, N.Y., Academic Press).

Diewert, W. E. 1971. "An Application of the Shephard Duality Theorem: Generalized Leontief Production Function," *Journal of Political Economy* vol. 79, no. 3, pp. 481–507.

———. 1973. "Functional Forms for Profit and Transformation Functions," *Journal of Economic Theory* vol. 6, no. 3, pp. 284–316.

———. 1974. "Applications of Duality Theory," in M. Intrilligator and D. A. Kendrick, eds., *Frontiers of Quantitative Economics*, vol. II (Amsterdam, North-Holland).

———. 1976. "Exact and Superlative Index Numbers," *Journal of Econometrics* vol. 4, no. 2, pp. 115–145.

———. 1978. "Superlative Index Numbers and Consistency in Aggregation," *Econometrica* vol. 46, no. 4, pp. 883–900.

———. 1980a. "Aggregation Problems in the Measurement of Capital," in D. Usher, ed., *The Measurement of Capital* (Chicago, University of Chicago Press).

———. 1980b. "Capital and the Theory of Productivity Measurement," *American Economic Review* vol. 70, no. 2, pp. 260–267.

———. 1981. "The Economic Theory of Index Numbers: A Survey," in A. Deaton, ed., *Essays in the Theory and Measurement of Consumer Behaviour in Honour of Sir Richard Stone* (London, Cambridge University Press).

———. 1985. "The Measurement of the Economic Benefits of Infrastructure Services," Discussion Paper No. 85–11 (Vancouver, British Columbia, Department of Economics, University of British Columbia).

Dogramaci, A., ed. 1981. *Productivity Analysis, A Range of Perspectives Studies in Productivity Analysis*, vol. 1 (Boston, Mass., Martinus Nijhoff).

Domar, E. 1961. "On the Measurement of Technological Change," *The Economic Journal*, vol. 71, pp. 709–729.

Douglas, P. H. 1976. "The Cobb-Douglas Production Function Once Again: Its History, Its Testing, and Some New Empirical Values," *Journal of Political Economy* vol. 84, no. 5, pp. 903–927.

Easter, D. W., M. E. Abel, and G. Norton. 1977. "Regional Differences in Agricultural Productivity in Selected Areas of India," *American Journal of Agricultural Economics* vol. 59, no. 2, pp. 257–265.

Eckstein, Z. 1984. "A Rational Expectations Model of Agricultural Supply," *Journal of Political Economy* vol.92, pp. 1–19.

Epstein, L. 1981. "Duality Theory and Functional Forms for Dynamic Factor Demands," *Review of Economic Studies* vol. 48, no. 151, pp. 81–95.

———, and M. G. S. Denny. 1983. "The Multivariate Flexible Accelerator Model: Its Empirical Restrictions and Application to U.S. Manufacturing," *Econometrica* vol. 51, no. 3, pp. 647–674.

———, and A. J. Yatchew. 1985. "The Empirical Determination of Technology and Expectations: A Simplified Procedure," *Journal of Econometrics* vol. 27, no. 2, pp. 235–258.

Evenson, R. E. 1968. "The Contribution of Agricultural Research and Extension to Agricultural Production" (Ph.D. dissertation, University of Chicago).

———, and Y. Kislev. 1975. *Agricultural Research and Productivity* (New Haven, Conn., Yale University Press).

Feder, G. 1982. "Adoption of Interrelated Agricultural Innovations: Complementarity and the Impacts of Risk, Scale and Credit," *American Journal of Agricultural Economics* vol. 64, no. 1, pp. 94–101.

Ferguson, C. E. 1971. *The Neoclassical Theory of Production and Distribution* (Cambridge, England, University Press).

Fisher, I. 1922. *The Making of Index Numbers* (Boston, Mass., Houghton Mifflin).

Forsund, F. R., C. A. K. Lovell, and P. Schmidt. 1980. "A Survey of Frontier Production Functions and of Their Relationship to Efficiency Measurement," *Journal of Econometrics* vol. 13, no. 1, pp. 5–25.

Fuss, M. and D. McFadden, eds. 1978. *Production Economics: A Dual Approach to Theory and Applications* (Amsterdam, North-Holland).

———, D. McFadden, and Y. Mundlak. 1978. "A Survey of Functional Forms in the Economic Analysis of Production, in M. Fuss and D. McFadden, eds., *Production Economics: A Dual Approach to Theory and Applications* (Amsterdam, North-Holland).

Green, J., and W. P. Heller. 1981. "Mathematical Analysis and Convexity with Applications to Economics," in K. J. Arrow and M. D. Intrilligator, eds., *Handbook of Mathematical Economics* vol. 1 (Amsterdam, North-Holland).

Griliches, Zvi. 1957. "Hybrid Corn: An Exploration in the Economics of Technological Change," *Econometrica* vol. 25, no. 4, pp. 501–523.

———. 1963. "Estimates of the Aggregate Agricultural Production Function from Cross-Sectional Data," *Journal of Farm Economics* vol. 45, no. 2, pp. 419–428.

———. 1964. "Research Expenditures, Education, and the Aggregate Agricultural Production Function," *American Economic Review* vol. 54, no. 6, pp. 961–974.

Hall, R. E. 1973. "The Specification of Technology with Several Kinds of Output," *Journal of Political Economy* vol. 81, no. 4, pp. 878–892.

Halter, A. N., H. O. Carter, and J. G. Hocking. 1957. "A Note on Transcendental Production Functions," *Journal of Farm Economics* vol. 39, no. 4, pp. 966–974.

Hanoch, G. 1971. "CRESH Production Functions," *Econometrica* vol. 39, no. 5, pp. 695–712.

Hansen, L. P., and T. J. Sargent. 1980. "Formulating and Estimating Dynamic Linear Rational Expectations Models," *Journal of Economic Dynamics and Control* vol. 2, no. 2, pp. 7–46.

―――― and K. J. Singleton. 1982. "Generalized Instrumental Variables Estimation of Nonlinear Rational Expectations Models," *Econometrica* vol. 50, no. 5, pp. 1269–1286.

Hatchett, S. A. 1984. "Dynamic Input Decisions: An Econometric Assessment of Crop Response to Irrigation" (Ph.D. dissertation, University of California, Davis).

Hayami, Y., and V. W. Ruttan. 1971. *Agricultural Development: An International Perspective* (Baltimore, Md., Johns Hopkins University Press).

Hazilla, M., and R. Kopp. 1984. "Industrial Energy Substitution: Econometric Analysis of U.S. Data, 1958–1974," EA-3462, Final Report to Electric Power Research Institute, Palo Alto, Calif.

――――, and R. Kopp. 1986a. "Testing for Separable Functional Structure Using Partial Equilibrium Models," *Journal of Econometrics* vol. 33, no. 1, pp. 119–142.

――――, and R. Kopp. 1986b. "Restricted Cost Function Models: Theoretical and Econometric Considerations," Discussion Paper QE86–05 (Washington, D.C., Resources for the Future).

Heady, E. O. 1952. "Use and Estimation of Input-Output Relationship or Productivity Coefficients," *Journal of Farm Economics* vol. 34, no. 5, pp. 775–786.

――――, and J. L. Dillon. 1962. *Agricultural Production Functions* (Ames, Iowa, Iowa State University Press).

――――, G. L. Johnson, and L. S. Hardin. 1956. *Resource Productivity, Returns to Scale, and Farm Size* (Ames, Iowa, Iowa State University Press).

Hicks, J. R. 1963. *The Theory of Wages* (New York, St. Martins).

Huffman, W. E. 1974. "Decision Making: The Role of Education," *American Journal of Agricultural Economics* vol. 56, no. 1, pp. 85–97.

Hulten, C. R. 1973. "Divisia Index Numbers," *Econometrica* vol. 41, no. 6, pp. 1017–1025.

――――. 1978. "Growth Accounting with Intermediate Inputs," *Review of Economic Studies* vol. 45, no. 141, pp. 511–518.

Jorgenson, D. 1984. "Econometric Methods for Modeling Producer Behavior," Discussion Paper 1086 (Cambridge, Mass., Harvard Institute for Economic Research).

――――, and Z. Griliches. 1967. "The Explanation of Productivity Change," *Review of Economic Studies* vol. 34, no. 99, pp. 249–283.

Kendrick, J. W. 1961. *Productivity Trends in the United States* (Princeton, N.J., Princeton University Press for the National Bureau of Economic Research).

――――. 1973. *Postwar Productivity Trends in the United States, 1948–1969* (New York, N.Y., National Bureau of Economic Research).

Kennedy, C., and A. Thirlwall. 1972. "Surveys in Applied Economics: Technical Progress," *The Economic Journal* vol. 82, no. 321, pp. 12–72.

Kohli, U. R. 1981. "Nonjointness and Factor Intensity in U.S. Production," *International Economic Review* vol. 22, no. 1, pp. 3–18.

Konüs, A. A. 1939. "The Problem of the True Index of the Cost of Living," *Econometrica* vol. 7, no. 11, pp. 10–29 (translation of a paper first published in *Ekonomicheskii Byulleten Konyunkturnovo Instituta* vol. 3, (1924), pp. 64–71).

Kopp, R. J. 1981. "The Measurement of Productive Efficiency: A Reconsideration," *Quarterly Journal of Economics* vol. 96, no. 3, pp. 477–504.

——— and W. E. Diewert. 1982. "The Decomposition of Frontier Cost Function Deviations into Measures of Technical and Allocative Efficiency," *Journal of Econometrics* vol. 19, no. 3, pp. 319–331.

Kuznets, S. 1952. "Long-Term Changes in the National Income of the United States of America Since 1870," *Income and Wealth,* series II (Cambridge, England, Bowes and Bowes).

Lau, L. J. 1972. "Profit Functions of Technologies with Multiple Inputs and Outputs," *Review of Economics and Statistics* vol. 54, no. 3, pp. 281–289.

———. 1978. "Applications of Profit Functions," in M. Fuss and D. McFadden, eds., *Production Economics: A Dual Approach to Theory and Applications,* vol. 1 (Amsterdam, North-Holland).

Lee, L., and M. Pitt. 1986. "Microeconomic Demand Systems with Binding Nonnegativity Constraints: The Dual Approach," *Econometrica* vol. 54, no. 5, pp. 1237–1242.

Lockheed, M. E., D. T. Jamison, and L. J. Lau. 1980. "Farmer Education and Farm Efficiency: A Survey," *Economic Development and Cultural Change* vol. 29, no. 1, pp. 37–76.

Lucas, Robert E., Jr. 1967. "Adjustment Costs and the Theory of Supply," *Journal of Political Economy* vol. 75, no. 4, pp. 321–334.

———, and T. J. Sargent. 1981. *Rational Expectations and Econometric Practice* (Minneapolis, Minn., University of Minnesota Press).

Magnus, Jan R. 1978. "Maximum Likelihood Estimation of the GLS Model with Unknown Parameters in the Disturbance Covariance Matrix," *Journal of Econometrics* vol. 7, no. 3, pp. 281–312.

Malmquist, S. 1953. "Index Numbers and Indifference Surfaces," *Trabajos de Estadistica* vol. 4, no. 2, pp. 209–242.

McFadden, D. 1978a. "Appendix A.3, Convex Analysis," in D. McFadden and M. Fuss, eds., *Production Economics: A Dual Approach to Theory and Applications* (Amsterdam, North-Holland).

———. 1978b. "Cost, Revenue, and Profit Functions," in D. McFadden and M. Fuss, eds., *Production Economics: A Dual Approach to Theory and Applications* (Amsterdam, North-Holland).

Mundlak, Y. 1963. "Specification and Estimation of Multiproduct Production Functions," *Journal of Farm Economics* vol. 45, no. 2, pp. 433–443.

———. 1968. "Elasticities of Substitution and the Theory of Derived Demand," *Review of Economic Studies* vol. 35, no. 102, pp. 225–236.

———, and A. Razin. 1969. "Aggregation, Index Numbers, and the Measurement of Technical Change," *Review of Economics and Statistics* vol. 51, no. 2, pp. 166–175.

Muth, J. F. 1960. "Optimal Properties of Exponentially Weighted Forecasts," *Journal of the American Statistical Association* vol. 55, no. 290, pp. 299–306.

Nadiri, Ishaq M. 1970. "Some Approaches to the Theory and Measurement of Total Factor Productivity: A Survey," *Journal of Economic Literature* vol. 8, no. 4, pp. 1137–1177.

Peterson, W., and Y. Hayami. 1977. "Technical Change in Agriculture," in L. Martin, ed., *A Survey of Agricultural Economics Literature*, vol. 1 (Minneapolis, Minn., University of Minnesota Press).

Petzel, T. E. 1978. "The Role of Education in the Dynamics of Supply," *American Journal of Agricultural Economics* vol. 60, no. 3, pp. 445–451.

Pindyck, R. S., and J. J. Rotemberg. 1983. "Dynamic Factor Demands and the Effects of Energy Price Shocks," *American Economic Review* vol. 73, no. 5, pp. 1066–1079.

Pollack, A. 1971. "The Theory of the Cost of Living Index," Research Discussion Paper No. 11 (Washington, D.C., U.S. Department of Labor, Bureau of Labor Statistics, Office of Prices and Living Conditions).

Pope, R. D. 1982. "To Dual or Not to Dual?" *Western Journal of Agricultural Economics* vol. 7, no. 2, pp. 337–351.

Revankar, N. S. 1971. "A Class of Variable Elasticity Production Functions," *Econometrica* vol. 39, no. 1, pp. 61–72.

Roumasset, J., J. M. Boussard, and I. Singh, eds. 1979. *Risk, Uncertainty, and Agricultural Development* (New York, Agricultural Development Council).

Rudin, W. 1964. *Principles of Mathematical Analysis*, 2nd edition (New York, McGraw-Hill).

Samuelson, P. A., and Swamy, S. 1974. "Invariant Economic Index Numbers and Canonical Duality: Survey and Synthesis," *American Economic Review* vol. 64, no. 4, pp. 566–593.

Schuh, G. E., and H. Tollini. 1979. "Costs and Benefits of Agricultural Research: The State of the Art," Staff Working Paper No. 360 (Washington, D.C., The World Bank).

Schultz, T. W. 1964. *Transforming Traditional Agriculture* (New Haven, Conn., Yale University Press).

———. 1975. "The Value of the Ability to Deal with Disequilibria," *Journal of Economic Literature* vol. 13, no. 3, pp. 827–846.

Shephard, R. W. 1953. *Cost and Production Functions* (Princeton, N.J., Princeton University Press).

———. 1970. *Theory of Cost and Production Functions* (Princeton, N.J., Princeton University Press).

Solow, R. M. 1955. "The Production Function and the Theory of Capital," *Review of Economic Studies* vol. 23, no. 61, pp. 101–108.

———. 1957. "Technological Change and the Aggregate Production Function," *Review of Economics and Statistics* vol. 39, no. 3, pp. 312–320.

Star, S., and R. E. Hall. 1976. "An Approximate Divisia Index of Total Factor Productivity," *Econometrica* vol. 44, no. 2, pp. 257–263.

Theil, H. 1968. "On the Geometry and the Numerical Approximation of Cost of Living and Real Income Indices," *De Economist* vol. 116, pp. 677–689.

Tornqvist, L. 1936. "The Bank of Finland's Consumption Price Index," *Bank of Finland Monthly Bulletin* no. 10, pp. 1–8.

USDA. See U.S. Department of Agriculture

U.S. Department of Agriculture, Economics, Statistics, and Cooperative Service. 1980. "Measurement of U.S. Agricultural Productivity: A Review of Current Statistics and Proposals for Change," Technical Bulletin No. 1614 (Washington, D.C., USDA).

Uzawa, H. 1962. "Production Function with Constant Elasticities of Substitution," *Review of Economic Studies* vol. 24, no. 81, pp. 291–299.

Varian, H. R. 1978. *Microeconomic Theory* (New York, Norton).

von Oppen, M., P. P. Rao, and K. V. Subba Rao. 1985. "Impact of Market Access on Agricultural Productivity in India," *Proceedings of the International Workshop on Agricultural Markets in the SAT* (Patancheru, A.P., India, International Crop Research Institute for the Semi-arid Tropics).

Walters, A. A. 1963. "Production and Cost Functions: An Econometric Survey," *Econometrica* vol. 31, no. 1, pp. 1–66.

Welch, F. 1970. "Education in Production," *Journal of Political Economy* vol. 78, no. 1, pp. 35–59.

Zellner, A. 1962. "An Efficient Method of Estimating Seemingly Unrelated Regressions and Tests for Aggregation Bias," *Journal of American Statistical Association* vol. 57, no. 298, pp. 348–368.

——— and N. S. Revankar. 1969. "Generalized Production Functions," *Review of Economic Studies* vol. 36, no. 107, pp. 241–250.

3

A REVIEW OF THE EVIDENCE ON AGRICULTURAL PRODUCTIVITY AND AGGREGATE TECHNOLOGY*

SUSAN M. CAPALBO AND TRANG T. VO

Productivity growth is often cited as one of the major factors contributing to the continued economic growth of the postwar agricultural sector. The main purpose of this chapter is to present the empirical evidence on productivity and technical change for the aggregate agricultural sector. Although the presentation is to some extent a selective literature review, we provide some historical insights by viewing the research that developed in light of the evolution of theories and approaches to productivity measurement. This in turn facilitates the comparisons among studies, as many of the empirical discrepancies are partly a result of differences in defining and measuring technical change and productivity.

To keep the review to a reasonable length, we focus in greater detail on the developments of the last decade and a half, and draw upon the surveys by Nadiri (1970), Kennedy and Thirlwall (1972), and Peterson and Hayami (1977) for the results of the earlier periods. In addition we limit our coverage of empirical results to selected U.S. and Canadian aggregate sector analyses.[1]

In general, the study of productivity encompasses measurement and identification of the phenomenon; analysis of impacts and distortions due to macro and sector policies; and the identification of the incidence of the benefits of productivity and its distributional impacts. This chapter reviews the descriptive information on agricultural productivity and aggregate technology, that is, the determination of changes in productivity. We do not address in any detail the factors responsible for these growth rates or the welfare effects of productivity changes.

Approaches to measurement of productivity and technical change may be grouped into two broad categories: (a) analysis of accounting data, which involves mainly the development of indexes of outputs and

---

*The comments by Stanley Johnson and anonymous reviewers on early versions of this paper are greatly appreciated.

[1] No effort is made to provide a statistical compendium of all results.

inputs and the computation of nonparametric factor productivity measures, and (b) analyses that estimate production relations with econometric or programming techniques. Classifying all studies into one of the two categories is somewhat arbitrary, as both approaches are frequently used within a given study. Furthermore, from the discussions in chapter 2 of alternative approaches to productivity measurement it is evident that even the nonparametric methods impose an implicit structure on the aggregate production technology.

Because of this overlap of methodological boundaries, we structure our review along historical lines. In the second section, we summarize the pre-1970 productivity and technical change literature. This literature, as reflected in the debates of the 1950s and 1960s, highlighted the problems of aggregation and identification in estimating the rate of technical progress as a residual component of the growth of output and laid the groundwork for much of the subsequent analyses. In the main, the empirical techniques for measurement developed during this period relied on index numbers and primal measures of total factor productivity (TFP) and technical change.

The methodological linkages between index numbers and production structures that were extensively delineated in the 1970s by Diewert and others were used in the empirical research that emerged in the mid-1970s and 1980s. A synthesis of these methodologies and empirical studies are presented in the third section. The chapter concludes with comparisons of the empirical evidence from the post-1970 studies.

HISTORICAL PERSPECTIVE: PRE-1970

Productivity has generally been thought of in terms of the efficiency with which inputs are transformed into useful output within the production process. The tendency to view productivity growth as "manna from heaven" was quickly dispelled in the post–World War II era. Since then, productivity growth has been recognized as resulting from the diversion of resources to activities that generate technical progress.

Among the earliest methods of quantifying agricultural productivity were the index number approaches, beginning with the research by Barton and Cooper (1948) and Loomis and Barton (1961). These productivity measures were ratios of an index of aggregate output to either a single factor—typically labor—or an index of all factors. The single factor, or partial productivity, index masked many of the factors accounting for observed productivity growth, such as the substitution among inputs. The use of partial indexes was replaced by the TFP measures in the late 1950s, when output and input indexes were constructed using either a linear aggregation with market prices as weights or a geometric aggregation with factor and revenue shares as weights.

The popularization of the arithmetic indexes for TFP measurement is often associated with the work of Kendrick (1961). The study by Loomis and Barton (1961) set the precedent for the use of the Laspeyres arithmetic index by the U.S. Department of Agriculture (USDA). The appropriateness of such an index has been the subject of much debate, as both the Laspeyres and Paasche indexes imply an underlying production function characterized by either an infinite elasticity of substitution between inputs or zero elasticity of substitution. Solow (1957) and Domar (1961) linked the Cobb-Douglas production function to the geometric index of inputs. Ruttan (1957), Chandler (1962), and Lave (1964) were among the earliest users of the geometric approach for studying productivity and technical change in U.S. agriculture.

It was recognized that the TFP indexes leave a large portion of the growth of output unexplained (Abramovitz, 1956; Solow, 1957; Fabricant, 1959). Partially as an attempt to minimize the unexplained residual, productivity was linked to shifts in the production relations. Solow's research associated the geometric index of productivity with technical change by explicitly utilizing the aggregate production function; he also adjusted the capital input for quality changes. Ruttan (1957) utilized the production framework and viewed technical change as an upward shift in the production function. Schultz (1953) stressed that by allowing the production function to shift over time economic growth is being treated as exogenous to the system.

Subsequent work with production functions was aimed at separating technical progress from other factors. Issues studied included the substitution of capital for labor, economies of scale, the effects of human capital investment, and errors in measurement. Jorgenson and Griliches (1967, 1972) looked at the determinants of TFP for the entire U.S. economy. By adjusting output and input data for aggregation errors, errors in capital prices, and utilization of capital and labor, they were able to reduce substantially the unexplained portion of the growth in output.

The work by Griliches (1960, 1963, 1964) on U.S. agriculture also reflected the idea that research should aim at minimizing the unexplained portion of output growth. Griliches used differences in education to adjust labor for quality and included research and extension expenditures as inputs. His results indicated that for the late 1940s and 1950s the amount of technical change or unexplained residual that could not be attributed to the quality adjusted inputs was small: economies of scale, labor quality changes, and R&D were major factors in explaining the output growth. Griliches also attempted to test the hypothesis that technical change is embodied in the quality changes of the conventional inputs by differentiating agricultural capital by age (older capital presumably is less productive than newer vintages). He found that the age of capital had little explanatory power in accounting for the differences in output.

Technical progress was characterized as neutral or biased. Based on Hicks's definitions, neutral technical change implied that the ratio of the marginal product of two inputs was left unchanged. Hicks argued that historically the accumulation of capital relative to labor tended to decrease capital's share of production. At the same time, the rise in the price of labor relative to capital tended to have an offsetting effect, and caused induced innovation to have a labor-saving bias.

Empirical tests of the induced innovation hypothesis were hampered by the difficulty of distinguishing movements along the production function from shifts in the production function. For example, to what extent should the increase in capital intensity be attributed to factor substitution as a result of the change in relative prices, and to what extent to a possible labor-saving bias in technical change? Stout and Ruttan (1958) argued that technical change could not have been neutral in U.S. agriculture: the rapid decline in farm employment from 1925 to 1955 could not have resulted only from the movement of the price of labor relative to other inputs. Hayami and Ruttan (1970, 1971) compared trends of factor prices and factor proportions in the long-term agricultural development in the United States and Japan and found the patterns to be consistent with the induced innovation theory; that is, changes in the level of factor prices influenced the direction of innovation activity.

The presence of economies of scale was also recognized as influencing the estimates of technological progress. To the extent that economies of scale exist, the rate of technological change is overestimated by using production models that impose linear homogeneity (see discussions in chapter 2).

In summary, by the late 1960s substantial progress was made in formulating the concept of TFP and its relation to technical change. The empirical evidence that emerged during this period on the level and sources of aggregate agricultural productivity growth was based on index number approaches that implied fairly restrictive types of production technologies, and single-equation estimates of an aggregate production function. Attempts to explain the aggregate residual by attributing the growth of agricultural productivity to changes in scale and the quality of inputs were limited and, as Peterson and Hayami (1977) concluded, best summarized by Griliches (1963, 1964): the unexplained growth in agricultural productivity can be attributed in fairly equal proportions to input quality changes, economies of scale, and public investment in research and extension.

HISTORICAL PERSPECTIVE: POST-1970

Methodological Developments

A general thrust of the methodological developments since 1970 has been to improve upon the techniques for separating the factors contributing to productivity growth rates and to test hypotheses raised in previous decades. These advances have been due largely to the development of the flexible functional form production models and related econometric procedures, and the development and use of the theory of exact index numbers.

We provide only a brief summary of the post-1970 developments in the theory of productivity measurement (for a detailed discussion, see chapter 2). First, Diewert's (1976) work on exact and superlative index numbers relaxed the rigid structure imposed *a priori* on the aggregate production function that resulted from using either the geometric or arithmetic indexing procedures. The Tornqvist-Theil superlative index, which is an exact index for the homogeneous translog approximation of the transformation function, is utilized as an alternative index of TFP. It is noted that the Tornqvist-Theil productivity index, corrected by a scale factor, is also equal to the mean of two Malmquist productivity indexes, the latter index being defined for production structures with arbitrary returns to scale, elasticities of substitution, and technical change biases (Caves, Christensen, and Diewert, 1982).

Second, the use of the Divisia indexing procedure (or the Tornqvist-Theil approximation) reflected efforts to adjust the conventional inputs for differences in quality. That is, since the Divisia procedure relies on current factor prices in constructing the weights and to the extent that quality improvements to the inputs are reflected in higher wage rates and rental prices, these adjustments are accounted for, at least conceptually.

Third, through the use of duality theory, technical change can be viewed as parametric shifts in the cost function. The conditions required for the equivalence of the primal and dual measures of technological change are discussed in chapter 2.

Fourth, since the early 1970s, many studies have estimated substitution and various output elasticities, economies of scale, and factor biases that have utilized flexible functional forms, such as the generalized Leontief, the translog, or the generalized Box-Cox (Berndt and Khaled, 1979). The joint econometric estimation of a system of factor demand and output supply equations derived explicitly from a flexible functional form for the production technology has shed new light on quantifying the technical characteristics of the production process. In conjunction with these econometric advances, the production models have relaxed the competitive equilibrium assumptions implicit in the growth accounting approach, and decomposed the rate of productivity growth into scale effects, nonmarginal cost pricing effects, and technological change (Denny, Fuss, and Waverman, 1981). Restricted cost and profit function models have been estimated for agriculture as well as for other sectors (Caves, Christensen, and Swanson, 1981). How important this is to the measurement of multifactor productivity is an empirical issue, dependent on the extent of the temporary disequilibrium.

Finally, researchers have recognized the role nonmarket factors and environmental regulations have had in influencing the TFP measures, and have attempted to adjust productivity measures accordingly (Pittman, 1984; Kopp and Smith, 1981). Methodologies for incorporating nonmarket effects range from additional quality adjustments to the conventional inputs to modeling the production process and the TFP measures in a dynamic context. For example, the (shadow) costs of adjusting to changing levels (or qualities) of the resources are incorporated into the cost function (Epstein and Denny, 1980). To date, few empirical studies have been based on these dynamic frameworks.

Empirical Results

Productivity Indexes. The use of Divisia indexing procedures for aggregating U.S. agricultural outputs and inputs was first adopted by Brown (1978). Subsequent updates and modification of these indexes have been performed by Ball (1985) for 1948–79, and by Capalbo, Vo, and Wade (1986) for 1948–83. These modifications incorporate the recommendations of a recent American Agricultural Economics Association (AAEA) task force that reviewed current USDA productivity statistics (USDA, 1980) and the recommendations and suggestions made by Shumway in chapter 4. In this section we report only the Capalbo, Vo, and Wade (CVW) indexes, as they include the most recent years. The CVW output and input indexes are provided in the appendix to this chapter.

In order to characterize the growth of U.S. agriculture over the post-World War II period, we have calculated the rates of growth of total revenue, aggregate real output, total cost, and aggregate real input for several subperiods (table 3-1). Revenue shares were used to aggregate the outputs; cost shares were used as weights in the input aggregations. Total revenue grew at an average rate of 4.33 percent per year over the 1950–82 period. Aggregate real output growth was somewhat slower, averaging only 1.76 percent over this same period. Since 1960, total revenue has grown more than three times as quickly as aggregate output, indicative of the substantial price increases that characterized this period.

Costs were calculated using a service price for capital and land. Both capital and land are factors that are purchased in one time period but deliver a flow of factor services over subsequent time periods. Given perfect rental markets for these factor services, rental prices would provide the correct service flow prices. Unfortunately, these markets do not exist, so we utilized a proxy for the unobservable flow price based on the work of Hall and Jorgenson (1967) and Christensen and Jorgenson (1969, 1970). In general the service price depends upon the asset price, the rate of return, the rate of physical deterioration, the rate of capital appreciation or depreciation, and the tax structure.

Total cost grew more rapidly than total revenue for all subperiods. The increases in the value of land accounted for the major portion of the growth rate of total costs. The substantially higher growth rates for total costs relative to growth rate of aggregate inputs are also indicative of the substantial increase in the factor prices over the entire period.

Tables 3-2 and 3-3 display the average annual growth rates for two aggregate output and nine aggregate input categories, respectively. Note that crops accounted for a substantially larger share of total revenue relative to livestock products (60 percent versus 40 percent). The rate of growth of livestock output, in constant dollars, declined continuously

TABLE 3-1. AVERAGE ANNUAL RATES OF GROWTH OF TOTAL REVENUE, AGGREGATE REAL OUTPUT, TOTAL COST, AND AGGREGATE REAL INPUT, 1950–1983

Period	Revenue (current dollars)	Output (constant dollars)	Cost (current dollars)	Input (constant dollars)
1950–60	0.74	1.66	2.22	0.14
1960–70	2.44	0.84	5.26	0.05
1970–82	8.89 (7.73)	2.60 (1.45)	11.28 (9.41)	0.34 (0.13)
1950–82	4.33 (4.01)	1.76 (1.33)	6.57 (5.97)	0.17 (0.09)

Note: Growth rates in parentheses are the average annual rates of growth over the given period using 1983 as the last year.

TABLE 3-2. RATES OF GROWTH OF OUTPUT, 1950–1982

Period	Livestock	Crops	All outputs
1950–60	1.34	1.91	1.66
1960–70	1.16	0.56	0.84
1970–82	0.84	3.72	2.60
(83)	(0.94)	(1.78)	(1.45)
1950–82	1.10	2.17	1.76
(83)	(1.13)	(1.45)	(1.33)

Note: Growth rates in parentheses are the average annual rates of growth over the given period using 1983 as the last year.

throughout the post–World War II years. A substantially different pattern exists in the rate of growth of all crops, most notably the high rate of growth for crops in the 1970–82 period.

The input aggregations reflect adjustments for changes in education and composition of the labor force; changes in the vintage of the equipment; nonhomogeneity of the land, on a state-by-state basis; and changes in active ingredients of the pesticides and fertilizers over time. The rates of growth of the aggregate input groups had distinct patterns. The growth rate of family labor was negative for all periods; hired labor followed a similar pattern from 1950–70. This decline in agricultural labor was less pronounced than previous estimates by the USDA, because the labor inputs were quality adjusted for changes in the composition of the agricultural labor force. The difference between an unadjusted and a quality-adjusted labor index is shown in figure 3-1. The unadjusted labor index declined nearly twice as fast as the quality-adjusted index.

In real terms, the growth rates for land and structures were nearly nil over the entire postwar period. The most dramatic growth rates were for agricultural chemicals (fertilizers and pesticides), especially during the first two decades. The apparent decline in the growth rates for chemicals in the 1970s and early 1980s should be interpreted cautiously, as these indexes were aggregated using total expenditures based on pounds of active ingredients as weights, which may not reflect the toxicity or marginal value product of the chemical.

The cost shares for selected years are provided in table 3-4. The shares for equipment and animal capital, for fertilizer and pesticides, and for energy have been fairly stable. For each of these input groups, the total expenditures kept pace with the rise in the total costs (table 3-1). This implies that the factor prices for equipment, animal capital, and energy showed a substantial increase over this period, whereas the factor prices for agricultural chemicals were fairly stable. The substantial rise in the

TABLE 3-3. RATES OF GROWTH OF INPUTS, 1950–1982

Period	Hired labor	Family labor	Equipment	Animal capital	Structures and land	Fertilizer	Pesticides	Energy	Other materials	All inputs
1950–60	−3.02	−4.07	0.74	1.59	0.38	5.91	4.77	3.27	2.23	0.14
1960–70	−2.24	−3.78	2.71	−0.22	0.00	7.60	10.81	2.72	0.60	0.05
1970–82	0.84	−1.70	2.56	−0.13	0.07	2.08	3.22	−0.75	0.84	0.34
(83)	(1.08)	(−1.86)	(1.96)	(−0.26)	(0.00)	(0.75)	(1.31)	(−1.14)	(0.47)	(0.13)
1950–82	−1.32	−3.09	2.04	0.38	0.10	5.01	6.07	1.58	1.20	0.17
(83)	(−0.70)	(−3.11)	(1.89)	(0.31)	(0.06)	(4.39)	(5.23)	(1.36)	(1.04)	(0.09)

Note: Growth rates in parentheses are the average annual rates of growth over the given period using 1983 as the last year.

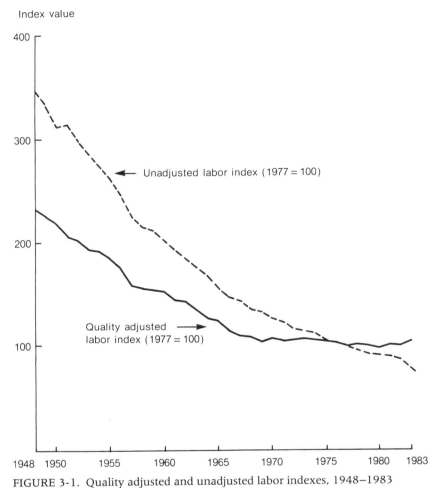

FIGURE 3-1. Quality adjusted and unadjusted labor indexes, 1948–1983

cost share for land and structures reflects the dramatic increases in per acre values (service prices) of land over this period.

The conventionally measured Divisia index of TFP growth was obtained as the difference in the proportional rates of growth of aggregate output and aggregate input (see discussion in chapter 2). The index of TFP from 1948 to 1983 is shown in tables 3-5 and 3-6. This index grew at an average annual rate of 1.57 (1.22) percent from 1950–1982 (1950–1983), with the most substantial growth occurring during the

TABLE 3-4. COST SHARES FOR 1960, 1970, AND 1980

Input	1960	1970	1980
Labor	0.24	0.20	0.11
Equipment and animal capital	0.25	0.23	0.21
Structures and land	0.16	0.26	0.41
Fertilizer and pesticides	0.04	0.05	0.06
Energy	0.04	0.03	0.04
Other materials	0.26	0.22	0.17

TABLE 3-5. TOTAL FACTOR PRODUCTIVITY FOR U.S. AGRICULTURE, 1948–1983

Year	TFP	Year	TFP
1948	70.33	1966	83.45
1949	68.52	1967	85.44
1950	67.00	1968	85.63
1951	68.87	1969	86.55
1952	71.89	1970	84.36
1953	74.33	1971	88.70
1954	74.21	1972	89.69
1955	73.51	1973	91.15
1956	79.11	1974	94.40
1957	77.19	1975	96.32
1958	78.62	1976	96.31
1959	75.84	1977	100.00
1960	77.99	1978	97.29
1961	78.64	1979	102.50
1962	80.17	1980	102.31
1963	81.54	1981	109.63
1964	83.41	1982	110.71
1965	82.62	1983	100.12

1970s and early 1980s. This TFP index, based on the CVW data, can be contrasted to the USDA index of TFP. Explanations for the differences in these two indexes can be traced to the differences in the measurement of the inputs and outputs. In general, the USDA labor index does not include age- and education-related adjustments; it does not distinguish between hired and family workers; it does not assign weights to the pesticide and fertilizer components of the combined indexes; it does not incorporate taxes on capital and land into the measure of the service prices for these inputs; and the land index does not account for non-homogeneous changes in quality, as reflected in disproportionate increases in land value on a state-by-state basis. The USDA aggregate input index has a slightly negative rate of growth of inputs over the period from 1950–82.

TABLE 3-6. ANNUAL RATES OF GROWTH OF TOTAL FACTOR
PRODUCTIVITY FOR U.S. AGRICULTURE, 1950–1982
(Percent)

Period	Annual rate of growth
1950–60	1.52
1960–70	0.79
1970–82 (83)	2.26 (1.32)
1950–82 (83)	1.57 (1.22)

Note: Growth rates in parentheses are the average annual rates of growth over the period using 1983 as the last year.

The USDA index of aggregate output grew at a faster average annual rate than the CVW aggregate output index for the period from 1950 to 1982 (1.95 percent versus 1.76 percent). This difference is partially attributed to the fact that the USDA output measure is net of farm production that is used as an input on farms—that is, seed and feed—whereas the CVW index reflects a fully gross output index.[2] The AAEA task force report cites the "dubious" nature of the data used to net out farm-produced feed and seed as a rationale for using a fully gross output approach (USDA, 1980). A second reason for the discrepancy in the rate of growth of the indexes may be the indexing procedure itself. The USDA output groups are aggregated arithmetically using base-period price weights. The bias due to changing relative prices is a perennial concern in using the arithmetic indexing procedure.[3]

As a result of these differences in the aggregate output and input indexes, the USDA index of TFP grew at a faster rate than the CVW Divisia index of TFP. From 1950 to 1960 the USDA index grew at 2.32 percent per annum compared to 1.52 percent per annum for the CVW index; from 1960 to 1970 the USDA index grew at 1.22 percent versus 0.79 percent; and from 1970 to 1982 the USDA rate of growth was identical to the CVW growth rate. However, for the 1950–82 period the rate of growth of aggregate output not explained by the growth of aggregate inputs was nearly 25 percent higher based on the USDA indexes (1.94 percent versus 1.57 percent).

The total factor productivity index and the land and labor productivity indexes based on the CVW data base are graphed in figure 3-2. The partial productivity measures are useful performance indicators if and

---

[2] To avoid double counting, CVW include feed and seed produced and used by the farm sector as an input.

[3] See the AAEA task force report (USDA, 1980) for further discussion of these biases in the context of the U.S. agricultural indexes of inputs and outputs.

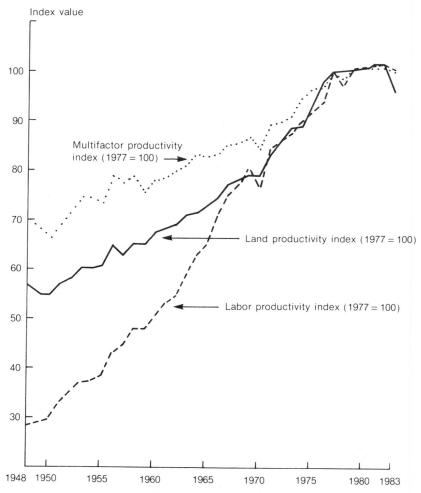

FIGURE 3-2. Indexes of multifactor, labor, and land productivity, 1948–1983

only if the amount of the other resources used remains unchanged. As evident from the preceding discussions, this has not occurred. The amounts of nonlabor inputs increased at an annual rate of 1.12 percent from 1950 to 1982. The labor productivity index thus does not measure outputs obtained from given resources, nor does it measure the attributes of labor as a productive resource. Higher output per labor hour was achieved by the increased use of fertilizer, machinery, and other capital equipment.

The rate of growth of an index of nonland inputs was fairly constant over this same period. For this reason we see a fairly close parallel between the growth rates of TFP and land productivity, especially through the early 1970s. Higher output per acre was achieved by the simultaneous increased use of nonlabor, nonland inputs and the decreased use of labor inputs. However, as with all productivity indexes, these partial indexes are void of any economic interpretation, because there is no information on the relationship between technological change, scale economies, and substitution among inputs.

Econometric Results. The results reviewed in this section are based on aggregate econometric production models of U.S. and Canadian agriculture. In general, these models used the cost or profit function and the corresponding set of factor demands and output supply functions in a joint estimation procedure. For each study, we briefly highlight the assumptions and specification of the model, the results of hypothesis tests on the structure of the production technology, and the data set. Cross-study comparisons of the factor demand and output supply elasticities, scale effects, elasticities of substitution, and technological change effects are presented in the following section of this chapter.

One of the earliest empirical studies on aggregate U.S. agriculture to use duality theory was Binswanger's (1974a,b) translog cost function model. Binswanger assumed that the production technology was homothetic, implying that the factor cost shares could be expressed as functions of factor prices and time; his measure of technological change bias depends only on these variables. If the technology is nonhomothetic, measures of technological change must take into account the output, or scale, effect on factor cost shares. By imposing the (untested) homotheticity restrictions, Binswanger's measures of technological change biases as well as factor demand elasticities may be misleading. Furthermore, by estimating only the set of factor share equations, he was unable to obtain an estimate of the rate of technical change as reflected in the intertemporal shift of the cost function.

Binswanger's model was applied to pooled cross-section, time series data for thirty-nine states for 1949, 1954, 1959, and 1964. He used the estimated parameters from the cost share model to decompose the observed changes in cost shares over the period from 1912 to 1968 into technological change effects and observed factor price changes. His results indicated that technical change had in general been labor-saving and machinery- and fertilizer-using.

Binswanger compared the time paths of the technological change biases with the time paths of factor prices in order to test the induced innovation hypothesis. There had been an overall machinery-using bias and an average rise in the relative machinery price. With respect to labor, there had been an overall factor-saving bias and an average rise in relative labor costs. These trends were especially pronounced from the mid-1940s. For fertilizer the opposite was true, with a substantial fertilizer-using bias and sharp decline in the relative price index for fertilizer appearing.

Closely related to the Binswanger analysis is the study by Brown and Christensen (1981). To characterize the structure of U.S. agriculture, Brown and Christensen specified a single-output, five-input translog restricted cost function. The quasi-fixed inputs were land and family labor. Hired labor, durable capital, and materials were treated as variable inputs. Linear homogeneity of the variable cost function was imposed with respect to the variable factor prices and constant returns to scale on the underlying production function.

The empirical results are suspect on three accounts: variable costs were increasing (decreasing) as output levels decrease (increase), although the coefficient was not statistically significant; there was technological regression; and variable costs increased when the level of the quasi-fixed input increased, holding all other factors constant.

More recently Hazilla and Kopp (1986) have reestimated the Brown-Christensen model imposing (a) the monotonicity restrictions for the arguments of the cost function and (b) the set of theoretical restrictions for concavity of the restricted cost function in input prices given fixed output and quasi-fixed input levels, and for convexity of the restricted cost function in output and quasi-fixed input levels, given input prices. In comparing their results with those of Brown and Christensen they were able to demonstrate the distortions to perceptions of production structures that result from a reliance on "fitted" models that are inconsistent with the underlying theory of firm behavior. Since the Brown-Christensen curvature violations were primarily associated with the quasi-fixed factors and output, the most severe discrepancies between their results and the Hazilla-Kopp results are with respect to the long-run elasticities of substitution and factor demand elasticities, and with respect to the cost elasticity. For example, the Hazilla-Kopp cost elasticity (inverse of short-run returns to scale elasticity) was approximately 1.046 at the point of approximation compared with the Brown-Christensen estimate of -0.32. Further comparisons are made in the next section.

Ray (1982) utilized a multiple-output, five-input translog cost function to characterize U.S. agricultural production over the 1939–1977 period. His cost function was restricted *a priori* to be Hicks neutral; he also imposed linear homogeneity in input prices and marginal cost pricing for the outputs. The two outputs were livestock and crop; the input categories were hired labor, farm capital (machinery and land), fertilizer, purchased intermediate inputs, and miscellaneous inputs.

Ray tested for constant returns to scale by testing for constant average costs; this hypothesis was strongly rejected. He also rejected homotheticity of the underlying technology. He concluded that Hicks-neutral technical change occurred at an annual rate of 1.8 percent. In this model technological change did not affect either the cost shares or the revenue shares; thus, any changes in these shares over time had to originate from changes in factor prices or output. His estimate of scale economies indicated that the U.S. agricultural sector was characterized by decreasing returns, although scale increased from 0.64 in 1939 to 0.83 in 1977.

One needs to be cautious in using the single-output cost function test for economies of scale for a multiple-output model, because average cost is not defined for multiple-output technologies. Economies of scale in an overall sense require that all outputs increase proportionately; economies of scale with respect to the ith output are not necessarily an indication of returns to scale in the overall case (unless the output-specific returns to scale measure exceeds unity).

Lopez (1980) analyzed the structure of Canadian agriculture for the 1946–1977 period using a modified generalized Leontief cost function; his set of estimating equations included only the factor demands expressed as input-output ratios. He tested for symmetry, a fixed coefficient production technology, constant returns to scale (and nonhomotheticity), and factor-augmenting technical change. The symmetry restrictions could not be rejected; the hypothesis of a fixed proportion Leontief production function was strongly rejected, indicating factor substitution responses to relative price changes; the hypothesis of constant return to scale was rejected. Because of Lopez's form of the stated hypothesis for constant returns to scale, rejection implied a nonhomothetic production function. Finally, the hypothesis of no factor-augmenting technical change could not be rejected. Lopez indicated that in light of this final test, the "observed decrease in agricultural labor demand and increased capitalization would reflect changes in relative prices and a nonhomothetic production function rather than (biased) technical change" (p. 43).

Chan and Mountain (1983) also provided evidence on returns of scale and technical change in Canadian agriculture utilizing a translog production model. The input categories were land, labor, machinery, and materials, and the data cover the 1952–1977 period.

Chan and Mountain specified a series of models in which they sequentially relaxed the assumptions regarding nonneutral technical change and nonconstant returns to scale. Based on the likelihood ratio test, the Hicks-neutral specification was not rejected. At the point of approximation, the rate of Hicks-neutral technical change was 0.28 percent, although the parameter was not statistically significant. Furthermore, under Hicks neutrality, there were increasing returns to scale of the magnitude of 2.5.

An alternative model was specified in which constant returns to scale were imposed. Based on the likelihood ratio test, constant returns to scale could be rejected at the 5 percent significance level. Finally, the joint hypothesis of no technical change and nonconstant returns to scale was not rejected at either the 5 percent or 10 percent level of significance. For this latter specification, the estimated scale parameter was approximately 2.5. Chan and Mountain concluded that if the increasing returns to scale measure was as dominant as their econometric results indicated, employing the conventional Tornqvist measure of growth overestimates technical change.[4]

The Chan and Mountain study involved estimating a set of equations that was nonlinear in the parameters, including the scale parameter. The estimated parameters from such a specification are likely to be less stable than those resulting from a linear specification. Although the returns to scale parameters in the nonconstant returns to scale models were always in excess of 2.0 and statistically significant, the alternative model that imposed constant returns to scale was not rejected. Such a large "acceptable" range for the returns to scale parameter is puzzling.

Antle (1984) utilized a normalized four-input, one-output translog profit function model to analyze technological change biases and the structure of the U.S. agricultural sector for 1910–78. He estimated only the set of normalized factor share equations and thus did not obtain an estimate of the rate of technological change. His input aggregates were farm labor, farm real estate, mechanical power and machinery, and chemicals. Farm labor and mechanical power and machinery were adjusted to reflect quality changes. The homotheticity restrictions were rejected for both the pre- and post-World War II periods. For the prewar period the factor biases were toward machinery and labor and against land. For the postwar period the biases were toward machinery,

---

[4] Chan and Mountain show that the conventional Tornqvist approximation to changes in TFP is appropriate only under constant returns to scale. The discrete approximation to the expression for $\partial \ln Q/\partial T$ based on the translog production function is

$$\text{TFP}_t = (\ln Q_t - \ln Q_{t-1}) - \theta \Sigma \frac{1}{2}(S_{it} + S_{i,t-1})(\ln X_{it} - \ln X_{i,t-1})$$

where S_i denotes the ith cost share and θ is the degree of homogeneity of the production function. Decreasing (increasing) returns would under(over)estimate technological change, defined in the narrow sense as changes in TFP.

chemicals, and land and against labor, indicating that technology has been labor-saving and capital- and chemical-using.

During the prewar period, only chemical inputs declined in response to a scale increase; during the postwar period, usage of chemicals and land increased whereas usage of labor declined with respect to scale increases. Antle reported that both output and inputs, especially labor, were less price responsive after World War II: changes in factor use were more a function of scale and technological change in this period relative to the earlier years.

Antle reestimated the model imposing the previously rejected homotheticity restrictions and compared the technical change biases. During the postwar period, the homothetic model had smaller biases associated with labor and chemicals relative to the nonhomothetic model, and had a significant bias against land. The land bias had been insignificant in the nonhomothetic model. The homothetic specification also resulted in larger factor demand elasticities.

Capalbo and Denny (1986) have developed a series of tests designed to evaluate alternative structures for shifts in technology over the long run for the agricultural sectors of the United States and Canada. Their analysis was based on a three-input translog production model, where the inputs were labor (unadjusted for quality changes), capital (land, structures, and equipment), and material inputs. The time period covered the decades of the 1960s and 1970s. Technological change had been occurring at an average annual rate of 1.9 percent and 1.4 percent in the Canadian and U.S. agricultural sectors, respectively, during the 1960s and 1970s, as measured at the point of approximation (1970). Furthermore, there was evidence in both countries that the long-run shifts had been Hicks neutral in models that use gross rather than net output measures.

The specification of a single aggregate output imposes an additional limitation due to the implied separability restrictions, that is, an aggregate output quantity index must exist. The use of multiple-output models—and in particular, multiple-output profit models—does not impose such severe separability restrictions, nor does it require allocation of the inputs to a specific output. The following three studies adopted a multiple-output production structure.

Based on a multiple-output profit model, Lopez (1984) separated the substitution and expansion effects for the inputs and outputs and obtained both the compensated and uncompensated sets of output supply and input demand elasticities. His empirical results pertained to the Canadian agricultural sector and were based on cross-sectional data for 1971. At the sample mean, the uncompensated elasticities based on a modified generalized Leontief profit function are inelastic except for hired labor and farm capital. Lopez also noted the difference in the responses of hired and family labor to price changes. The demand for family labor was highly inelastic; the demand for hired labor was elastic. The cross elasticities indicated that higher wages for hired labor affected livestock production and had almost no effect on crop production. Higher returns to family labor (which is primarily an opportunity cost to the farmer) had less effect on altering the composition of the output.

A recent attempt to incorporate agricultural policies into an aggregate production model was the study by Huffman and Evenson (1986). They provided measures of biases of technology caused by agricultural research, extension, and farm programs, as well as evidence on the shadow prices or returns to the fixed factors and government policies. Their analysis was based on a multiple-output, multiple-input restricted quadratic profit model fitted to pooled time series cross-sectional data for grain farms in forty-two states for the six agricultural census years between 1949 and 1974.

Huffman and Evenson found that an increase in agricultural research and extension increased variable profits but reduced total profits, including economic rents to the fixed factor (land) and government policies. Agricultural research resulted in a bias in favor of fertilizer, fuel, and labor inputs, as well as the production of wheat, but had a neutral effect on the machinery input and a bias against soybean production. In contrast, farmers' education (measured as a weighted average number of years of schooling) induced the opposite biases. Huffman and Evenson also note that the estimated bias effects associated with farm programs were weak. Although additional research is needed better to understand the interactions among the conventional inputs and outputs and the public agricultural research and policy variables, this research represents a major effort to quantify nonmarket impacts.

In a regional study with methodological implications for aggregate production models, Shumway (1983) specified a multiple-output, multiple-input restricted quadratic profit function for the Texas agricultural sector. Family labor and land were fixed variables and government policies, weather, and technical change were exogenous factors. Shumway constructed a series of tests for nonjointness in production and homothetic separability. The tests for homothetic separability translate into tests for weak separability and homotheticity (Lau, 1978). Weak separability of the technology was tested by examining the normalized profit function for evidence of homotheticity and weak separability in the set of normalized prices.

On the basis of these latter two sets of hypothesis tests, Shumway concluded that only three outputs—cotton, sorghum, and corn—could be consistently aggregated for the Texas agricultural sector; that is, for this output group neither weak separability nor homotheticity were rejected. He compared the parameter estimates based on a fully disaggregated system (six output supply functions and three variable input demand functions) to the estimates based on a partially aggregated output supply system (four output supply functions and three variable input demand functions), and concluded that the empirical results obtained from the former model were more plausible than the results obtained from the latter specification, even when the aggregation was based on the technology tests for nonjointness and homothetic separability.

In a study by Thirtle (1985), the technical progress parameters estimated for a crop production model for U.S. agriculture were used to test the induced innovation theory. In particular, Thirtle estimated a nested CES/Cobb-Douglas production model that included neutral shift parameters for biological (primarily fertilizer) and mechanical innovations for each field crop.

His results supported the induced innovation hypothesis. The production of corn and wheat, both of which have high land-labor ratios relative to cotton, were much more responsive than cotton to biological innovations. His evidence on mechanical innovations also appeared consistent with the induced innovation hypothesis. Wheat and soybeans, which are also much less labor-intensive than cotton, showed far lower rates of mechanical (or labor-saving) technical change. More generally, Thirtle's results showed that the more labor-intensive (relative to land) was the crop production function, the greater was the labor-saving bias of technical change.

SOME COMPARATIVE REMARKS

As indicated in chapter 2, direct comparisons of the technological change biases and other structural characteristics of the production process from the econometric studies would require adjustments to account for differences in the functional specifications and the data. Even without such adjustments, however, it is useful to make qualitative comparisons of the empirical results. By and large these comparisons pertain to the estimated rates and biases of technological change and the substitution and input demand elasticities as reflected by the post-1970 contributions.

Technological Change

The manner in which technological change is specified in an econometric model is critical for the interpretation of other structural parameters. For example, under Hicks neutrality, cost share equations exclude a time trend variable; all variation in shares over time is attributed to changes in input prices or scale. If technical change is nonneutral, the observed changes in shares are a result of technical change biases, as well as changes in factor prices and scale. Thus, misspecifying the form of technical change biases the parameter estimates of the model and the estimates of the substitution elasticities and the factor demand elasticities.

The assumptions made with respect to homotheticity or input-output separability of the underlying production relation are also crucial in estimating the rate and biases of technical change. If the technology is nonhomothetic or nonseparable, changes in factor cost shares stem from factor price changes and scale (output) changes as well as from biased technical change. Furthermore, the classifications of technological change biases obtained from profit function and cost function models are not comparable unless one accounts for the scale effects (see chapter 2).

If the econometric model does not include the cost, profit, or production function in the set of estimating equations, the total effects of technological change cannot be measured; by estimating the set of share equations, only the factor biases can be measured. Of the studies we reviewed, Ray; Brown and Christensen; Hazilla and Kopp; Chan and Mountain; and Capalbo and Denny reported estimates of the effects of technological change on the production technology. Ray reported a rate of Hicks-neutral technological progress of approximately 1.8 percent per year for the period 1939–77 for the United States; Brown and Christensen indicated technological regression; Hazilla and Kopp indicated that technical progress occurred at an average annual rate of 0.06 percent; Chan and Mountain reported a rate of 0.28 percent per year (unadjusted for scale effects) for Canadian agriculture; and Capalbo and Denny reported a rate of 1.9 percent and 1.3 percent for Canada and the United States, respectively, when evaluated at the point of approximation. The 95 percent confidence intervals for both the Ray and the Capalbo and Denny measures of technological progress for the United States include the nonparametric estimate of technical change using the CVW data reported in table 3-5.

Many of the studies discussed in this survey have tested for and subsequently rejected Hicks-neutral technical change. Exceptions are Chan and Mountain, who did not reject Hicks-neutral technical change for a translog production model for Canadian agriculture; Lopez, who accepted Hicks-neutral technical change within a nonhomothetic cost structure of Canadian agriculture; Capalbo and Denny, who accepted Hicks-neutral technical change within a gross output production model for U.S. and Canadian agriculture for the 1960–1980 period; and Ray,

TABLE 3-7. TECHNOLOGICAL CHANGE BIASES FOR SELECTED STUDIES

Research study	Biases			
	Labor[a]	Capital[b]	Fertilizer[c]	Land[d]
Antle (1984)	Saving	Using	Using	Using
Binswanger (1974)	Saving	Using	Using	—
Brown and Christensen (1981)	Saving	Using	Saving	—

Note: Blank spaces indicate that no estimate was reported.
[a] Defined as hired labor in Brown and Christensen.
[b] Defined as machinery in Antle and Binswanger.
[c] Defined as materials in Brown and Christensen and as chemicals in Antle.
[d] Treated as a fixed input in Brown and Christensen.

who imposed Hicks neutrality on his aggregate model for the United States without testing for its validity.

The direction of the factor bias can be determined in the nonneutral technological change translog cost and profit models by the sign of coefficient on the cross term between time and the logarithm of the factor price.[5] A comparison of the biases of technical change since World War II is presented in table 3-7. Among the studies that did not impose Hicks neutrality, the evidence has been strongly in favor of capital- and land-using biases, and a labor-saving bias. The evidence for fertilizer is mixed, which may be due to the differences in the procedure for aggregating the chemical inputs.

Input Substitution and Factor Demand Elasticities

Although both factor demand elasticities and elasticities of substitution provide information on the relationships among inputs, one should be cautious in comparing the results obtained from a profit function model with those obtained from a cost function model. With the cost function, output is fixed and factor demand elasticities reflect movements along a given isoquant (Hicksian elasticities). The elasticities of substitution from the cost function parameters are related to the curvature of this isoquant. With a profit function model, output is allowed to vary. The factor demand elasticities derived from the profit function parameters correspond to the uncompensated, or Marshallian, elasticities, and the relationships among the inputs reflect gross substitutes or complements.

Comparisons of the elasticities of substitution for the cost function models are provided in table 3-8. The Brown and Christensen estimates

[5] To obtain measures of the Hicksian bias in the nonhomothetic models, one would need to adjust the bias measure for scale effects, as discussed by Antle and Capalbo in chapter 2 of this volume.

TABLE 3-8. ESTIMATED ELASTICITIES OF SUBSTITUTION FROM VARIOUS COST FUNCTION MODELS

Research study	Sample/model	Elasticities of substitution[a]					
		Labor[b]/capital[c]	Labor[b]/fertilizer[d]	Labor[b]/land[e]	Capital[c]/fertilizer[d]	Capital[c]/land[e]	Land[e]/fertilizer[d]
Binswanger (1974)	Pooled: annual, U.S. states, 1949, 1954, 1959, 1964; translog cost model	0.85 (0.25)	−1.62 (0.53)	0.20 (0.24)	−0.67 (1.72)	1.22 (0.46)	2.99 (0.93)
Brown and Christensen[f] (1981)	Time series: annual, U.S., 1947–74; restricted translog cost model	0.32	1.24	—	0.01	—	—
Lopez (1980)	Time series: annual, Canada, 1946–77; generalized Leontief cost model	1.78 *	0.88 **	0.11 ***	1.56 *	0.23 ****	0.99 **
Ray (1982)	Time series: annual, U.S., 1939–77; translog cost model	0.75	5.73	—	1.21	—	—

Note: Blank spaces indicate that no estimate was reported.

[a] Standard errors are given in parentheses.
[b] Defined as hired labor in Brown and Christensen and in Ray.
[c] Defined as machinery in Binswanger and as farm real estate and capital in Ray.
[d] Defined as materials in Lopez and in Brown and Christensen.
[e] Defined as land and structures in Lopez. Treated as fixed input in Brown and Christensen.
[f] Short-run elasticities.

* Denotes that the coefficient is statistically significant from one.
** Denotes that the coefficient is not statistically different from one.
*** Denotes that the coefficient is not statistically different from zero.
**** Denotes that the coefficient is statistically significant from zero.

are short-run elasticities of substitution; all other reported elasticities are long-run elasticities. The Binswanger elasticities pertain to pooled time series/cross-section data.

From the reported elasticities of substitution the following conclusions can be made: farm capital, fertilizer, and other purchased materials substitute for hired labor in varying degrees in the time series models of U.S. agriculture. The higher degree of substitutability between hired labor and fertilizer relative to hired labor and capital was consistent with the observed decline in hired labor and the large increase in fertilizer use in the United States. Only the Binswanger estimates, based on time series/cross-section data, indicated a statistically significant complementary relationship between labor (family and hired) and fertilizer for U.S. agriculture. For the Canadian agricultural sector, more substitutability between labor and capital has occurred, perhaps as a result of the inclusion of family and owner operators in the Canadian labor category. Finally, in both the U.S. and Canadian studies, capital and land have been substitutes for all inputs. Note that the complementary relationship reported by Binswanger is not statistically significant.

Chalfant (1984) compared the estimated elasticities of substitution resulting from the class of Box-Cox flexible functions (Berndt and Khaled, 1979) and the logarithmic Fourier flexible form (Gallant, 1982). The class of Box-Cox flexible functional forms includes the generalized Leontief, the generalized square-root quadratic, and the translog. Chalfant's results suggest that the translog and generalized Leontief cost functions are less appropriate for modeling agricultural production, since these models do not result in negative own-elasticities of substitution for all inputs. The estimates resulting from the use of the Fourier flexible form also failed to satisfy the negative own elasticities for all of the factors. While Chalfant's study points out the differences in the elasticities of substitution obtained from alternative functional forms, the absence of standard errors precludes making any comparisons based on the statistical significance of these elasticities.

With respect to the factor demand elasticities reported in table 3-9, the following conclusions can be drawn for U.S. agriculture in the post-World War II era: (a) the own-price elasticities of demand for all inputs, with the exception of capital in Binswanger's study, have been less than 1 in absolute value; (b) Binswanger's pooled time series/cross-section cost model produced consistently higher own-price demand elasticities for labor, capital, and chemicals than the time series studies on U.S. agriculture; (c) imposing the curvature restrictions altered the reported own-price elasticities of demand for hired labor, capital, and purchased materials—the Hazilla and Kopp estimates were more elastic than the

TABLE 3-9. SUMMARY OF OWN-PRICE AND CROSS-PRICE ELASTICITIES OF DEMAND FOR INPUTS USED IN U.S. AGRICULTURE

Research study	Sample/model	Price of			
		Labor[a]	Capital[b]	Chemicals[c]	Land[d]
(a) Demand for labor					
Antle (1984)	Time series: annual, U.S., 1910–46, 1947–78; translog profit model, single output	−0.01 (0.22)	0.06 (0.10)	0.11 (0.08)	−0.19 (0.09)
Binswanger (1974)	Pooled time series/cross-section: U.S. states, 1949, 1954, 1959, 1964; translog cost model, single output	−0.91 (0.06)			
Brown and Christensen[e] (1981)	Time series: annual, U.S., 1947–74; restricted translog cost model, single output	−0.66			
Hazilla and Kopp[e] (1986)	Time series: annual, U.S., 1947–74; restricted translog cost model, single output	−0.96			
Lopez (1980)	Time series: annual, Canada, 1946–77; generalized Leontief cost model, single output	−0.52			
Lopez (1984)	Cross-section: Canada, 1971; generalized Leontief profit model, multiple outputs				
Uncompensated		−1.24 (0.65)	−0.03 (0.02)		0.26 (0.26)
Compensated		−0.38	0.21		0.45
Ray (1982)	Time series: annual, U.S., 1939–77; translog cost model, multiple outputs	−0.85			
Shumway (1983)	Time series: annual, Texas field crops, 1957–79; restricted quadratic profit model, single output	−0.43	−0.01	0.20	
(b) Demand for capital					
Antle (1984)	Time series: annual, U.S., 1910–46, 1947–78; translog profit model, single output	0.09 (0.10)	−0.25 (0.11)	−0.22 (0.06)	−0.08 (0.10)

Study	Description				
Binswanger (1974)	Pooled time series/cross-section, U.S. states, 1949, 1954, 1959, 1964; translog cost model, single output		−1.09 (0.18)		
Brown and Christensen[e] (1981)	Time series: annual, U.S., 1947–74; restricted translog cost model, single output		−0.05		
Hazilla and Kopp[e] (1986)	Time series: annual, U.S., 1947–74; restricted translog cost model, single outputs		−0.52		
Lopez (1980)	Time series: annual, Canada, 1946–77; generalized Leontief cost model, single output		−0.35		
Lopez (1984)	Cross-section: Canada, 1971; generalized Leontief profit model, multiple outputs				
Uncompensated		−0.05 (0.05)	−1.58 (0.62)		0.08 (0.03)
Compensated		0.53	−1.48		0.19
Ray (1982)	Time series: annual, U.S., 1939–77; translog cost model, multiple outputs		−0.53		
Shumway (1983)	Time series: annual, Texas field crops, 1957–79; restricted quadratic profit model, multiple outputs	−0.01	−0.37	−0.15	

(c) Demand for chemicals

Study	Description				
Antle (1984)	Time series: annual, U.S., 1910–46, 1947–78; translog profit model, single output	0.49 (0.34)	0.01 (0.25)	−0.25 (0.20)	
Binswanger (1974)	Pooled time series/cross-section: U.S. states, 1949, 1954, 1959, 1964; translog cost model, single output			−0.95 (0.16)	
Brown and Christensen[e] (1981)	Time series: annual, U.S., 1947–74; restricted translog cost model, single output			−0.20	
Hazilla and Kopp[e] (1986)	Time series: annual, U.S., 1947–74; restricted translog cost model, single output			−0.54	−1.02 (0.21)

TABLE 3-9. (continued)

Research study	Sample/model	Labor[a]	Capital[b]	Chemicals[c]	Land[d]
Lopez (1980)	Time series: annual, Canada, 1946–77; generalized Leontief cost model, single output			−0.41	
Ray (1982)	Time series: annual, U.S., 1939–77; translog cost model, multiple outputs			−0.40	
Shumway (1983)	Time series: annual, Texas field crops, 1957–79; restricted quadratic profit model, multiple outputs	0.39	−0.15	−0.70	
(d) Demand for land					
Antle (1984)	Time series: annual, U.S., 1910–46, 1947–78; translog profit model, single output	−0.07 (0.12)	0.10 (0.12)	−0.29 (0.06)	−0.18 (0.13)
Binswanger (1974)	Pooled time series/cross-section, U.S. states, 1949, 1954, 1959, 1964; translog cost model, single output				−0.34 (0.09)
Lopez (1980)	Time series: annual, Canada, 1946–77; generalized Leontief cost model, single output				−0.42
Lopez (1984)	Cross-section: Canada, 1971; generalized Leontief profit model, multiple outputs				
Uncompensated		0.26 (0.04)	0.04 (0.01)		−0.36 (0.09)
Compensated		0.43	0.09		−0.48

Notes: Standard errors, when reported, are in parentheses. A blank space indicates that no estimation was reported or attempted.

[a] Defined as hired labor in Lopez (1984), Brown and Christensen, Hazilla and Kopp, and Ray.
[b] Defined as machinery in Antle and Binswanger and as farm real estate and capital in Ray.
[c] Defined as fertilizer in Shumway, Lopez (1984), Binswanger, and Ray; defined as materials in Lopez (1980) and Brown and Christensen.
[d] Defined as land and structures in Lopez (1980, 1984). Treated as a fixed input in Brown and Christensen, Hazilla and Kopp, Shumway, and Ray.
[e] Short-run or restricted price elasticities.

Brown and Christensen estimates; and (d) based on a profit function model, only the own-price elasticity for capital was statistically significant. Moreover, the results reported from the profit function model indicate that the changes in factor use are more responsive to technological change and scale changes than to input prices.

The two studies by Lopez (1980, 1984) on Canadian agriculture offer a means for comparisons of the relative magnitudes of the compensated and uncompensated factor price elasticities.[6] The set of compensated input demand elasticities obtained from the estimated parameters of his profit function model (1984) are comparable to elasticities obtained from his cost function model (1980). Differences in the compensated elasticities from the profit model and the elasticities reported for the cost function also highlight the differences in elasticities that may result from utilizing cross-section versus time series data.

For labor, there is a little variability between the compensated own-price elasticities from the 1980 and 1984 studies: the compensated elasticity from the cross-section profit model is only slightly smaller (in absolute value) than the compensated own-price elasticity for labor in the time series cost model. This is not true, however, for capital. The compensated own-price elasticity based on the profit model is substantially greater than the corresponding elasticity estimated directly from the cost function (-1.48 versus -0.35).

Comparisons of the compensated and uncompensated elasticities of demand are provided in Lopez (1984). For both labor and capital, the uncompensated own-price elasticities are higher relative to the compensated elasticities, implying that the compensated factor demand function cuts the uncompensated factor demand function from below. When output levels are allowed to adjust to factor price changes, the change in factor use is smaller than if output levels had remained fixed. This is consistent with the results for the U.S. models—the cost function models produced larger own-price elasticities relative to Antle's profit function model. However, additional research is needed on the alternative specifications before more definitive statements can be made with respect to model-dependent results.

Conclusions

From this review of the aggregate econometric models we can conclude that (a) the U.S. and Canadian agricultural sectors during the post–

---

[6] One should also note that, as the generalized Leontief is not a self-dual functional form, the generalized Leontief cost and profit models provide different second-order approximations to the underlying technology.

World War II period are best characterized by a nonhomothetic, nonneutral specification of the aggregate technology and (b) with the exception of Binswanger's and Lopez's estimates of the factor demand elasticities for capital, there is a general consistency with regard to factor demand inelasticity in the U.S. and Canadian agricultural sectors since the end of World War II. This is in spite of differences in aggregation, data, and specification of the production technology.

A third conclusion pertains to the comparisons of the measures of TFP from the nonparametric (index) and parametric (econometric) approaches. Over a comparable period, 1948–75, the nonparametric measures of TFP growth fall within the 95 percent confidence interval of the measures of technical change produced by the econometric models of U.S. agriculture. On the one hand, this suggests that the distortions to the TFP index due to returns to scale, nonhomogeneity, and non-marginal cost pricing may have had offsetting effects and the TFP index may be fairly representative of technological progress. However, since the measures of TFP are based on different data sets, one should be cautious in drawing such conclusions. Further discussions and comparisons of these two measures of technological progress are found in chapter 5. Capalbo's econometric models are estimated using the CVW data set, and the measures of technological change are thus directly comparable to the nonparametric measures discussed in this chapter.

This review has exposed some of the limitations of the existing research. For example, it is not clear what should be done with empirical models that violate theoretical properties. One solution is to impose the correct curvature restrictions *a priori*, as exemplified in the Hazilla-Kopp study. Apart from the financial considerations associated with the econometric estimation of a considerably more involved model, the merits of molding the data into such a restrictive model are questionable. An alternative would be to check the statistical significance of the curvature restrictions for the model at the point of approximation. Statistically significant violations of these restrictions may signal the need for better data development as opposed to a more constraining specification.

The Hayami and Ruttan (1970, 1971), Binswanger (1974a,b), and Thirtle (1985) studies have provided evidence supporting the induced innovation theory, without providing a statistical test of this hypothesis. In chapter 12 of this volume, Antle suggests a statistical test of the induced innovation hypothesis: if the production process can be modeled as a static phenomenon, the final-form factor demand equations with lagged prices can be used to test the induced innovation hypothesis. Antle's results indicate some violations of the induced innovation hypothesis based on the static formulations. However, if the production process is dynamic, the final-form equations for models of exogenous technological change include lagged prices. Therefore, it becomes necessary to identify the structural parameters in a dynamic model to test the induced innovation theory.

Suggestions for further research on the aggregate measurement of TFP include the following. First, refinements to the data base should include determining the service price or shadow value of land net of the capitalized values of government programs or effects of nonagricultural land use (see Alston, 1986; Burt, 1986; Melichar, 1979; and Phipps, 1984). Second, for the econometric analyses, it is reasonable to assume that some of the factors of production are not instantaneously or "costlessly" adjusted. Recognition of quasi-fixity of resource stocks, such as land, makes the estimation of full equilibrium models inappropriate. To date, the success with estimating restricted cost or profit functions and the use of dynamic models to specify the aggregate technology and production process for the agricultural sector is limited.

APPENDIX 3-A.
U.S. AGRICULTURAL DATA BASE

The data employed for the empirical analyses consist of annual data over the period 1948–1983 on the prices and quantities of agricultural outputs and inputs. Gross agricultural output is aggregated into two groups, crops and livestock products. Crops consist of small grains (wheat, barley, oats, rye, and rice); coarse grains (corn and sorghum); other field crops; vegetables; and tree crops. The livestock group includes dairy and animal products. The commodities are aggregated using the Tornqvist approximation to the Divisia indexing procedure, where the weights are the value shares of each commodity in the aggregation group. Table 3A-1 gives the Divisia quantity and implicit price indexes for the aggregate output groups. The 1970 revenue shares for the aggregate outputs are 0.54 for all crops (0.04 for small grains, 0.16 for coarse grains, 0.16 for field crops, 0.03 for fruits and nuts, 0.15 for vegetables) and 0.46 for livestock.

The four major inputs are labor, nonland capital, land, and intermediate inputs. Labor consists of hired workers and self-employed/unpaid family labor; capital consists of farm durable equipment, nonresidential structures, animal stock, and animal replacement inventories; land is mea-

TABLE 3A-1. OUTPUT QUANTITY AND PRICE INDEXES
(1977 = 100)

	Small grains		Coarse grains		Field crops	
Year	Implicit price	Quantity	Implicit price	Quantity	Implicit price	Quantity
1948	5,632.21	.745000	15,870.8	.416000	14,762.9	.598000
1949	5,230.19	.631000	15,186.8	.395000	13,072.6	.569000
1950	5,839.30	.635000	18,173.3	.392000	15,154.3	.539000
1951	6,101.22	.608000	20,034.9	.371000	15,822.8	.612000
1952	6,131.25	.704000	18,650.2	.403000	15,695.6	.625000
1953	5,822.24	.658000	17,995.1	.419000	14,374.2	.662000
1954	5,799.80	.659000	17,371.1	.432000	14,002.6	.609000
1955	5,265.00	.650000	16,126.6	.453000	13,696.2	.641000
1956	5,486.49	.622000	15,683.9	.471000	13,506.5	.646000
1957	5,198.19	.628000	13,462.5	.514000	12,624.8	.598000
1958	4,844.21	.833000	13,630.6	.536000	12,698.3	.681000
1959	4,944.82	.663000	12,631.9	.582000	13,006.6	.658000
1960	4,813.41	.766000	12,067.7	.615000	12,977.4	.693000
1961	5,159.80	.696000	13,426.8	.572000	13,483.2	.749000
1962	5,434.06	.665000	13,656.7	.612000	13,346.7	.768000
1963	5,114.59	.676000	13,516.9	.675000	13,934.7	.750000
1964	4,303.39	.710000	14,241.3	.620000	13,900.6	.762000
1965	4,293.21	.737000	14,023.0	.695000	13,848.7	.817000
1966	4,882.42	.723000	14,944.5	.719000	14,134.8	.766000
1967	4,403.23	.787000	12,526.3	.792000	13,974.1	.748000
1968	4,017.74	.855000	13,122.8	.751000	13,444.8	.808000
1969	3,986.00	.809000	14,155.5	.787000	13,255.3	.831000
1970	4,248.21	.762000	16,111.0	.717000	14,386.1	.802000
1971	4,268.27	.857000	13,241.1	.914000	15,956.0	.849000
1972	5,453.52	.794000	19,053.5	.900000	18,303.3	.897000
1973	11,437.30	.853000	30,831.2	.931000	24,981.9	.954000
1974	11,958.30	.865000	36,805.7	.828000	31,058.1	.869000
1975	10,281.00	1.020000	30,989.4	.933000	28,882.9	.940000
1976	8,392.02	1.000000	26,265.9	.974000	33,100.5	.906000
1977	7,323.75	1.000000	24,583.3	1.000000	30,326.7	1.000000
1978	8,593.86	.909000	27,364.1	1.060000	31,811.2	1.010000
1979	10,778.70	1.030000	30,738.2	1.110000	33,640.8	1.140000
1980	11,759.40	1.110000	37,985.1	.971000	40,039.8	.999000
1981	10,624.60	1.320000	30,560.9	1.140000	36,656.4	1.200000
1982	10,014.30	1.310000	32,717.4	1.140000	36,438.6	1.200000
1983	10,350.20	1.130000	41,050.1	.708000	42,439.0	1.000000

TABLE 3A-1. OUTPUT QUANTITY AND PRICE INDEXES (*continued*)
(1977 = 100)

	Fruits		Vegetables		Animal products	
Year	Implicit price	Quantity	Implicit price	Quantity	Implicit price	Quantity
1948	1,838.76	.546000	18,798.9	.836000	32,545.3	.697000
1949	1,644.71	.577000	17,257.4	.771000	27,538.4	.731000
1950	1,964.66	.583000	16,423.0	.738000	27,636.8	.758000
1951	1,699.45	.619000	17,670.0	.787000	32,795.0	.781000
1952	1,882.43	.597000	18,851.5	.835000	30,905.7	.787000
1953	1,928.69	.615000	19,197.3	.860000	27,869.7	.802000
1954	1,933.71	.603000	18,421.8	.826000	25,843.1	.826000
1955	2,000.23	.634000	17,259.0	.774000	24,606.5	.844000
1956	1,934.36	.657000	18,409.6	.829000	24,373.8	.911000
1957	2,035.16	.617000	18,181.7	.812000	25,998.0	.838000
1958	2,141.54	.657000	17,158.2	.771000	28,011.6	.847000
1959	2,085.65	.687000	16,631.1	.745000	26,369.9	.872000
1960	2,362.01	.649000	16,871.7	.760000	26,416.9	.867000
1961	1,994.11	.734000	16,457.5	.739000	26,339.5	.899000
1962	2,309.45	.632000	16,060.1	.721000	26,513.0	.906000
1963	2,431.32	.679000	15,735.5	.707000	25,808.9	.923000
1964	2,298.77	.724000	16,200.1	.719000	25,133.7	.943000
1965	2,051.18	.756000	15,549.2	.688000	27,151.1	.918000
1966	2,042.92	.799000	16,388.1	.715000	30,849.9	.925000
1967	2,791.88	.670000	16,520.7	.727000	29,658.3	.944000
1968	2,564.88	.793000	16,233.3	.719000	30,869.0	.941000
1969	2,297.26	.866000	16,268.8	.709000	34,057.5	.942000
1970	2,621.70	.804000	15,651.2	.686000	34,949.2	.974000
1971	2,729.18	.877000	15,543.6	.667000	34,355.7	.979000
1972	3,180.18	.806000	16,725.8	.718000	38,569.8	1.001000
1973	3,623.08	.930000	17,542.3	.738000	49,633.3	1.003000
1974	3,457.49	.948000	25,629.3	1.070000	48,300.4	.989000
1975	3,355.62	.993000	25,310.1	1.050000	50,734.8	.948000
1976	3,770.39	.974000	23,217.7	.959000	53,359.8	.987000
1977	4,813.72	1.000000	24,211.2	1.000000	53,150.2	1.000000
1978	5,631.60	.975000	24,691.8	1.010000	62,974.8	1.002000
1979	5,509.43	1.130000	25,786.6	1.060000	73,039.9	1.025000
1980	5,373.62	1.610000	26,444.1	1.070000	73,114.1	1.062000
1981	5,800.21	1.040000	28,294.8	1.130000	74,364.0	1.084000
1982	5,432.04	1.190000	28,182.6	1.140000	74,239.7	1.078000
1983	5,326.43	1.110000	27,573.8	1.100000	73,220.5	1.100000

TABLE 3A-1. OUTPUT QUANTITY AND PRICE INDEXES (*continued*)
(1977 = 100)

Year	Aggregate output	
	Implicit price	Quantity
1948	88,108.9	.670078
1949	78,340.5	.652255
1950	81,940.1	.649691
1951	91,029.7	.674694
1952	89,566.1	.701437
1953	84,748.5	.718498
1954	80,813.4	.712636
1955	76,595.7	.717031
1956	77,262.0	.754224
1957	76,552.5	.725589
1958	77,677.9	.751458
1959	74,844.2	.748183
1960	74,785.4	.767090
1961	75,812.0	.773757
1962	76,198.9	.778148
1963	75,365.9	.791226
1964	74,722.8	.796025
1965	76,024.2	.807621
1966	82,570.4	.811439
1967	79,185.2	.829702
1968	79,785.2	.838346
1969	83,842.2	.846447
1970	87,758.7	.834117
1971	85,926.2	.880955
1972	101,272.0	.899146
1973	138,463.0	.927367
1974	156,874.0	.932693
1975	149,783.0	.964824
1976	148,052.0	.962839
1977	144,409.0	1.000000
1978	161,030.0	1.006960
1979	179,257.0	1.070240
1980	194,316.0	1.054310
1981	186,120.0	1.136290
1982	186,755.0	1.139920
1983	198,478.0	1.007720

sured as the total land in farms; and intermediate inputs include energy, pesticides, fertilizer, feed and seed, and other miscellaneous inputs. The Divisia quantity and implicit price indexes for these inputs are shown in table 3A-2. The cost shares (1970) are 0.20 for labor, 0.28 for nonland capital, 0.30 for intermediate input, and 0.22 for land.

The agricultural labor data are based on the Gollop and Jorgenson (1983) labor data. The hired and self-employed/unpaid family labor are quality-adjusted to reflect the sex, age, education level, and occupation characteristics of the workforce. This cross-classification is in terms of numbers of workers, average hours worked, and average compensation per hour. A Divisia index of labor is obtained by summing across all classifications using the adjusted average hourly compensation for each classification as weights.

For the capital inputs—durable equipment, nonresidential structures, and animal stock—we computed a service price using the method outlined in Christensen and Jorgenson (1969, 1970). The service prices of durable equipment and nonresidential structures are equal to the sum of the opportunity cost of the respective capital plus depreciation and taxes. In absence of a reliable *ex ante* estimate of capital gains, the opportunity cost reflects only current returns. A similar procedure is used to construct the service price and quantity indexes of each group of animal capital (breeding hogs, milk cows, beef cows, bulls, and stocker sheep). Taxes on animal stock are also assumed to be zero.

An index of total land in farms is constructed by aggregating the state-level farmland data using the Divisia indexing procedure. The weights are the average per acre farmland price in each state. A service price for land is calculated using the Christensen and Jorgenson method. Capital gains are set equal to zero.

The intermediate and material inputs consist of livestock feed, seed, pesticides, fertilizer, energy, and miscellaneous inputs. Livestock feed includes commercial feedstuffs and grains, and other farm products. Seed usage is adjusted to the current calendar year based on the crop production cycle. The fertilizer data are aggregated to reflect the proportion of the primary nutrient contents of nitrogen, phosphorus, and potassium. The aggregate index of fertilizer combines the indexes for fertilizer and lime.

The expenditure on pesticides estimated by the USDA is deflated by a translog price index provided by the U.S. Environmental Protection Agency. Energy consists of gasoline, diesel fuel, liquefied petroleum, natural gas, and electricity. The miscellaneous input group includes grazing fees, machine hire and custom work, and veterinary fees.

TABLE 3A-2. INPUT QUANTITY AND PRICE INDEXES
(1977 = 1.00000)

Year	Family labor		Hired labor		Land		Structures		Other capital[a]	
	Implicit price	Quantity	Implicit price	Quantity	Implicit price	Quantity	Implicit price	Quantity	Implicit price	Quantity
1948	3,535.98	2.79300	1,882.06	1.59400	1,653.50	1.18600	1,518.77	.37600	8,608.31	.690955
1949	3,252.84	2.73300	1,900.81	1.48200	1,690.18	1.18900	1,506.05	.40200	9,752.25	.712570
1950	3,363.99	2.59900	1,791.11	1.57500	1,652.92	1.18900	1,448.12	.42700	8,911.34	.733897
1951	3,684.23	2.46700	1,978.41	1.48200	1,902.16	1.18800	1,511.64	.44900	11,032.8	.777396
1952	3,937.87	2.39800	2,029.73	1.41300	2,159.73	1.18700	1,615.04	.47200	13,903.0	.806263
1953	4,107.59	2.33300	2,046.20	1.34200	2,431.24	1.18700	1,699.86	.49100	10,942.1	.810574
1954	4,262.59	2.30400	2,053.59	1.26900	2,319.16	1.18100	1,631.43	.50700	10,264.8	.805533
1955	3,465.72	2.24600	2,073.43	1.25300	2,484.81	1.17900	1,636.47	.52100	10,133.7	.812952
1956	3,586.99	2.12100	2,252.89	1.15900	2,834.84	1.15700	1,774.51	.53600	10,173.3	.802942
1957	3,835.46	1.96300	2,454.71	1.10400	3,520.55	1.16000	1,999.47	.54800	10,625.7	.785882
1958	4,201.20	1.83900	2,552.89	1.12500	3,652.22	1.14900	1,981.74	.55800	11,762.0	.802402
1959	4,282.77	1.84600	2,728.28	1.08200	4,397.50	1.14500	2,057.61	.58000	13,199.4	.829797
1960	3,934.65	1.72900	2,657.22	1.16400	4,783.69	1.13600	2,056.03	.59500	12,380.0	.833923
1961	4,397.55	1.63500	2,728.43	1.18200	4,936.75	1.13300	2,050.23	.61100	11,844.0	.832701
1962	4,978.96	1.56800	2,805.58	1.18300	5,101.11	1.12400	2,050.23	.63000	11,384.7	.825616

TABLE 3A-2. (continued)

	Energy		Fertilizer		Pesticides		Miscellaneous		Total inputs	
Year	Implicit price	Quantity	Implicit price	Quantity	Implicit price	Quantity	Implicit price	Quantity	Implicit price	Quantity
1948	2,619.85	.413000	4,807.18	.169000	1,296.70	.910000	1,210.43	.706000	36,364.5	.958335
1949	2,772.73	.440000	4,972.74	.180000	1,376.24	.101000	1,294.44	.616000	35,983.9	.958790
1950	2,763.68	.457000	4,626.46	.191000	1,243.06	.144000	1,381.26	.674000	36,262.6	.976393
1951	2,819.71	.477000	4,749.72	.218000	1,611.57	.121000	1,388.21	.814000	40,959.5	.986552
1952	2,723.53	.510000	4,798.78	.238000	1,544.72	.123000	1,430.96	.852000	44,308.8	.983594
1953	2,740.60	.532000	4,803.01	.254000	1,396.40	.111000	1,484.82	.850000	42,311.9	.975995
1954	2,722.12	.547000	4,729.59	.266000	1,284.62	.130000	1,521.52	.841000	41,479.7	.970394
1955	2,729.87	.559000	4,646.87	.278000	1,307.19	.153000	1,575.26	.892000	39,521.2	.984773
1956	2,730.77	.572000	4,496.46	.276000	1,299.52	.207000	1,619.94	.900000	40,907.7	.963949
1957	2,780.49	.574000	4,400.08	.292000	1,204.97	.161000	1,640.61	.970000	43,091.0	.952743
1958	2,676.22	.593000	4,463.42	.301000	1,183.25	.191000	1,733.81	1.046000	45,190.6	.966230
1959	2,607.78	.617000	4,408.79	.341000	1,238.09	.231000	1,775.49	1.290000	48,755.6	.993983
1960	2,619.87	.634000	4,358.49	.345000	1,250.00	.232000	1,807.52	1.336000	48,057.4	.991126
1961	2,643.96	.646000	4,405.64	.365000	1,213.24	.272000	1,863.84	1.325000	49,436.3	.991867
1962	2,581.46	.669000	4,371.42	.395000	1,202.61	.306000	2,116.51	1.207000	50,939.6	.980209
1963	2,514.98	.701000	4,285.42	.446000	1,114.71	.340000	2,079.90	1.260000	52,032.2	.979764

Year								
1963	5,019.27	1.45300	2,904.69	1.17500	5,376.06	1.11600	2,069.02	.64700	11,583.6	.840977
1964	5,400.14	1.38200	3,215.82	1.07500	5,847.40	1.10900	2,129.78	.66300	11,398.5	.841009
1965	5,825.26	1.36200	3,464.46	1.02700	6,375.46	1.10100	2,197.86	.68000	11,947.8	.881305
1966	6,968.86	1.25200	3,754.40	.96500	7,763.45	1.09100	2,433.33	.69800	13,208.4	.899351
1967	6,385.12	1.22300	4,068.39	.89200	8,944.41	1.08100	2,657.84	.72100	13,722.4	.909792
1968	7,101.62	1.23000	4,419.69	.85300	10,555.1	1.07300	2,959.18	.74000	14,369.3	.902773
1969	7,698.87	1.18600	4,978.96	.80800	12,535.1	1.06700	3,377.42	.75900	15,603.4	.907470
1970	8,183.28	1.18400	4,435.48	.93000	14,694.9	1.05700	3,985.19	.78200	17,403.2	.914818
1971	8,353.94	1.13300	4,653.04	.90500	14,343.1	1.05200	4,120.93	.80300	17,708.1	.920354
1972	8,049.43	1.13300	4,794.93	.90700	15,031.5	1.04800	4,324.36	.81600	18,349.3	.926767
1973	8,969.14	1.13400	5,506.98	.93100	17,807.6	1.04700	4,679.80	.84500	20,251.4	.961424
1974	9,624.89	1.08500	6,033.23	.99300	24,972.6	1.03900	5,668.16	.87600	23,070.2	.986296
1975	10,798.7	1.07300	6,392.51	1.01400	28,978.5	1.03500	6,673.64	.91400	23,952.0	.997363
1976	12,446.3	1.02400	7,096.62	1.03500	32,990.5	1.00700	6,838.48	.95400	25,245.3	.996623
1977	12,820.0	1.00000	7,794.00	1.00000	37,344.7	1.00000	6,644.73	1.00000	26,827.3	1.000000
1978	13,041.8	1.00500	7,805.32	1.05300	44,549.5	.99700	7,369.42	1.05000	30,797.6	1.035670
1979	14,793.6	1.00300	8,778.00	1.01800	56,990.1	.99600	8,875.40	1.08000	38,818.7	1.044100
1980	15,830.0	1.00000	9,346.10	.99100	83,575.2	.99500	11,591.3	1.11000	46,556.4	1.061020
1981	15,039.3	1.01800	8,758.50	1.03500	106,996.0	.99600	14,331.8	1.12000	51,501.9	1.064870
1982	17,011.4	.96500	9,907.70	1.02900	106,708.0	.99600	15,065.4	1.12000	52,315.5	1.044770
1983	13,754.8	.93000	7,856.60	1.25500	86,578.4	.98100	13,734.7	1.12000	50,027.6	1.010000

Year								
1964	2,616.06	.685000	4,210.70	.490000	1,083.56	.371000	2,342.99	1.155000	54,365.4	.965888
1965	2,437.58	.745000	4,155.80	.515000	1,104.89	.429000	2,462.85	1.151000	57,049.0	.985557
1966	2,524.32	.740000	4,104.59	.582000	1,135.35	.495000	2,608.74	1.112000	64,268.7	.980750
1967	2,512.52	.759000	4,009.28	.648000	1,182.36	.669000	2,774.21	1.076000	66,426.8	.979307
1968	2,442.89	.788000	3,533.37	.699000	1,205.54	.686000	3,043.19	1.060000	72,098.7	.985431
1969	2,453.87	.813000	3,285.55	.719000	1,504.98	.602000	3,324.54	1.004000	80,298.4	.984527
1970	2,421.87	.832000	3,293.56	.738000	1,403.51	.684000	3,441.73	.978000	88,650.4	.992486
1971	2,383.10	.864000	3,452.82	.784000	1,411.11	.810000	3,760.70	.960000	88,668.2	.996156
1972	2,272.63	.906000	3,543.75	.787000	1,341.51	1.019000	3,910.74	.975000	96,091.0	1.003490
1973	2,545.66	.898000	3,903.54	.820000	1,447.29	.977000	4,136.96	1.028000	112,874.0	1.016540
1974	3,537.69	.902000	6,813.90	.877000	1,571.13	.963000	4,237.94	1.082000	140,530.0	.993181
1975	4,032.99	.970000	8,938.14	.805000	1,756.65	1.015000	4,883.14	1.031000	154,889.0	1.004610
1976	4,814.36	1.002000	6,746.25	.957000	1,850.75	1.139000	5,424.61	1.014000	166,913.0	.999390
1977	5,425.00	1.000000	6,693.34	1.000000	1,938.00	1.000000	6,142.80	1.000000	176,879.0	1.000000
1978	5,875.62	1.021000	6,636.73	.930000	2,164.17	1.227000	6,292.34	1.068000	199,170.0	1.027270
1979	7,241.30	.978000	7,084.22	1.010000	2,657.65	1.150000	6,580.41	1.179000	242,589.0	1.034830
1980	10,948.80	.859000	9,453.19	1.040000	2,929.35	1.129000	7,166.28	1.047000	321,305.0	1.025970
1981	13,076.00	.789000	10,120.80	1.050000	3,200.76	1.120000	7,396.99	1.144000	375,379.0	1.030740
1982	12,940.80	.760000	9,978.39	.948000	3,591.74	1.007000	7,716.89	1.179000	381,791.0	1.026050
1983	13,297.10	.717000	9,399.16	.814000	432.41	.811000	8,176.86	1.048000	333,754.0	1.006160

Note: [a] Includes durable equipment, livestock, and livestock inventory.

REFERENCES

Abramovitz, M. 1956. "Resource and Output Trends in the United States Since 1870," *American Economic Review* vol. 46, no. 1, pp. 5–23.

Alston, Julian M. 1986. "An Analysis of Growth of U.S. Farmland Prices, 1963–82," *American Journal of Agricultural Economics* vol. 68, no. 1, pp. 1–9.

Antle, J. M. 1984. "The Structure of U.S. Agricultural Technology, 1910–1978," *American Journal of Agricultural Economics* vol. 66, no. 4, pp. 414–421.

Ball, V. Eldon. 1985. "Output, Input and Productivity Measurement in U.S. Agriculture: 1948–79," *American Journal of Agricultural Economics* vol. 67, no. 3, pp. 475–486.

Barton, G. T., and M. R. Cooper. 1948. "Relation of Agricultural Production to Inputs," *Review of Economics and Statistics* vol. 30, no. 1, pp. 117–126.

Berndt, E., and M. Khaled. 1979. "Parametric Productivity Measurement and Choice Among Flexible Functional Forms," *Journal of Political Economy* vol. 87, no. 6, pp. 1220–1243.

Binswanger, H. P. 1974a. "A Cost Function Approach to the Measurement of Factor Demand Elasticities and of Elasticities of Substitution," *American Journal of Agricultural Economics* vol. 56, no. 2, pp. 377–386.

———. 1974b. "The Measurement of Technical Change Biases with Many Factors of Production," *American Economic Review* vol. 64, no. 6, pp. 964–976.

Brown, R. S. 1978. "Productivity, Returns and the Structure of Production in the U.S. Agriculture, 1947–74" (Ph.D. dissertation, University of Wisconsin, Madison).

———, and L. R. Christensen. 1981. "Estimating Elasticities of Substitution in a Model of Partial Static Equilibrium: An Application to U.S. Agriculture, 1947–74," in E. Berndt and B. Field, eds., *Modeling and Measuring Natural Resource Substitution* (Cambridge, Mass., MIT Press).

Burt, Oscar R. 1986. "Econometric Modeling of the Capitalization Formula for Farmland Prices," *American Journal of Agricultural Economics* vol. 86, no. 1, pp. 10–26.

Capalbo, S. M., and M. Denny. 1986. "Testing Long-Run Productivity Models for the Canadian and U.S. Agricultural Sectors," *American Journal of Agricultural Economics* (forthcoming).

———, T. T. Vo, and J. C. Wade. 1985. "An Econometric Data Base for the U.S. Agricultural Sector," Discussion Paper No. RR85 01, National Center for Food and Agricultural Policy (Washington, D.C., Resources for the Future).

Caves, D. W., L. R. Christensen, and W. E. Diewert. 1982. "Multilateral Comparisons of Output, Input and Productivity Using Superlative Index Numbers," *Economic Journal* vol. 92, no. 365, pp. 73–86.

———, L. R. Christensen, and J. A. Swanson. 1981. "Productivity Growth, Scale Economies, and Capacity Utilization in U.S. Railroads, 1955–1974," *American Economic Review* vol. 17, no. 5, pp. 994–1002.

Chalfant, J. A. 1984. "Comparison of Alternative Functional Forms with Application to Agricultural Input Data," *American Journal of Agricultural Economics* vol. 66, no. 2, pp. 216–220.

Chan, M. W. L., and D. C. Mountain. 1983. "Economies of Scale and the Tornqvist Discrete Measure of Productivity Growth," *Review of Economics and Statistics* vol. 65, no. 4, pp. 663–667.

Chandler, C. 1962. "The Relative Contribution of Capital Intensity and Productivity to Changes in Output and Income," *Journal of Farm Economics* vol. 44, no. 2, pp. 335–348.

Christensen, L. R., and D. W. Jorgenson. 1969. "The Measurement of U.S. Real Capital Input 1929–1967," *Review of Income and Wealth* vol. 15, no. 4, pp. 293–320.

———, and D. W. Jorgenson. 1970. "U.S. Real Product and Real Factor Input 1929–1967," *Review of Income and Wealth* vol. 16, no. 1, pp. 19–50.

Denny, M., M. Fuss, and L. Waverman. 1981. "The Measurement and Interpretation of Total Factor Productivity in Regulated Industries, with an Application to Canadian Telecommunications," in T. G. Cowing and R. E. Stevenson, eds., *Productivity Measurement in Regulated Industries* (New York, Academic Press).

Diewert, W. E. 1974. "Applications of Duality Theory," in M. Intrilligator and D. A. Kendrick, eds., *Frontiers of Quantitative Economics* vol. II (Amsterdam, North-Holland).

———. 1976. "Exact and Superlative Index Numbers," *Journal of Econometrics* vol. 4, no. 2, pp. 115–145.

Domar, E. 1961. "On the Measurement of Technological Change," *Economic Journal* vol. 71 (December), pp. 709–729.

Epstein, L., and M. Denny. 1980. "Endogenous Capital Utilization in a Short-Run Production Model: Theory and an Empirical Application," *Journal of Econometrics* vol. 12, no. 2, pp. 189–207.

Fabricant, S. 1959. *Basic Facts on Productivity Change* (New York, Columbia University Press for the National Bureau of Economic Research).

Gallant, A. R. 1982. "Unbiased Determination of Production Technologies," *Journal of Econometrics* vol. 20, no. 2, pp. 285–323.

Gollop, F., and D. Jorgenson. 1983. "Sectoral Measures of Labor Cost for the United States, 1948–1978," Discussion Paper no. 963 (Cambridge, Mass., Harvard Institute of Economic Research).

Griliches, Zvi. 1960. "Measuring Inputs in Agriculture: A Critical Survey," *Journal of Farm Economics* vol. 42, no. 5, pp. 1411–1433.

———. 1963. "The Sources of Measured Productivity Growth: United States Agriculture, 1940–1960," *Journal of Political Economy* vol. 71, no. 4, pp. 331–346.

———. 1964. "Research Expenditures, Education, and the Aggregate Agricultural Production Function," *American Economic Review* vol. 54, no. 6, pp. 961–974.

Hall, R. E., and D. W. Jorgenson. 1967. "Tax Policy and Investment Behavior," *American Economic Review* vol. 57, no. 3, pp. 391–414.

Hayami, Y., and V. W. Ruttan. 1970. "Agricultural Productivity Differences Between Countries," *American Economic Review* vol. 60, no. 5, pp. 895–911.

———, and V. W. Ruttan. 1971. *Agricultural Development: An International Perspective* (Baltimore, Md., Johns Hopkins University Press).

Hazilla, M., and R. J. Kopp. 1986. "Restricted Cost Function Models: Theoretical and Econometric Considerations," Discussion Paper QE86-05 (Washington, D.C., Resources for the Future).

Hicks, J. 1932. *The Theory of Wages* (London, Macmillan).
Huffman, W. E., and R. E. Evenson. 1986. "Supply and Demand Functions for Multi-Product U.S. Cash Grain Farms: Biases Caused by Research and Other Prices, 1948–1974," Staff Paper Series No. 155 (Ames, Iowa, Department of Economics, Iowa State University).

Jorgenson, D. W., and Z. Griliches. 1967. "The Explanation of Productivity Change," *Review of Economic Studies* vol. 34, no. 99, pp. 249–283.
———, and Z. Griliches. 1972. "Issues in Growth Accounting: A Reply to Edward F. Denison," *Survey of Current Business* vol. 52 (May), part II, pp. 65–94.

Kendrick, J. W. 1961. *Productivity Trends in the United States* (Princeton, N.J., Princeton University Press for National Bureau of Economic Research).
Kennedy, C., and A. Thirlwall. 1972. "Surveys in Applied Economics: Technical Progress," *The Economic Journal* vol. 82, no. 325, pp. 12–72.
Kopp, R. J., and V. K. Smith. 1981. "Productivity Measurement and Environmental Regulation: An Engineering-Econometric Analysis," in T. G. Cowing and R. E. Stevensen, eds., *Productivity Measurement in Regulated Industries* (New York, N.Y., Academic Press).

Lau, L. J. 1978. "Applications of Profit Functions," in M. Fuss and D. McFadden, eds., *Production Economics: A Dual Approach to Theory and Applications* (Amsterdam, North-Holland).
Lave, L. B. 1964. "Technological Change in U.S. Agriculture: The Aggregation Problem," *Journal of Farm Economics* vol. 46, no. 1, pp. 200–217.
Loomis, R. A., and G. T. Barton. 1961. *Productivity of Agriculture* (Washington, D.C., U.S. Department of Agriculture, Economic Research Service).
Lopez, R. E. 1980. "The Structure of Production and the Derived Demand for Inputs in Canadian Agriculture," *American Journal of Agricultural Economics* vol. 62, no. 1, pp. 38–45.
———. 1984. "Estimating Substitution and Expansion Effects Using a Profit Function Framework," *American Journal of Agricultural Economics* vol. 66, no. 3, pp. 358–367.

Martin, L., ed. 1977. *A Survey of Agricultural Economics Literature* vol. 1 (Minneapolis, Minn., University of Minnesota Press).
Melichar, E. 1979. "Capital Gains versus Current Income in the Farming Sector," *American Journal of Agricultural Economics* vol. 61, no. 5, pp. 1085–1092.

Nadiri, Ishaq M. 1970. "Some Approaches to the Theory and Measurement of Total Factor Productivity: A Survey," *Journal of Economic Literature* vol. 8, no. 4, pp. 1137–1177.

Peterson, W., and Y. Hayami. 1977. "Technical Change in Agriculture," in L. Martin, ed., *A Survey of Agricultural Economics Literature*, vol. 1 (Minneapolis, Minn., University of Minnesota Press).
Phipps, T. 1984. "Land Prices and Farm-Based Returns," *American Journal of Agricultural Economics* vol. 66, no. 4, pp. 422–429.
Pittman, R. W. 1983. "Multilateral Productivity Comparisons with Undesirable Outputs," *The Economic Journal* vol. 93, no. 372, pp. 883–891.

Ray, S. C. 1982. "A Translog Cost Function Analysis of U.S. Agriculture, 1939–1977," *American Journal of Agricultural Economics* vol. 64, no. 3, pp. 490–498.

Ruttan, V. W. 1957. "Agricultural and Nonagricultural Growth in Output per Unit of Input," *Journal of Farm Economics* vol. 39, no. 5, pp. 1566–1575.

Schultz, T. W. 1953. *Economic Organization of Agriculture* (New York, N.Y., McGraw-Hill).

Shumway, C. R. 1983. "Supply, Demand and Technology in a Multiproduct Industry: Texas Field Crops," *American Journal of Agricultural Economics* vol. 65, no. 4, pp. 748–760.

Solow, R. M. 1957. "Technical Change and the Aggregate Production Function," *Review of Economics and Statistics* vol. 39, no. 3, pp. 312–320.

Stout, T., and V. W. Ruttan. 1958. "Regional Patterns of Technological Change in American Agriculture," *Journal of Farm Economics* vol. 40, no. 2, pp. 196–207.

Thirtle, C. G. 1985. "Induced Innovation in United States Field Crops, 1939–1978," *Journal of Agricultural Economics* vol. 36, no. 1, pp. 1–14.

USDA. See U.S. Department of Agriculture.

U.S. Department of Agriculture, Economic Research Service. 1980. "Measurement of U.S. Agricultural Productivity: A Review of Current Statistics and Proposals for Change," Technical Bulletin No. 1614 (Washington, D.C., USDA).

4

THE STATISTICAL BASE FOR AGRICULTURAL PRODUCTIVITY RESEARCH:
A REVIEW AND CRITIQUE*

C. RICHARD SHUMWAY

A rich statistical data base exists for productivity research in agriculture. Several agencies have been involved in collection, tabulation, and interpretation of these data, with responsibility for most data collection and summarization residing with the U.S. Department of Agriculture (USDA).

The USDA has long been concerned with sectoral productivity. An early innovator, it has for more than two decades been the sole government agency regularly to prepare and publish total factor productivity (TFP) indexes. Other agencies, in particular the Department of Labor, have continued to emphasize the much less useful, and less theoretically palatable, partial factor productivity (PFP) indexes. For this innovativeness, the USDA is to be praised.

Although innovative in many ways, the USDA has been resistant to change in others. Almost as soon as USDA began to publish TFP indices, Griliches (1960) challenged the quality of some of the data and a number of procedures used by USDA to measure agricultural productivity. Since that time, other economists have criticized particular aspects of the productivity series and its statistical underpinnings. Few of their suggestions for improvement have been implemented.

Although the USDA has been reluctant to modify the official productivity series, useful data have been collected and reside in unpublished data files. These files provide the interested researcher a substantial data base for conducting investigations into agricultural production relationships, including productivity measurement, product supplies, input demands, and technical change. Some of the data are deep and sound; others are shallow and suspect. Although one must often dig deep into the statistical foundations and adjustment procedures used to construct

*This chapter is Texas Agricultural Experiment Station Technical Article No. 19744.

particular series to be sure they mean what the user assumes they mean, much information is available in the critiques and in certain USDA publications.

The purpose of this paper is to review the statistical base available for productivity research on agriculture and to consolidate suggestions for improving or better using that base. In developing this information, I shall draw liberally on the excellent report of an American Agricultural Economics Association (AAEA) task force chaired by Bruce Gardner (1980). The status of agricultural productivity data and an evaluation of needed improvements perceived up to that time are well documented in that report. My contribution will be limited to extracting key issues, identifying institutional responses and recent developments (as of the summer of 1984), and offering my own perceptions of additional needs.

A DESCRIPTION OF THE PRODUCTIVITY DATA BASE

Productivity indexes are published annually by the USDA for the United States and ten farm production regions comprising the contiguous forty-eight states in *Economic Indicators of the Farm Sector: Production and Efficiency Statistics*. The 1984 issue included annual indexes of farm production by twelve commodity categories (nine crop and three livestock) for the years between 1939 and 1982; farm inputs by seven input categories plus purchased and nonpurchased inputs; and TFP and several PFPs (output per hour of labor, crop production per acre, livestock production per breeding unit, and farm production per hour of labor by enterprise group). The publication also provides much useful physical data, including acreage of cropland used for crops and harvested for specified purposes; animal units of breeding livestock; quantity of fertilizer used in agricultural production; farm machinery inventory; hours of labor required in farmwork by enterprise group; and number of persons supplied farm products by one farmworker.

The procedures used to develop this publication are reported in *Major Statistical Series of the U.S. Department of Agriculture* (USDA, 1970) and elsewhere (for example, Loomis and Barton, 1961; Lambert, 1973; Barton and Durost, 1960). However, as noted in the AAEA report,

> ... the details of the procedures used—sources for each element of basic data, assumptions used to fill in the many data gaps, techniques for estimation between benchmark surveys and extrapolation beyond the latest benchmark, criteria for revising data and changing procedures—have never been available in print (AAEA, p. 13).

Responsibility for consolidating the data, making necessary adjustments, and preparing the report resided in 1984 with the Economic Indicators and Statistics Branch, National Economics Division, Economic Research Service (ERS) of USDA. Most of the hard data underlying the report come from the Statistical Reporting Service (SRS) of USDA and other agencies that conduct surveys contractually for the ERS.

The AAEA report outlines in detail the procedures involved and time sequence associated with construction of the USDA productivity statistics. I shall limit my discussion of procedures to highlights and recent developments.

All data used in the computation of official productivity indexes are annual time series. Most are maintained by region since 1939; some national series go back to 1870. Data are frequently revised in response to each new census and as new information becomes available on production practices or data errors. However, although USDA continues to publish data beginning with 1939 in each annual publication, some revisions affect only a portion of the series. This leads to a problem of consistency when the entire series is used, since major differences in assumptions and procedures may affect different years.

At all levels of aggregation, both inputs and outputs are aggregated by means of modified Laspeyres-type indexes. Base period prices are used as weights to aggregate items into larger categories. To permit some comparison across years served by different base periods, index values for each major input group are spliced in the year when the base period changes. This equates the values for that year under both old and new weights.

Economic Research Service Input Series

Labor. The labor data are estimates of hours required for farmwork and are based on state and regional estimates of the time required to perform various production activities on an enterprise basis. A constant proportion (recently, 15 percent) is added for overhead labor. Until 1978 the labor data were not based on surveys of actual labor usage but were obtained as expert opinions from state agricultural experiment station and extension personnel. In 1978, labor requirements were reestimated based on survey data. The 1974 cost of production surveys of the Firm Enterprise Data System were used to provide a new benchmark. These surveys were conducted by SRS for ERS. Most estimates since that time have been based on extrapolation and further subjective judgment.

No differentiation is made between operator, family, and hired labor in the published series. Nor is any account made for labor available in agriculture but idle or partially idle during certain periods.

Real Estate. Service flows from real estate are estimated as the sum of the following components:

1. Net cash rent (after property taxes) multiplied by the equity portion of the constant-dollar (base period) value of land and service buildings (excluding dwellings) on farms.
2. Base period (currently 1976–1978) mortgage interest rate multiplied by the debt portion of the constant-dollar value of land and service buildings.
3. Depreciation on constant-dollar value of service buildings (the depreciation rate currently used is 2 percent).
4. Grazing fees paid for public lands.
5. Accidental damage and cost of repairs to service buildings.

The acreage of various types of land (cropland, pasture, and other) is based on Census of Agriculture surveys (1978 was the most recent Census for which data were available in 1984). The estimate of total land in farms is updated annually based on SRS data. Implicit service flows from land diverted from current production under the commodity programs are not excluded from the real estate input series.

Machinery and Mechanical Power. Service flows from machinery and mechanical power include depreciation on the stock of capital (based on a declining balance method); opportunity cost of funds invested in capital equipment (using the base period farm mortgage interest rate); repairs; parts; tires; fuel and oil; electricity; custom work; small hand tools; blacksmithing; and hardware. Data series are maintained on each category. However, the statistical data base underlying many of these series is not deep.

For example, the last Census of Agriculture benchmark used for the stock of machinery was for 1949. Annual changes in the stock since that time have been estimated based upon trade sources and the SRS farm production expenditure surveys of 1955, 1971, and subsequent annual surveys.

The percent of truck and automobile use for farm purposes has been modified in recent years based on the SRS farm production expenditure surveys but the rates have varied so much as to make the series suspect. Before 1955, the rates were 78 and 40 percent for trucks and automobiles, respectively. Since that time, the rates have ranged from a low of 73 and 20 percent to a high of 90 and 46 percent, respectively. Estimates for the other categories of expense are periodically revised based on the SRS farm production expenditure surveys.

Chemicals. The agricultural chemicals category includes fertilizers, lime, herbicides, insecticides, and fungicides. Fertilizer nutrients (nitrogen, phosphorus, and potassium) are price weighted (using base period prices) back to 1965. Prior to 1965, the data series consisted of unweighted tonnage of nutrients. Annual data on tonnage of fertilizer nutrients are from SRS and are published by the USDA (1982). Annual tonnage of lime is obtained from the National Lime Institute.

Pesticide expenditure data based on the SRS farm production expenditure surveys and the annual publication *Pesticide Review* (USDA, 1982) are deflated for each year since 1965 by the SRS index of prices paid for agricultural chemicals (see USDA, 1983a). Prior to 1965, price indexes were developed internally by ERS.

Feed, Seed, and Livestock. Feed, seed, and livestock services include the nonfarm sector's contribution to the use of feed, seed, and livestock as an input. Expenditures on feed are estimated by ERS from the Census of Agriculture and data provided by the American Feed Manufacturer's Association. Census of Manufacturing data are used to estimate the margin between the farm value of feeds and the value of manufactured feeds.[1] Expenditure data are deflated by the SRS index of prices paid by farmers for feed.

The nonfarm sector's contribution to purchased seeds is estimated by the differences between prices paid by farmers for seeds and prices received for the respective crops in the base period, deflated by the SRS index of prices paid by farmers for seed. Because of a lack of data, the entire value of hybrid seeds is attributed to the nonfarm sector.

The nonfarm value added in livestock and transfers of livestock products covers expenditures on baby chicks and turkeys, milk hauling, and feeder livestock marketing charges.

Taxes and Interest. The taxes and interest category includes real estate and personal property tax payments and forgone returns to funds in-

---

[1] Data from the 1969 *Census of Manufacturing* were still being used for this purpose in 1984.

vested in livestock, operating capital, and other nonreal estate, nonmachinery assets. Livestock inventories are the average of each year's SRS beginning and ending inventory estimates. Crop inventories are one half the SRS January stock estimate. Operating capital is estimated by ERS as farmers' demand deposits deflated by the SRS index of prices paid by farmers for factors of production. The service flow is estimated by using the base period mortgage interest rate. The additional service flow from nonreal estate assets, due to the positive differential between short-term and mortgage interest rates, is added.

With the intent of reflecting the service flow of such intangible inputs as education, roads, and research, taxes on real estate other than dwellings and taxes on personal property are estimated by ERS. State sales and income taxes are not included. Real estate tax expenditures are deflated by the SRS index of taxes (see USDA, 1983a), and personal property tax expenditures are deflated by the Bureau of Labor Statistics (BLS) index of purchases of goods and services by state and local governments.

Miscellaneous. The miscellaneous input category includes insurance, irrigation operating and maintenance charges, veterinary expenses, dairy supplies, cotton ginning, telephone, containers, and binding materials. Most expenditure data are ERS estimates based on a variety of information, including the SRS farm production expenditure surveys and data from the Agricultural Stabilization and Conservation Service and the Dairy Herd Improvement Association. Price deflators are mainly from SRS but also include a price index for irrigation operating and maintenance charges from the Bureau of Reclamation.

Estimated service flow data are tabulated and maintained for detailed input categories by ERS. For a listing of these categories and their 1967 deflated values, see AAEA (1980, pp. 50–51).

ERS Output Series

Farm output is a gross measure, excluding only farm-produced intermediate products used as inputs in agriculture. It is maintained by ERS for twelve output categories (three livestock and nine crop). Annual data are from the Crop Reporting Board of SRS. The series are designed to measure output in the calendar year in which it is produced. The series are comprehensive; only some very minor crops, livestock by-products, and farm forest products are excluded. Base period prices are used to aggregate individual commodity production into the output categories and into aggregate output.

Other Government Input Series

Labor data are collected for the farming sector in two separate surveys that are not used in developing the ERS labor series for productivity measurement. They are the recently interrupted SRS quarterly sample survey of number of operator, family, and hired workers on farms and the Census Bureau's Current Population Survey. Both give independent estimates of labor use (or at least availability) on farms.

The Bureau of Labor Statistics (BLS), which is responsible for maintaining relevant data on productivity of the nonagricultural sector, also collects some data relevant to agriculture. Of particular importance are the BLS price indices for machinery. Quality adjustments in machinery are explicitly taken into account in their series to a much greater extent than in the SRS price indices.

EVALUATION OF THE DATA BASE

USDA has been lauded frequently by economists for important longstanding contributions to sectoral productivity research. For more than twenty years, the USDA has collected necessary data to support published TFP measures (Brown, 1978, p. 111; AAEA, 1980, p. 42) with output measured in gross rather than net (value added) terms (Christensen, 1975, pp. 912–913; AAEA, 1980, p. 42). In addition to publishing these series, USDA has continued to publish several PFP series, including farm production per hour of labor, crop production per acre, and livestock production per breeding unit. The first of these PFP measures facilitates the important tasks of conducting intersectoral comparisons since the only productivity measures reported for most sectors of the U.S. economy are the BLS PFP statistics.[2]

These contributions by the USDA are basic and extremely important. We turn now, however, to criticisms and suggestions for improvement in the USDA published indexes and underlying data base.

Many economists have suggested ways in which the quality of information available for productivity research on the agricultural sector could be improved. In 1960, Griliches articulated the need for several modifications in data collection and inference used in the USDA productivity series. Others (including Christensen, 1975; Penson, Hughes, and Nelson, 1977; Brown, 1978; Gollop and Jorgenson, 1980; the AAEA task force, 1980; and Gardner, 1981) have reinforced Griliches's basic tenets and proposed further changes. Their recommendations are sum-

---

[2] BLS has recently announced plans to begin publishing TFP indexes by sector.

marized briefly below under seven topics: aggregation, definition of the farming sector, input quality, capital stocks and service flows, data inadequacies and gaps, data use, and concepts and information.

Aggregation Procedures

Griliches (1963, p. 333) noted that aggregation procedures implicitly maintain hypotheses about elasticity of substitution. An arithmetic index, for example, implies infinite elasticity of substitution; a geometric index implies an elasticity of one. For this reason, the modified Laspeyres arithmetic index procedure used by the USDA to aggregate both inputs and outputs has come under repeated criticism (Christensen, 1975; Brown, 1978; AAEA, 1980).

In the Laspeyres index, base period prices are used to weight individual quantities for aggregation into larger categories. The base periods currently used are

1910 to 1940: 1935–1939 base period
1940 to 1955: 1947–1949 base period
1955 to 1965: 1957–1959 base period
1965 to 1975: 1967–1969 base period
1975 to present: 1976–1978 base period

As noted in chapter 2, the Laspeyres index of inputs is exact for a linear or a Leontief production function. That is, for the Laspeyres index to give an exact representation of the production function, inputs must be perfect substitutes. As there is evidence that inputs used in agricultural production are imperfect substitutes, the use of the Laspeyres index constitutes a potentially serious abstraction from reality. Each of the above-mentioned critics has recommended that the USDA use a superlative index of inputs and of outputs, that is, one which is exact for a production function that provides a second-order approximation to an arbitrary twice differentiable linear homogeneous production function. Most often recommended is the Tornqvist discrete approximation to the Divisia index. The Divisia index is exact for the homogeneous translog function that has the above approximation property.

The recommendation to change indexing procedures, or at least to shorten the period over which any base period prices are used (Brown, 1978, p. 140), is particularly important for inputs. Because relative output prices have not changed as greatly as input prices, this is a less serious concern for output aggregation (AAEA). Gardner (1981, p. 44), for example, found that the Laspeyres price index of the basic farm crops resulted in a bias relative to the Divisia price index of "less than ½ of 1 percent per year in the estimated rate of change of farm prices in the 1970s."

Another problem with the USDA aggregation procedure is that it maintains the hypothesis of separability of outputs and inputs. Since empirical tests of separability of agricultural outputs and inputs have generally been rejected (for example, Saez, 1983, and Shumway, 1983), this is an issue of great practical importance. Diewert (1976) proposed an index of technological change that does not maintain the separability hypothesis. Brown (1978, pp. 136–138) estimated the Diewert index and noted substantial differences in subperiod rates of productivity growth relative to the conventional index. One potential problem with the Diewert index, however, is that it is not invariant to the item selected for the numéraire output or input.

Definition of Farming Sector

In order to make sense out of sectoral productivity measures, it is essential to have a clear definition of what a sector constitutes. Sectors can be defined by establishment or by product. The establishment definition means that the conventional units of production, for example, farms, comprise the sector. All outputs of such establishments are then included in the measure of output and their inputs in the measure of input. The problem with this definition is that the activities of the establishments have changed over time. Some agricultural products—for example, broilers—are now produced largely by nonfarm firms. Consequently, "adherence to an establishment definition for productivity measurement leads to shifts in productivity merely due to changes in the activities encompassed by the establishments" (AAEA, 1980, p. 26).

The product definition, which identifies the sector according to specified production activities, regardless of the nature of the establishment performing the activities, is conceptually preferred. The problem in implementing the product definition, however, is that many of the data needed to measure agricultural productivity are collected only on an establishment basis. Therefore, it may be preferable to use the establishment definition exclusively rather than combine the two measures, as is now the case.

A second major problem associated with definition of the sector is treatment of that portion of technological progress attributable to such things as increased education, public research, and so forth. Because the quality of inputs used in agricultural production has improved over time, it is generally thought appropriate to attribute the quality improvement to nonagricultural sectors rather than to agriculture. However, much of the cost of public schooling in rural areas is supported by the agricultural producing industry, as is part of the cost of public research on agriculture and much of the cost of implementing research findings. This gives rise to a dilemma in determining how much of the quality change to attribute to the agricultural sector, that is, how much of any change in expenditure (or revenue) is due to price change and how much to quantity change.

Input Quality

Once the sector is defined, the data must be collected and necessary adjustments made to develop a quality-constant series of inputs used by the sector and a quality-constant series of outputs produced by it. The quality adjustment should be limited to those quality changes due to other sectors.

The USDA has been strongly and frequently criticized for failing to adjust the input series for quality. In 1960 Griliches (p. 1412) charged:

> The USDA, instead of recognizing that it is a difficult problem with no perfect solution, but one worth fighting for, has taken the position that nothing *should* be done about it, that it is not *desirable* to hold quality constant when pricing items bought by farmers, and thus has made the quality change problem much more serious for the USDA Prices Paid Index than is the case for similar indexes in other sectors of the economy.

Christensen (1975, p. 914) and the AAEA task force (1980, p. 45) have more recently voiced the same concern.

The criticisms for failure to adjust for quality changes have been focused most sharply on labor. Griliches (1960, pp. 1411–1416) argued that recent labor quantities had been underestimated largely because the farm labor force had become more highly educated over time. The AAEA task force (1980, pp. 28–29) concurred and suggested that quality has also been affected by a changing age, sex, and family versus hired labor structure. None of these factors is considered in the official USDA labor series used in the measurement of productivity.

Griliches (1960, p. 1415) attempted an early adjustment of the labor series based on sex and years of schooling. Gollop and Jorgenson (1980) performed a much more involved and sophisticated analysis of labor quality in many sectors, including agriculture, covering the period 1947–1973. In a recent working paper, they have extended the quality series to 1978. Both reports considered changes over time in wages and such attributes as age, sex, education, employment class (family or hired labor), occupation class (manager, laborer, and so forth), and industry.

Griliches (1960, p. 1416) recommended that the USDA account for changes in fertilizer quality by using prices to weight nitrogen, phosphorus, and potassium content in the fertilizer, since nutrient ratios have changed substantially over time. Brown (1978, pp. 117–120) made the same plea.

Because the BLS machinery price index has been computed for standardized machinery quality, the AAEA task force (1980, p. 46) recommended that it be used to compute the machinery input series from expenditure data. The currently-used SRS machinery price index does not adequately compensate for quality changes that have taken place in this input. Consequently, unless appropriate adjustments are made in this index, it is judged to be an inferior series to the BLS index.

Capital Stocks and Service Flows

No aspect of the USDA productivity data series has been subject to as many varied criticisms as the capital stock and service flow series. Concerns over stock measurement have focused on machinery and structures; criticisms of the service flow measurement have included land as well.

Griliches (1960, pp. 1417–25), Brown (1978, pp. 117–120), and Penson and coauthors (1977, pp. 77–81) have charged that the USDA depreciation rates based on a declining balance asset value are too high to measure correctly the productive capacity (or stock) of machinery and buildings. If service flows are to be measured as a constant proportion of the stock, they argue, the productive value of the stock of capital provides the best measure of the base from which these services flow. Griliches proposed a fifteen-year one-hoss-shay asset value pattern (that is, original value after years one through fourteen, zero after year fifteen) as potentially more relevant than the USDA declining balance (convex) pattern. Penson and coauthors (1977) found that a concave pattern of the productive value of tractors was most consistent with engineering data. Penson and coauthors (1981) showed that such a pattern best explained farmers' investment behavior. The USDA's depreciation rates performed worst of all the patterns studied.

Repairs should be added to the capital stock and depreciated over several years rather than being charged as a service flow of capital in the year when made (Brown, 1978, pp. 117–120). Conservation practices on land should be treated similarly (AAEA, 1980, p. 32). The AAEA task force (1980, p. 44) recommended that a separate data series be maintained on diverted land. However, because of the highly variable way in

THE STATISTICAL BASE FOR AGRICULTURAL PRODUCTIVITY RESEARCH   149

which diverted land has been treated under the commodity programs (consider, for example, planting restrictions), the task force did not recommend that diverted land be subtracted from the total available for production.

Rental price of capital should be used to measure service flows where possible (Griliches, 1963, p. 343; AAEA, 1980, pp. 32–33; Brown, 1978, p. 119). A shadow price for public grazing land is preferred to cash rent because the latter substantially underestimates the marginal value product of that asset (AAEA, 1980, p. 46).

Data Inadequacies and Gaps

Several data series used for productivity measurement represent constructed data and provide no independent information. Such series are based on questionable data sources, arbitrary construction rules, or both. These series include labor, capital repairs, veterinary services, and dairy supplies. Two labor surveys, the Census Bureau and the SRS surveys, could be used as alternatives to the ERS hours required series. Neither is perfect, but the Census Bureau data, based on a retrospective annual survey of farm workers (or the SRS data if based on a monthly survey of employees), are recommended by the AAEA task force (1980, p. 46). The Census Bureau series reports the number of workers on farms quarterly and distinguishes them by age, sex, and worker class (self-employed, family, and hired). The SRS series also reports the number of workers on farms by work status (operator, family, and hired). These classifications are important improvements over the published ERS series on hours required, which does not differentiate type of labor.[3] However, neither the Census Bureau nor the SRS series reports actual labor used; they only indicate the amount of labor available.

The series on stock and depreciation of service buildings has been charged with being highly tenuous (AAEA, 1980, pp. 36–37; Brown, 1978, p. 119; Grove, 1969, p. 185). Until recently, no survey data existed on service buildings. Adequate survey detail was not available on the machinery stock. Many arbitrary assumptions have been involved in construction of that series, and little information has been available on machinery quality. Gardner (1981, p. 36), however, recently noted that the data on equipment and structures investment have been improved greatly since 1970 because of the greater use of farm production expenditure surveys.

The only data collected on fertilizer quantities deal with total disappearance. The amount used in agriculture is not distinguished from that

---

[3] The unpublished ERS labor data are differentiated by type.

used for forestry, residential, and other purposes. The quantity of new chemicals being used by the livestock industry is not collected either and so it is not included in the productivity data series. Neither is hybrid seed production. Seed used on the farm where grown is not subtracted from output.

Breeding livestock inventories appear to be undervalued (using slaughter values), but inadequate data are being collected to accurately determine inventory values. No data are being collected on the stock, repair, and maintenance of irrigation equipment, or on legal services, soil testing, accounting services, or office space in dwellings used by the agricultural production industry.

Several nonconventional inputs are having an increasingly important effect on agricultural production. These include irrigation water, the environment, management services, public infrastructure, insurance, and governmental activity (AAEA, pp. 33–45). Perhaps the most important issue that must be addressed with regard to these is to decide which should be measured precisely enough to attempt measurement at all.

Brown (1978, p. 112) recommended reorganization of the input categories to be more consistent with the capital, labor, energy, and materials classification used in analysis of many nonagricultural sectors (see, for example, Berndt and Wood, 1975). Although such reorganization should have little effect on estimates of total productivity changes, it could facilitate other uses of the data.

Some interest exists for computing productivity measures by geographic and commodity divisions. The USDA already reports both data and productivity measures for ten farm production regions comprising the contiguous forty-eight states. The data are not currently amenable to state or commodity productivity measurement. The AAEA task force (1980, p. 37) recommended that the USDA not devote additional resources for commodity productivity measurement. The data problems would be horrendous for an industry that produces so many commodities. Most firms produce multiple products and either do not or cannot maintain records on the allocation of all inputs among commodities. To attempt productivity measurement by commodity would entail extreme arbitrariness in the specification of input usage.

Data Use

There is some evidence that the rate of total productivity growth in agriculture reported by the USDA is overestimated. Griliches (1960) alleged that two problems with the USDA productivity data—namely, failure to account for quality improvements in inputs and overestimation of the rate of depreciation of the capital stock—biased the productivity growth estimates upward.

Based on data revisions more appropriately to account for input quality changes and service flows, Brown's (1978, p. 124) recomputation of the total productivity index for agriculture resulted in a reduction of productivity growth between 1947 and 1974 of 12 percent relative to the USDA estimate. Even using the USDA productivity data, Lu's (1975, p. 73) estimate of productivity growth between 1939 and 1972 based on a Cobb-Douglas production function was 34 percent lower than the USDA estimate.[4]

Although each of these estimates assumed neutral technical change, it is possible to assume otherwise both with the production function and with index numbers. Consequently, agricultural productivity measures need not suffer from this unrealistic maintained hypothesis, as all evidence indicates that technical change in agriculture has been labor-saving relative to capital over an extended period. The challenge is to determine what the magnitudes of the technical biases are. Production function and related econometric approaches are normally used for that purpose.

Other problems that appear with the productivity data stem from trying to use the same data for several purposes, for example, for growth accounting, income statistics, and productivity measurement (1980, AAEA, pp. 33–34). Data that were appropriately collected for one purpose may be quite inadequate—and even misleading—when used for another. The same data set may be quite useful for a wide variety of uses if it is maintained for electronic recall in sufficient detail and with sufficient documentation that the potential user can judge propriety for his or her purpose.

Concepts and Information

Two additional recommendations of the AAEA task force (1980, pp. 37–42) are particularly important and need no elaboration.

1. More resources should be devoted by the USDA to conceptual work on agricultural productivity as opposed to strict number generation.
2. More information needs to be provided by the USDA on how data series are generated. More information on both data collection and, perhaps even more importantly, assumptions and procedures in-

---

[4] Ball, however, finds some evidence that the official figures actually underestimate the true productivity growth rate in the postwar period.

volved in processing the raw data into each series, should be provided.

AGENCY RESPONSE TO SUGGESTIONS

Although an early innovator in sectoral productivity research, the USDA has been reluctant to implement many of its critics' suggestions. Of Griliches' recommendations made nearly twenty-five years ago, only two appear to have been incorporated into the USDA productivity series:

1. Since 1965 nutrient prices have been used to weight fertilizer nutrient quantities into a fertilizer quantity series. Unfortunately for intertemporal comparisons, the data prior to 1965 have not been correspondingly revised. Rather, those earlier data are simple sums of the quantities of the three major nutrients.
2. Capital service flow and other input series are now more firmly rooted in periodic surveys since the initiation of the SRS farm production expenditure surveys on an annual basis in the 1970s.

Perhaps because he was not a part of the USDA-State Experiment Station complex, Brown's work appears to have received little official attention within the USDA. Even the AAEA task force was apparently unaware of his work, which represents an important contribution to the literature on agricultural productivity data and estimation. The recent work of Eldon Ball has incorporated the significant suggestions of Brown in further ERS data construction.

In response to the AAEA task force, an interbranch committee of the National Economics Division of ERS was formed to determine which recommendations should be pursued by the USDA. A serious and well-reasoned response was prepared by the committee in an internal document (Teigen and coauthors, 1982). With only a single member objecting, this USDA committee concurred with all of the AAEA recommendations, encouraged implementation, and discussed interagency cooperation and resource requirements necessary to effect them. The only substantive difference between the recommendations of the USDA committee and the AAEA task force dealt with aggregation procedure. In that regard the committee recommended to

> ... continue using a Laspeyres index to aggregate all inputs. The Divisia index is a conceptually superior formula, and more frequent updating of factor share weights is desirable. However, strong pragmatic reasons dictated our recommendation: (a) considerable resources are required for

an annual updating of factor share weights; (b) there needs to be a consistency between measures of both inputs and outputs and the price indexes for those inputs and outputs (the price and crop production indexes reported by SRS are Laspeyres indexes); (c) the indexes reported by the preponderance of government agencies are Laspeyres indexes. It is more important to devote resources to measuring the conceptual content of inputs and outputs than to worry about the aggregation process (Teigen and coauthors, 1982, pp. 14–15).

Unfortunately, it appears that few of the AAEA task force or the USDA committee recommendations have yet affected either the data collection or construction procedures used by ERS in the official productivity data series.

FURTHER CONCERNS

I find myself in basic agreement with the AAEA task force and USDA committee about changes needed in the productivity data collection and construction. However, since these data are used for so many purposes other than productivity measurement, I am a little concerned about two additional issues:

1. Although Gardner found little difference caused by aggregation procedure in productivity growth of outputs, the choice of aggregation procedure for outputs may not be a trivial issue when the data are used for such things as supply analysis.
2. I agree with the AAEA task force recommendation not to devote significant resources to the development of commodity or commodity-group productivity measures. However, better data on input allocations among commodities are urgently needed. Most agricultural firms produce more than one product. Recent research on model specification of the multiple-product firm (for example, Just and coauthors, 1983; Shumway and coauthors, 1984) document the need for accurate data on inputs that are clearly allocated among products. It is not necessary to arbitrarily allocate inputs that are not allocated in practice. But simply lumping all inputs into generic categories does little to contribute to an understanding of individual product supplies or the cause of changes in input demands. Because of these secondary needs for the productivity data, I hope that the USDA will devote some modest resources to improving the quality of input allocation data collection and summarization.

Many raw and partially processed data, many of which are of very good quality, are currently maintained by the USDA and can be used by their analysts and others to conduct more appropriate productivity and other research. With limited resources, the USDA should concentrate its efforts on improving the quality of the basic data, describing procedures and assumptions used in constructing detailed data series, and facilitating easy electronic access to the data.

It is essential that more serious analysis of the agricultural production sector be conducted using the wealth of data already maintained by the USDA. It is perhaps even more important that pressure be kept on the USDA to implement basic reforms that will improve the quality of data available for all purposes.

REFERENCES

American Agricultural Economics Association Task Force on Measuring Agricultural Productivity. 1980. *Measurement of U.S. Agricultural Productivity: A Review of Current Statistics and Proposals for Change,* ESCS Technical Bulletin No. 1614 (Washington, D.C., U.S. Department of Agriculture).

Ball, V. E. 1984. *Measuring Agricultural Productivity, A New Look,* Staff Report No. AGES 840330 (Washington, D.C., U.S. Department of Agriculture, Economic Research Service).

Barton, G. T., and D. D. Durost. 1960. "The New USDA Index of Inputs," *Journal of Farm Economics* vol. 42, no. 5, pp. 1398–1410.

Berndt, E. R., and D. O. Wood. 1975. "Technology, Prices, and the Derived Demand for Energy," *Review of Economics and Statistics* vol. 57, no. 3, pp. 259–268.

Brown, R. S. 1978. "Productivity, Returns, and the Structure of Production in U.S. Agriculture, 1947–1974" (Ph.D. dissertation, University of Wisconsin, Madison).

Christensen, L. R. 1975. "Concepts and Measurement of Agricultural Productivity," *American Journal of Agricultural Economics* vol. 57, no. 5, pp. 910–915.

Diewert, W. E. 1976. "Exact and Superlative Index Numbers," *Journal of Econometrics* vol. 4, no. 2, pp. 115–145.

Gardner, B. L. 1981. "Changes in the Quality of Agricultural Statistics—Inputs, Farm Income, Output, and Prices," in R. K. Perrin and E. Reinsel, eds., *Economic Statistics for Agriculture: Current Directions, Changes and Concerns.* Proceedings of a Symposium at the American Agricultural Economics Association meetings, Clemson, S.C., July.

Gollop, F. M., and D. W. Jorgenson. 1980. "U.S. Productivity Growth by Industry, 1947–73," in J. W. Kendrick and B. N. Vaccara, eds., *New Developments in Productivity Measurement and Analysis*, chapter 1 (Chicago, Ill., University of Chicago Press).

———. "Sectoral Measures of Labor Cost for the United States, 1948–1978." Department of Economics Working Paper, Boston College, undated.

Griliches, Z. 1960. "Measuring Inputs in Agriculture: A Critical Survey," *Journal of Farm Economics* vol. 42, no. 5, pp. 1411–1427.

———. 1963. "The Sources of Measured Productivity Growth: United States Agriculture, 1940–60," *Journal of Political Economy* vol. 71, no. 4, pp. 331–346.

Grove, E. W. 1969. "Econometricians and the Data Gap: Comment," *Journal of Farm Economics* vol. 51, no. 1, pp. 184–187.

Just, R. E., D. Zilberman, and E. Hochman. 1983. "Estimation of Multicrop Production Functions," *American Journal of Agricultural Economics* vol. 65, no. 4, pp. 770–780.

Lambert, L. D. 1973. "Regional Trends in the Productivity of American Agriculture" (Ph.D. dissertation, Michigan State University).

Loomis, R. A., and G. T. Barton. 1961. *Productivity in Agriculture*, Technical Bulletin No. 1238 (Washington D.C., U.S. Department of Agriculture, Economic Research Service).

Lu, Y. C. 1975. "Measuring Productivity Change in U.S. Agriculture," *Southern Journal of Agricultural Economics* vol. 6, no. 2, pp. 69–75.

Penson, J. B., D. W. Hughes, and G. L. Nelson. 1977. "Measurement of Capacity Depreciation Based on Engineering Data," *American Journal of Agricultural Economics* vol. 59, no. 2, pp. 321–329.

———, R. F. J. Romain, and D. W. Hughes. 1981. "Net Investment in Farm Tractors: An Econometric Analysis," *American Journal of Agricultural Economics* vol. 63, no. 4, pp. 629–635.

Saez, R. R. 1983. "Multiproduct Agricultural Supply Response and Factor Demand Estimation in the U.S.: The Profit Function Approach" (Ph.D. dissertation, Texas A&M University).

Shumway, C. R. 1983. "Supply, Demand, and Technology in a Multiproduct Industry: Texas Field Crops," *American Journal of Agricultural Economics* vol. 65, no. 4, pp. 748–760.

———, R. D. Pope, and E. K. Nash. 1984. "Allocatable Fixed Inputs and Jointness in Agricultural Production: Implications for Economic Modeling," *American Journal of Agricultural Economics* vol. 66, no. 1, pp. 72–78.

Teigen, L. D., A. G. Smith, C. Cobb, V. E. Ball, and R. Simunek. 1982. "The USDA Productivity Indicators and Recommended Improvements," Economic Research Service (Washington, D.C., U.S. Department of Agriculture).

USDA. See U.S. Department of Agriculture.

U.S. Department of Agriculture. 1970. *Major Statistical Series of the U.S. Department of Agriculture* vol. II, HB-365 (Washington, D.C., USDA).
———. 1982. *Commercial Fertilizer Consumption for Year Ended June 30, 1982* (Washington, D.C., USDA, Economic Research Service).
———. 1983a. *Agricultural Prices, Annual Summary 1982* (Washington, D.C., USDA, Economic Research Service).
———. 1983b. *Pesticide Review 1982* (Washington, D.C., USDA, Agricultural Stabilization and Conservation Service).
———. 1984. *Economic Indicators of the Farm Sector: Production and Efficiency Statistics, 1982* (Washington, D.C., USDA, Economic Research Service).

PART II
MEASURING AGRICULTURAL PRODUCTIVITY AND TECHNICAL CHANGE

5

A COMPARISON OF ECONOMETRIC MODELS OF U.S. AGRICULTURAL PRODUCTIVITY AND AGGREGATE TECHNOLOGY*

SUSAN M. CAPALBO

INTRODUCTION

Over the past decade and a half, cost and profit functions have gained popularity and prominence as tools for estimating the structure of production. This has been due in part to the advent of duality theory and to the use of flexible functional forms that avoid the strong separability restrictions implicit in the Cobb-Douglas and CES functional forms. However, because in general these flexible functional forms are not self-dual—that is, the production and cost functions are not members of the same family of functional forms—the use of a production, cost, or profit function model does not incorporate the same maintained hypotheses regarding the production technology.[1] This is certainly true if the flexible functional form is viewed as an exact representation of the true function; it is not clear how important this becomes when the flexible function is viewed as a second-order local approximation to the production technology.

These considerations suggest that it would be useful to assess the sensitivity of the empirical results to the choice among flexible functional forms for production, cost, and profit models. Burgess (1975) compared the results obtained by representing the technology of the U.S. manufacturing sector as a translog production model and as a translog cost model. He reported that substantial differences exist even when the

---

* Comments by John Antle and two anonymous reviewers, and the research assistance of Trang Vo are gratefully acknowledged.

[1] Houthaker (1960) first proposed the concept of self-duality with respect to functional forms. The concept has been further developed by Samuelson (1965) and Christensen, Jorgenson, and Lau (1973). Diewert (1974) and the chapters in Fuss and McFadden (1978) provide extensive discussion of flexible functional forms and duality.

translog cost and translog production functions are treated as second-order approximations to the "true" functional form.

Analyses of the aggregate production structure for agriculture have used production, cost, and profit function models and various functional forms. However, as discussed by Capalbo and Vo in chapter 3, comparisons of the empirical findings from these studies have been hampered by the differences in the data sets. The intent in this chapter is to provide cross-comparisons of the results from econometrically estimated flexible form production, cost, and profit models of the U.S. agricultural sector. The functional forms used to specify the primal and dual models are viewed as second-order local approximations and each model is estimated using the same set of data. Using the results of this analysis, it is possible to assess the sensitivity of the inferences regarding technical change and structural measures of technology (factor price elasticities, and measures of scale) to the choice of the production model.

In the next two sections, full static equilibrium and partial static equilibrium production models are specified and estimated using aggregate data for U.S. agriculture. The characteristics of these models are displayed in table 5-1. Models 1 through 5 are static in the sense that inputs are assumed to be in equilibrium at each point in time. This occurs if all factors adjust instantaneously to changes in exogenous variables, or if the data reflect only equilibrium situations. For these models, the estimated elasticities of substitution, and the own- and cross-price elasticities of factor demand and output supplies correspond to the full equilibrium (long-run) measures.

Models 6 through 8 allow for quasi-fixed inputs and are referred to as partial static equilibrium models. The factors are quasi-fixed because the cost of adjusting to exogenous changes during a given period is sufficiently high. The firm or sector is in equilibrium with respect to a subset of the inputs (variable inputs) conditional on the observed levels of the other inputs (quasi-fixed inputs). Thus, there is no substitution between the quasi-fixed factors and the variable factors. A discussion of the theoretical properties of the restricted cost and profit functions are found in Diewert (1974, 1985) and Lau (1976). For these models, the Le Chatelier principle implies that the own-price elasticities of the variable factors must be less (in absolute terms) in the short run than the corresponding long-run elasticities, when evaluated at some given point in input space.

Unlike the series of models tested by Berndt and Khaled (1979) or Capalbo and Denny (1986), the specifications estimated in this chapter do not form a series of nested hypothesis tests. Therefore, the choice of a "best" model cannot be based on the likelihood ratio test. Alternative procedures might include using nonnested testing procedures, as pro-

Table 5-1. Model Characteristics

Model number	Functional form	Outputs	Variable inputs	Fixed inputs	Technical change	Scale
1	Translog production function	Aggregate output	Hired and family labor; land and structures; equipment, animal capital, and replacement inventories; materials	None	Nonneutral	Nonhomothetic
2	Translog cost function	Aggregate output	Hired and family labor; land and structures; equipment, animal capital, and replacement inventories; materials	None	Nonneutral	Nonhomothetic
3	Generalized Leontief cost function	Aggregate output	Hired and family labor; land and structures; equipment, animal capital, and replacement inventories; materials	None	Nonneutral	Nonhomothetic
4	Translog cost function	Crops, livestock	Hired and family labor; land and structures; equipment, animal capital, and replacement inventories; chemicals; other materials	None	Hicks neutral	Nonhomothetic
5	Translog cost function	Crops, livestock	Hired and family labor; land, structures, equipment, animal capital, and replacement inventories; materials	None	Nonneutral	Nonhomothetic
6	Translog variable cost function	Aggregate output	Hired labor; chemicals; other materials	Family labor; land, structures, equipment, animal capital, and replacement inventories	Nonneutral	Constant returns to scale in variable inputs
7	Translog restricted profit function	Aggregate output	Hired and family labor; chemicals; other materials	Land, structures, equipment, animal capital, and replacement inventories	Nonneutral	Nonhomothetic
8	Translog restricted profit function	Crops, livestock	Hired labor; chemicals; other materials	Family labor; land, structures, equipment, animal capital, and replacement inventories	Nonneutral	Nonhomothetic

posed by Pesaran and Deaton (1978) and others, or the nonparametric tests for theoretical consistency conditions developed by Diewert and Parkan (1982). Employing such tests would be valuable but these tests go well beyond the scope and intent of this chapter.

To simplify the presentation of the econometric results, the following procedure is adopted. First, the set of estimating equations, the list of variables, and the parameter estimates are summarized in a separate table for each model. Second, the econometric results are briefly discussed in the text and the results of the tests for monotonicity, curvature, and other structural hypotheses are noted. Readers unfamiliar with the specification and interpretation of the parameters for these types of models should review the material in appendixes A and B to chapter 2. The data, taken from Capalbo, Vo, and Wade (1985), pertain to the post-World War II agricultural sector in the United States and are reported in appendix A of chapter 3.

In the following section, we provide a perspective on results derived from the different econometric models, all of which purport to measure the production structure of the aggregate agricultural sector in the United States. Tables 5-2–5-14 illustrating the models appear on the following pages. We also attempt to identify a "best" model using criteria based on theoretical consistency at the point of approximation. Model-by-model comparisons of the estimated price elasticities, the measures of total factor productivity, and the factor biases from these econometric models are provided. In the final section, the conclusions are noted, along with the limitations and areas of future research.

FULL STATIC EQUILIBRIUM MODELS

Production Function Model (Model 1)

Model 1 is a translog production function with four aggregate inputs: labor (hired and family), land and structures, durable equipment and animal capital, and purchased materials (feed and seed, energy, fertilizer, chemicals, and miscellaneous). The translog production model, the set of share equations, and the results of estimating this model are given in table 5-2. The tests for monotonicity were satisfied at all data points. The curvature conditions were violated at the point of approximation (1970); the demand for durable equipment and animal capital was positively sloped.

For model 1, the estimate of returns to scale at the point of approximation, $\sum_i \alpha_i$, was approximately 0.8. The time trend parameters, α_t and ϕ_{tt}, indicate the direction of the shift in the production function and the

TABLE 5-2. MODEL 1: TRANSLOG PRODUCTION FUNCTION WITH ONE OUTPUT, FOUR INPUTS

Model: $\ln Q = \alpha_0 + \alpha_t T + \sum_i^4 \alpha_i \ln X_i + \frac{1}{2}\sum_i^4 \sum_j^4 \gamma_{ij} \ln X_i \ln X_j + \sum^4 \phi_{it} \ln X_i T + \frac{1}{2}\phi_{tt} \cdot T^2$

$S_i = \alpha_i + \sum_j^4 \gamma_{ij} \ln X_j + \phi_{it} \cdot T, \quad i = 1, \ldots, 4$

Restrictions: Symmetry
Homotheticity

Variables:
Q: aggregate output index
T: time trend
X_1: quantity index of family and hired labor
X_2: quantity index of land and structures
X_3: quantity index of durable equipment, animal capital, replacement inventories
X_4: quantity index of material inputs
S_i: ith input cost divided by total revenue $(P_i X_i/PQ)$

Empirical results: U.S. agricultural sector, 1948–1982
(standard errors in parentheses)

α_0	−0.052 (0.005)	γ_{11}	−0.106 (0.024)	γ_{24}	−0.024 (0.047)	ϕ_{1t}	−0.005 (0.001)
α_1	0.155 (0.004)	γ_{22}	−0.137 (0.101)	γ_{13}	−0.100 (0.024)	ϕ_{2t}	0.013 (0.001)
α_2	0.185 (0.016)	γ_{33}	0.179 (0.069)	γ_{41}	−0.063 (0.020)	ϕ_{3t}	−0.006 (0.001)
α_3	0.177 (0.008)	γ_{44}	−0.058 (0.033)	γ_{43}	0.029 (0.034)	ϕ_{4t}	0.001 (0.001)
α_4	0.277 (0.003)	γ_{21}	0.269 (0.033)			ϕ_{tt}	−0.0002 (0.0001)
α_t	0.014 (0.001)	γ_{23}	−0.107 (0.069)				

rate of change of this shift, respectively, at the point of approximation. Technical progress occurred at an average annual rate of 1.40 (measured at the point of approximation), and this rate was increasing slowly over time. The alternative hypotheses of Hicks-neutral technological change and constant returns to scale were tested for and rejected at the 5 percent level of significance in favor of the maintained hypothesis of nonneutral technological change and nonconstant returns to scale.

When the sample was restricted to the 1960–82 period, the estimated production function was well behaved, that is, monotonic and quasi-concave at the point of approximation. The measure of returns to scale increased to approximately 0.9, and the rate of technological change remained at 1.40 percent when measured at the point of approximation. The alternative hypotheses of Hicks-neutral technological change and constant returns to scale were also rejected at the 5 percent level of significance in favor of the nonneutral, nonconstant returns to scale hypotheses. However, the constant returns to scale hypothesis was not rejected at the 1 percent level of significance.

Cost Function Models (Models 2 Through 5)

The single-output and multiple-output translog cost functions estimated for U.S. agriculture are similar to the model discussed in appendix A to chapter 2. A generalized Leontief cost function was also estimated following the specification used by Lopez (1980).

Single Output Translog Total Cost Function (Model 2). The set of estimating equations, the restrictions on the model, and the parameter estimates for a nonhomothetic, nonHicks-neutral translog cost function are presented in table 5-3 (model 2). The inputs were aggregated into four categories—hired and family labor, land and structures, durable equipment and animal capital, and chemicals and other purchased inputs. The fitted cost function satisfied the regularity conditions that it be nondecreasing and concave in the input prices at the point of approximation. In addition, the cost function was increasing in output, although this parameter was not statistically significant. The measure of returns to scale for model 2—the inverse of the partial of the log of total cost with respect to the log of output $(\partial \ln C / \partial \ln Q)^{-1}$—was not statistically significant for this set of parameters.

The time trend parameters ϕ_t and ϕ_{tt} indicate direction and rate of change of the shift in the cost function independent of prices and quantities.[2] At the point of approximation, $\partial \ln C/\partial T = \phi_t + \phi_{tt}$, because by construction input prices, output, and the time trend have been normalized to 1.0. In table 5-3, $\phi_t = -0.001$, indicating that the cost function was shifting inward independent of changes in factor prices and output, although no statistical significance could be attached to this time shift. The parameter ϕ_{tt} also was not significantly different from zero. The parameters ϕ_{it}, $i = 1, \ldots, 4$, were significant and suggested that technological change was labor- and land-saving, and capital- and chemical-using.

[2] The translog cost function can be differentiated with respect to time, T,

$$\partial \ln C/\partial T = \phi_t + \phi_{tt} T + \phi_{tQ} \ln Q + \sum_i \phi_{it} \ln W_i$$

where C denotes total cost, Q denotes output, and W_i denotes input prices. This equation includes the parameters that involve time effects. The parameters ϕ_{it} indicate the effect of the time trend proxy for technological change on the shares holding all other variables constant.

TABLE 5-3. MODEL 2: TRANSLOG TOTAL COST FUNCTION WITH ONE OUTPUT, FOUR INPUTS

Model:
$$\ln TC = \alpha_0 + \alpha_Q \ln Q + \sum_{i}^{4} \alpha_i \ln W_i + \tfrac{1}{2}\gamma_{QQ}(\ln Q)^2$$
$$+ \tfrac{1}{2}\sum_{i}^{4}\sum_{j}^{4} \gamma_{ij} \ln W_i \ln W_j + \sum_{i}^{4} \rho_{iQ} \ln Q \ln W_i + \phi_t T + \tfrac{1}{2}\phi_{tt} T^2$$
$$+ \phi_{tQ} \ln Q \cdot T + \sum_{i}^{4} \phi_{it} \ln W_i \cdot T$$

$$S_i = \alpha_i + \sum_{j}^{4} \gamma_{ij} \ln W_j + \rho_{iQ} \ln Q + \phi_{it} T, \quad i = 1, \ldots, 4$$

Restrictions: Symmetry
Linear homogeneity in input prices

Variables:
Q: aggregate output index
T: time trend
W_1: price index of family and hired labor
W_2: price index of land and structures
W_3: price index of capital (durable equipment, animal capital, replacement inventories)
W_4: price index of material inputs
TC: total cost $\left(\sum_{i}^{4} W_i X_i\right)$
S_i: ith input cost divided by total cost

Empirical results: U.S. agricultural sector, 1948–82
(standard errors in parentheses)

α_0	10.958 (0.111)	γ_{12}	−0.070 (0.006)	γ_{34}	−0.065 (0.006)	ϕ_{tQ}	−0.011 (0.055)
α_1	0.220 (0.007)	γ_{13}	−0.045 (0.014)	γ_{44}	0.119 (0.012)	ϕ_t	−0.0005 (0.005)
α_2	0.320 (0.003)	γ_{14}	−0.017 (0.010)	γ_{QQ}	0.705 (3.55)	ϕ_{tt}	0.002 (0.001)
α_3	0.292 (0.005)	γ_{22}	0.157 (0.012)	ρ_{1Q}	0.497 (0.125)	ϕ_{1t}	−0.014 (0.002)
α_4	0.167 (0.005)	γ_{23}	−0.050 (0.005)	ρ_{2Q}	−0.050 (0.050)	ϕ_{2t}	0.002 (0.001)
α_Q	0.073 (0.291)	γ_{24}	−0.036 (0.009)	ρ_{3Q}	−0.271 (0.084)	ϕ_{3t}	0.005 (0.002)
γ_{11}	0.133 (0.022)	γ_{33}	0.160 (0.011)	ρ_{4Q}	0.276 (0.080)	ϕ_{4t}	0.012 (0.001)

The own price elasticities of factor demands, evaluated at the point of approximation, indicated that all factor demands were inelastic. Most inputs were substitutes in the production process; capital and materials were the only inputs that exhibited a statistically significant complementary relation.

A five-input translog cost function model was also estimated (but not reported) utilizing the same specifications on technological change and returns to scale. Chemicals were separated from other material inputs and comprised the fifth input category. For this specification both ϕ_t and α_Q were positive but not statistically significant; the other estimated parameters were fairly consistent with the results for the four-input translog cost model.

Based on these results, we conclude that the collinearity between the single-output index and the proxy for technological change in the translog cost function model precluded attempts successfully to separate the scale effects from the technical change effects. In addition, if the cost function is not included in the set of estimating equations, that is, in the Binswanger model, the parameter estimates for α_Q, ϕ_t, γ_{QQ}, and γ_{tt} are not obtained. Without the α_Q and γ_{QQ} parameters, the theoretical consistency of the estimated cost function cannot be determined.

Generalized Leontief Total Cost Function (Model 3). An adaptation of Diewert's generalized Leontief cost function was fitted to U.S. agricultural data in a manner similar to that used by Lopez (1980) in his study of the Canadian agricultural sector. The set of derived input demands are expressed in terms of input-output ratios. These equations, given in table 5-4, were derived from the following cost function:

$$C = Q \sum_i \sum_j \beta_{ij}(W_i W_j)^{1/2} + Q^2 \sum_i W_i \alpha_i + QT \sum_i \gamma_i W_i + \xi$$

where ξ includes all terms not involving input prices. The parameter estimates for the derived demand equations are presented in table 5-4. The estimated demand functions implied that underlying cost function satisfied the conditions for nonnegativity and monotonicity and concavity in input prices. Whenever the estimates for β_{ij} were statistically significant, they were positive.

We tested the hypotheses of fixed proportions technology, constant returns to scale, and Hicks-neutral technological change using the model with symmetry imposed. Using the likelihood ratio test, for the three alternative hypotheses the calculated values are higher than the critical χ^2 values at the 5 percent level of significance; thus, all three hypotheses are rejected.

A negative value for α_i in the factor demand equation indicates that as scale of production is expanded, the efficiency in the use of the factors increases, that is, there is an increase in the average product for the ith input (or alternatively, a decrease in the ith input/output coefficient). Our results indicated that there was increased efficiency for all inputs except labor. Furthermore, since the α_i's associated with all input de-

TABLE 5-4. MODEL 3: MODIFIED GENERALIZED LEONTIEF COST FUNCTION WITH ONE OUTPUT, FOUR INPUTS

Model: $R_i = \Sigma \beta_{ij} (W_j/W_i)^{1/2} + \alpha_i Q + \gamma_i T, \quad i = 1, \ldots, 4$

Restrictions: Symmetry

Variables:
- Q: aggregate output
- T: time trend
- W_1: price index of labor (family and hired)
- W_2: price index of land and structures
- W_3: price index of capital (durable equipment, animal capital, and inventories)
- W_4: price index of materials
- R_i: ith input divided by output

Empirical results: U.S. agricultural sector, 1948–1982 (standard errors in parentheses)

β_{11}	−2.159 (0.535)	β_{22}	1.679 (0.107)	α_1	2.717 (0.592)	γ_1	−0.092 (0.009)
β_{12}	0.086 (0.054)	β_{23}	−0.005 (0.033)	α_2	−0.596 (0.085)	γ_2	−0.008 (0.002)
β_{13}	0.155 (0.086)	β_{24}	−0.004 (0.023)	α_3	−0.706 (0.150)	γ_3	0.008 (0.002)
β_{14}	0.411 (0.106)	β_{33}	1.778 (0.200)	α_4	−0.987 (0.159)	γ_4	0.016 (0.002)
		β_{34}	0.063 (0.046)				
		β_{44}	1.619 (0.221)				

mand equations were significantly different from zero, the underlying production technology was nonhomothetic; that is, the input-output ratios were not independent of the output levels. A further indication of nonhomotheticity was the differences in the input-output ratio elasticities, $\epsilon_{iQ} = \partial \ln (X_i/Q)/\partial \ln Q$. These elasticities reflect the movements along an expansion path. A negative (positive) value would imply increased (decreased) efficiency in the use of that factor as scale increases. At the sample mean, labor decreased in efficiency while the other three inputs increased in efficiency. This is different from the results Lopez obtained for Canadian agriculture, in which all inputs were increasing in efficiency as scale increased.

The coefficients for the time trend variable in each equation were all significantly different from zero, supporting the earlier rejection of the Hicks-neutral technological change hypothesis. Furthermore, the technological biases were toward equipment and materials, and away from labor and land. Lopez found that under a nonhomothetic structure, the Hicks neutral technological change hypothesis could not be rejected for

Canadian agriculture. Using model 3, the results suggest that the U.S. agricultural sector was characterized by a nonhomothetic, nonHicks-neutral aggregate technology.

Finally, the own price elasticities for labor and materials evaluated at the point of approximation were inelastic and of the theoretically correct sign. For land and capital, the own price elasticities were positive, but not statistically significant.

Multiple-Output Translog Cost Function Models (Models 4 and 5). First, a two-output, five-input Hicks-neutral translog cost model was estimated using the cost function, four of the five cost share equations, and the two revenue share equations (model 4). The set of estimating equations and the parameter estimates are presented in table 5-5. The theoretical restrictions for symmetry ($\delta_{ij} = \delta_{ji}, \gamma_{rs} = \gamma_{sr}$) and linear homogeneity in input prices ($\Sigma\beta_r = 1, \sum_s \gamma_{sr} = \sum_i \rho_{ir} = 0$ for all r) were imposed *a priori*, along with the restrictions for Hicks neutrality. This specification is similar to the model used by Ray (1982).

The input categories were family and hired labor, land and structures, durable equipment and animal capital, agricultural chemicals, and other material inputs; the outputs were livestock products and crops. All factor demands were negatively sloped at the point of approximation ($\beta_i^2 - \beta_i + \gamma_{ii} < 0$ for $i = 1, \ldots, 5$). However, the estimated parameters violated the curvature conditions for both outputs ($\alpha_i^2 - \alpha_i + \delta_{ii} > 0$, for $i = 1, 2$), indicating that the supply functions were also negatively sloped at the point of approximation. The rate of technological change was 1.7 percent per year and statistically significant.

Following Caves, Christensen, and Swanson (1981), the measure of returns to scale for this multiple output cost model is $(1/(\sum_i \partial \ln C / \partial \ln Q_i)$. At the point of approximation, this simplifies to $1/(\Sigma\alpha_i)$, as all variables are normalized to 1.0. The estimated measure of returns to scale at the point of approximation was 0.74 and was statistically significant.

Given the violation of the curvature conditions with respect to the outputs in model 4, a less restrictive multiple output translog cost function was specified and tested for Hicks neutrality (table 5-6). To accommodate the additional parameters, we limited the number of input categories and specified a three-input, two-output translog cost function (model 5). Only the restrictions for symmetry and linear homogeneity of the cost function in factor prices were imposed *a priori*. The inputs were family and hired labor, land and capital, and intermediate materials; outputs were livestock products and crops. The cost function, two of the three share equations, and the two revenue share equations comprised

TABLE 5-5. MODEL 4: TRANSLOG TOTAL COST FUNCTION WITH TWO OUTPUTS, FIVE INPUTS

Model:
$$\ln TC = \alpha_0 + \sum_i^2 \alpha_i \ln Q_i + \tfrac{1}{2} \sum_i^2 \sum_j^2 \delta_{ij} \ln Q_i \ln Q_j + \sum_r^5 \beta_r \ln W_r$$
$$+ \tfrac{1}{2} \sum_r^5 \sum_s^5 \gamma_{rs} \ln W_r \ln W_s + \sum_i^2 \sum_r^5 \rho_{ir} \ln Q_i \, W_r + \phi T$$

$$S_r = \beta_r + \sum_s^5 \gamma_{rs} \ln W_s + \sum_i^2 \rho_{ir} \ln Q_i, \qquad r = 1, \ldots, 5$$

$$R_i = \alpha_i + \sum_j^2 \delta_{ij} \ln Q_j + \sum_r^5 \rho_{ir} \ln W_r, \qquad i = 1, 2$$

Restrictions: Symmetry
Linear homogeneity in input prices
Hicks neutrality

Variables:
Q_1: quantity index of crop products
Q_2: quantity index of livestock products
W_1: price index of family and hired labor
W_2: price index of land and structures
W_3: price index of capital (durable equipment, animal capital, and inventories)
W_4: price index of chemicals (pesticides and fertilizer)
W_5: price index of materials (energy, feed and seed, miscellaneous inputs)

TC: total cost $\left(W_r X_r \Big/ \sum_r^5 W_r X_r \right)$

S_r: rth input cost divided by total cost, $r = 1, \ldots, 5$
R: ith output revenue divided by total cost, $i = 1, 2$
T: time trend

Empirical results: U.S. agricultural sector, 1948–1982
(standard errors in parentheses)

α_0	11.0836 (0.007)	β_4	0.054 (0.003)	γ_{14}	−0.031 (0.004)	ρ_{12}	−0.112 (0.027)
α_1	0.818 (0.028)	β_5	0.254 (0.006)	γ_{15}	−0.004 (0.010)	ρ_{13}	−0.041 (0.023)
α_2	0.535 (0.013)	γ_{11}	0.062 (0.011)	γ_{23}	−0.024 (0.006)	ρ_{14}	0.020 (0.013)
δ_{11}	−0.756 (0.132)	γ_{22}	0.128 (0.008)	γ_{24}	0.013 (0.004)	ρ_{15}	−0.050 (0.030)
δ_{22}	−0.240 (0.100)	γ_{33}	0.173 (0.007)	γ_{25}	−0.028 (0.009)	ρ_{21}	−0.045 (0.023)
δ_{12}	−0.559 (0.067)	γ_{44}	0.004 (0.003)	γ_{34}	−0.041 (1.004)	ρ_{22}	0.050 (0.020)
β_1	0.192 (0.006)	γ_{55}	0.146 (0.016)	γ_{35}	−0.108 (0.008)	ρ_{23}	−0.054 (0.018)
β_2	0.249 (0.002)	γ_{12}	−0.090 (0.008)	γ_{45}	−0.006 (0.005)	ρ_{24}	−0.048 (0.012)
β_3	0.250 (0.005)	γ_{13}	0.001 (0.007)	ρ_{11}	−0.041 (0.031)	ρ_{25}	0.097 (0.035)
						ϕ	−0.017 (0.001)

TABLE 5-6. MODEL 5: TRANSLOG TOTAL COST FUNCTION WITH TWO OUTPUTS, THREE INPUTS

Model:

$$\ln TC = \alpha_0 + \sum_i^2 \alpha_i \ln Q_i + \frac{1}{2} \sum_i^2 \sum_j^2 \delta_{ij} \ln Q_i \ln Q_j$$

$$+ \sum_r^3 \beta_r \ln W_r + \frac{1}{2} \sum_r^3 \sum_s^3 \gamma_{rs} \ln W_r \ln W_s + \sum_i^2 \sum_r^3 \rho_{ir} \ln Q_i \ln W_r + \phi_t T$$

$$+ \phi_{tt} \cdot T^2 + \sum_i^2 \delta_{it} T \cdot \ln Q_i + \sum_r^3 \gamma_{rt} T \ln W_r$$

$$S_r = \beta_r + \sum_s^3 \gamma_{rs} \ln W_s + \sum_i^2 \rho_{ir} \ln Q_i + \gamma_{rt} T \qquad r = 1, \ldots, 3$$

$$R_i = \alpha_i + \sum_j^2 \delta_{ij} \ln Q_j + \sum_r^3 \rho_{ir} \ln W_r + \delta_{it} T \qquad i = 1, 2$$

Restrictions: Symmetry
Linear homogeneity in input prices

Variables:
Q_1: quantity index of crop products
Q_2: quantity index of livestock products
W_1: price index of family and hired labor
W_2: price index of land, structures, durable equipment, animal capital, and inventories
W_3: price index of materials (energy, feed and seed, chemicals, and miscellaneous inputs)
TC: total cost
S_r: rth input cost divided by total cost, $r = 1, 2, 3$
R_i: ith product revenue divided by total cost, $i = 1, 2$
T: time trend

Empirical results: U.S. agricultural sector, 1948–82
(standard errors in parentheses)

α_0	11.060 (0.014)	γ_{11}	0.043 (0.011)	ρ_{21}	0.118 (0.029)
α_1	0.617 (0.038)	γ_{22}	0.219 (0.007)	ρ_{23}	−0.075 (0.032)
α_2	0.482 (0.013)	γ_{33}	0.093 (0.016)	ϕ_t	−0.022 (0.002)
δ_{11}	0.978 (0.256)	γ_{12}	−0.085 (0.007)	ϕ_{tt}	−0.0008 (0.0003)
δ_{22}	0.433 (0.139)	γ_{13}	0.042 (0.012)	δ_{1t}	−0.034 (0.005)
δ_{12}	0.192 (0.106)	γ_{23}	−0.134 (0.008)	δ_{2t}	−0.017 (0.003)
β_1	0.184 (0.005)	ρ_{11}	0.068 (*)	γ_{1t}	−0.007 (0.001)
β_2	0.499 (0.003)	ρ_{22}	−0.043 (*)	γ_{2t}	0.003 (0.001)
β_3	0.316 (0.006)	ρ_{12}	−0.029 (0.024)	γ_{3t}	0.004 (0.001)
		ρ_{13}	−0.039 (0.043)		

the set of equations.[3] The monotonicity conditions were satisfied over the entire sample; furthermore, the curvature conditions for the estimated factor demands and output supply functions were satisfied at the point of approximation.

The alternative hypothesis of Hicks neutrality was strongly rejected on the basis of the χ^2 test. For model 5, technical change occurred at an average annual rate of 2.2 percent ($\phi_t = 0.022$), measured at the point of approximation, and decreased slightly over time ($\phi_{tt} = -0.0008$). Both parameters were statistically significant.

The sensitivity of the parameters associated with the output indexes in tables 5-6 and 5-7 should be noted. Under Hicks neutrality (model 4, table 5-5), δ_{11}, δ_{12}, and δ_{22} were all negative, which resulted in the curvature violations noted earlier; furthermore, all variations in the revenue shares were attributed to changes in input prices and/or scale. Under the nonneutral technical change specification (model 5, table 5-6), these parameters switched signs. Furthermore, the parameters on the cross terms between output and time (δ_{1t} and δ_{2t}) were negative and statistically significant; that is, variations in the revenue shares were inversely related to time and directly related to scale.

In summary, there appears to be strong collinearity between time and an aggregate output index. This hampers attempts to separate scale from technological change using single-output cost function models. For the multiple-output specifications, enough variations exist among the output indexes and time to separate the scale effects from bias effects.

PARTIAL STATIC EQUILIBRIUM MODELS

Restricted Cost Function Model (Model 6)

In estimating a single output, one quasi-fixed input restricted cost function model, the multicollinearity problems that plagued the single-output full static equilibrium cost model remained evident. The scale and technological change effects could not be successfully separated; the

---

[3] The estimation procedure was Zellner's SUR procedure. Potentially, there is an endogeneity problem involving output, because the profit maximization conditions are used in deriving the revenue share equations (Fuss and Waverman, 1981). However, given the relatively limited sample size, and in the absence of any obvious instruments, not much can be gained from a 2SLQ or a 3SLQ procedure. The model was also estimated without using the revenue share equations. However, including the revenue share equations led to more acceptable estimates; that is, the cost function was better behaved when we used the additional information provided by the revenue shares.

TABLE 5-7. MODEL 6: TRANSLOG VARIABLE COST FUNCTION WITH ONE OUTPUT, THREE VARIABLE INPUTS, TWO QUASI-FIXED INPUTS

Model:

$$\ln CV = \alpha_0 + \alpha_Q \ln Q + \sum_i^3 \alpha_i \ln W_i + \sum_i^2 \beta_i \ln Z_i + \tfrac{1}{2}\gamma_{QQ}(\ln Q)^2$$

$$+ \tfrac{1}{2}\sum_i^3\sum_j^3 \gamma_{ij} \ln W_i \ln W_j + \tfrac{1}{2}\sum_i^2\sum_j^2 \delta_{ij} \ln Z_i \ln Z_j + \sum_i^3 \rho_{iQ} \ln Q \ln W_i$$

$$+ \sum_i^3\sum_j^2 \rho_{ij} \ln W_i \ln Z_j + \sum_i^2 \Pi_i \ln Q \ln Z_i + \phi_t T + \tfrac{1}{2}\phi_{tt} T^2 + \phi_{tQ} \ln Q \cdot T$$

$$+ \sum_i^3 \phi_{it} \ln W_i \cdot T + \sum_i^2 \phi_{Z_i \ln Z_i \cdot T}$$

$$S_i = \alpha_i + \rho_{Qi} \ln Q + \sum_j^3 \gamma_{ij} \ln W_j + \sum_j^2 \rho_{ij} \ln Z_j + \phi_{it} t, \quad i = 1, 2, 3$$

Restrictions: Symmetry
Linear homogeneity in input prices
Constant returns to scale

Variables:
Q: aggregate output index
T: time trend
W_1: price index of hired labor
W_2: price index of chemicals (fertilizer and pesticides)
W_3: price index of materials (feed and seed, miscellaneous inputs)
Z_1: quantity index of capital (land, structures, equipment, livestock, inventory)
Z_2: quantity index of family labor
S_i: ith input cost divided by total variable cost
CV: variable cost

Empirical results: U.S. agricultural sector, 1948–82
(standard errors in parentheses)

α_0	10.022 (0.022)	γ_{22}	0.056 (0.009)	ρ_{Q2}	−0.029 (0.091)	Π_2	−2.329 (9.376)
α_Q	0.024 (0.714)	γ_{33}	0.123 (0.016)	ρ_{Q3}	−0.147 (0.119)	ϕ_t	−0.008 (0.018)
α_1	0.171 (0.003)	γ_{13}	−0.064 (0.011)	ρ_{1Z_1}	−0.189 (0.042)	ϕ_{tt}	0.005 (0.014)
α_2	0.154 (0.004)	γ_{21}	0.002 (0.007)	ρ_{2Z_1}	0.016 (0.061)	ϕ_{tQ}	−0.152 (0.528)
α_3	0.675 (0.006)	γ_{23}	−0.058 (0.011)	ρ_{3Z_1}	0.172 (0.077)	ϕ_{t1}	−0.004 (0.002)
β_1	1.472 (0.531)	δ_{11}	5.664 (6.181)	ρ_{1Z_2}	0.013 (0.036)	ϕ_{t2}	0.006 (0.002)
β_2	−0.496 (0.344)	δ_{22}	−3.211 (4.573)	ρ_{2Z_2}	0.013 (0.044)	ϕ_{t3}	0.009 (0.003)
γ_{QQ}	−7.111 (18.723)	δ_{12}	−0.882 (5.608)	ρ_{3Z_2}	−0.025 (0.059)	ϕ_{tZ_1}	0.036 (0.299)
γ_{11}	0.062 (0.010)	ρ_{Q1}	0.177 (0.064)	Π_1	−4.782 (10.452)	ϕ_{tZ_2}	0.116 (0.256)

parameters associated with time and output were unstable as evidenced by sign reversals in response to slight permutations in the sample period or input aggregations.

Limited success was achieved by further disaggregating the restricted cost function. A translog variable cost function that has one output, three variable inputs (hired labor, fertilizer and pesticides, and other purchased inputs), and two quasi-fixed inputs (family labor and capital) was specified (table 5-7, model 6). Symmetry was imposed on the γ_{ij}'s and δ_{ij}'s, and we assumed linear homogeneity of the variable cost function in input prices and constant returns to scale on the underlying production structure. This structure is similar to the Brown and Christensen (1981) model.

At the point of approximation, the fitted variable cost function was nondecreasing in the prices of the variable factors and nonincreasing in the level of one of the quasi-fixed inputs—family labor. It was, however, increasing in the level of capital. The coefficient on output in the cost function was of the correct sign (positive) but was not statistically significant. The fitted variable cost function exhibited technological progress—ϕ_t was negative but not statistically significant. There were significant biases toward materials and agricultural chemicals and against hired labor.

Restricted Profit Function Models (Models 7 and 8)

As noted in chapter 2, at the levels of the variable inputs and outputs that maximize profits, the derivatives of the restricted profit function with respect to output and input prices and the level of the fixed factor provides the output supply, variable input demand, and shadow price functions, respectively.

Model 7 is a single-output translog restricted profit function. Hired and family labor, materials, and equipment and animal capital comprised the three variable factors; land and structures were treated as a single quasi-fixed factor. The restrictions for linear homogeneity of profits in prices given the fixed factor levels were imposed *a priori*. The set of equations and the estimated parameters are presented in table 5-8.

The fitted restricted profit function satisfied the monotonicity conditions at all data points. At the point of approximation, the diagonal elements of the Hessian matrix were of the correct sign. Furthermore, the Hessian consisting of the elements associated with the input and output prices was positive semidefinite, and the Hessian associated with the fixed factor was negative semidefinite. These, in turn, implied that the

TABLE 5-8. MODEL 7: TRANSLOG RESTRICTED PROFIT FUNCTION WITH ONE OUTPUT, THREE VARIABLE INPUTS, ONE QUASI-FIXED INPUT

Model:

$$\ln \Pi = \alpha_0 + \sum_i^3 \alpha_i \ln W_i + \alpha_Q \ln P_Q + \beta_Z \ln Z + \alpha_t T + \tfrac{1}{2} \sum_i \sum_j \gamma_{ij} \ln W_i \ln W_j$$

$$+ \tfrac{1}{2} \gamma_{QQ} \ln P_Q^2 + \tfrac{1}{2} \beta_{ZZ} \ln Z^2 + \alpha_{tt} T^2 + \sum_i \gamma_{iQ} \ln W_i \ln P_Q$$

$$+ \sum_i^3 \rho_{Zi} \ln Z \ln W_i + \rho_{ZQ} \ln Z \ln P_Q + \sum_i^3 \rho_{ti} \ln W_i T$$

$$+ \rho_{tQ} \ln P_Q T + \rho_{tZ} \ln Z T$$

$$\Pi_i = \alpha_i + \sum_j^3 \gamma_{ij} \ln W_j + \gamma_{iQ} \ln P_Q + \rho_{Zi} \ln Z + \rho_{ti} T \quad i = 1, 2, 3$$

$$\Pi_Q = \alpha_Q + \gamma_{QQ} \ln P_Q + \sum_j^3 \gamma_{iQ} \ln W_i$$

Restrictions: Symmetry
Linear homogeneous in prices

Variables: Π: restricted profits $\left(P_Q Q - \sum_i W_i X_i \right)$

W_i: ith variable input price index: 1 (labor), 2 (chemicals), 3 (other purchased materials)
P_Q: output price index
Z: quasi-fixed factor (land, structures, durable equipment, and animal capital)
T: time trend
Π_i: share of ith input cost in restricted profits, $-W_i X_i / \Pi$
Π_Q: share of revenue in restricted profits, $P_Q Q / \Pi$

Empirical results: U.S. agricultural sector, 1948–82
(standard errors are in parentheses)

α_0	10.616 (0.011)	γ_{11}	−0.244 (0.012)	γ_{1Q}	0.466 (0.031)	ρ_{ZQ}	−0.116 (0.035)
α_1	−0.331 (0.004)	γ_{12}	−0.049 (0.006)	γ_{2Q}	0.217 (0.015)	ρ_{t1}	0.009 (0.001)
α_2	−0.085 (0.002)	γ_{13}	−0.173 (0.037)	γ_{3Q}	0.227 (0.053)	ρ_{t2}	−0.002 (0.001)
α_3	−0.499 (0.012)	γ_{22}	−0.041 (0.008)	β_{ZZ}	−0.642 (19.800)	ρ_{t3}	0.0002 (0.001)
α_Q	1.917 (0.013)	γ_{23}	−0.127 (0.015)	ρ_{Z1}	0.157 (0.177)	ρ_{tQ}	−0.008 (0.001)
β_Z	1.083 (0.499)	γ_{33}	0.073 (0.063)	ρ_{Z2}	−0.094 (0.100)	ρ_{tZ}	0.039 (0.121)
α_t	0.025 (0.002)	γ_{QQ}	−0.911 (0.065)	ρ_{Z3}	0.042 (0.228)	α_{tt}	0.001 (0.001)

restricted profit function was a convex function of prices for each fixed factor, and a concave function of the fixed factor for each fixed price, as required by neoclassical production theory.

A more disaggregated restricted translog profit model—two outputs, three variable inputs, and two quasi-fixed factors—was also estimated; the model and the result are reported in table 5-9 (model 8). The re-

TABLE 5-9. MODEL 8: RESTRICTED TRANSLOG PROFIT FUNCTION WITH TWO OUTPUTS; THREE VARIABLE INPUTS, TWO QUASI-FIXED INPUTS

Model:
$$C_i = \alpha_i + \sum_{h=1}^{5} \alpha_{ih} \ln P_h + \sum_{j=1}^{2} \rho_{ij} \ln Z_j + \phi_{it} T, \quad i = 1, \ldots, 5$$

$$R_j = \beta_j + \sum_{h=1}^{2} \beta_{jh} \ln Z_h + \sum_{i}^{4} \rho_{ij} \ln P_i + \theta_{jt} T, \quad j = 1, 2$$

Restrictions: Symmetry
Linear homogeneity in variable input prices
Linear homogeneity in fixed inputs

Variables: P_1: price index of crop output
P_2: price index of livestock output
P_3: price index of hired labor
P_4: price index of chemicals (fertilizer and pesticides)
P_5: price index of materials (energy, feed and seed, miscellaneous inputs)
Z_1: quantity index of family labor
Z_2: quantity index of capital (land, structures, durable equipment, animal capital, and replacement inventories)

C_i: share of net output i in variable profit $\left(P_i X_i \bigg/ \sum_{i}^{5} P_i X_i\right)$

R_1: share of family labor in the total cost of fixed inputs $\left(W_1 Z_1 \bigg/ \sum_{i}^{2} W_i Z_i\right)$

T: time trend

Empirical Results: U.S. agricultural sector, 1948–1982
(standard errors are in parentheses)

α_1	0.842 (0.007)	α_{22}	0.330 (0.055)	α_4	−0.068 (0.002)	ϕ_{1t}	−0.003 (0.0015)
α_{11}	0.118 (0.046)	α_{23}	0.033 (0.014)	α_{44}	−0.024 (0.006)	ϕ_{2t}	−0.008 (0.002)
α_{12}	−0.428 (0.038)	α_{24}	0.041 (0.011)	ρ_{11}	−0.009 (0.030)	ϕ_{3t}	0.002 (0.001)
α_{13}	0.048 (0.014)	α_3	−0.077 (0.002)	ρ_{21}	−0.076 (0.036)	ϕ_{4t}	−0.002 (0.001)
α_{14}	0.046 (0.012)	α_{33}	−0.039 (0.008)	ρ_{31}	0.027 (0.013)	β_1	0.203 (0.007)
α_2	0.647 (0.008)	α_{34}	−0.026 (0.005)	ρ_{41}	−0.001 (0.010)	β_{11}	−0.015 (0.047)
						θ_{1t}	−0.014 (0.002)

TABLE 5-10. PARTIAL EQUILIBRIUM OR RESTRICTED FACTOR PRICE ELASTICITIES

Elasticity	Model		
	6	7[a]	8
ϵ_{HH}	−0.474	−0.594	−0.604
ϵ_{HC}	0.168	0.067	0.173
ϵ_{HM}	0.314	0.211 (N)	−0.126
ϵ_{CH}	0.186	0.485	0.243
ϵ_{CC}	−0.473	−0.606	−0.700
ϵ_{CM}	0.279	0.993	0.185
ϵ_{MH}	0.079	0.014 (N)	−0.028
ϵ_{MC}	0.063	0.169	0.037
ϵ_{MM}	−0.144	−1.645	−0.803

Note: (N) indicates that the estimated elasticity is not statistically significant at the 5 percent level. All elasticities are evaluated at the point of approximation. H denotes hired labor, M denotes purchased materials, C denotes fertilizers and pesticides.

[a] H denotes both hired and family labor.

strictions for symmetry, linear homogeneity in variable prices, and linear homogeneity in fixed factors were imposed *a priori* on the set of estimating equations, which included four of the five variable share equations and one of the fixed factor share equations.[4]

The two outputs were crops and livestock products; the variable inputs were hired labor, purchased materials, and equipment and animal capital. The two quasi-fixed factors were family labor and land and structures. The estimated shares satisfied Diewert's sufficiency conditions: the shares for the variable inputs were negative; those for the two outputs and the fixed inputs were positive. The partial own elasticities were also of the theoretically correct signs for all of the statistically significant elasticities (table 5-10).

Although the Hicksian rates and biases of technological change have historically been calculated for single-output specifications, a correspondence does exist in the multiple-output case under certain restrictions. Input and output separability is a convenient, although not a necessary, restriction. In the absence of separability, shifts in factor shares reflect both technological change and output mix changes. One would need to separate these two effects in order to calculate the Hicksian biases (see discussion in chapter 2). Although input-output

---

[4] In the econometric estimation, the prices in the model are normalized by the price of the excluded variable factor; likewise the fixed quantity is normalized by the quantity of the excluded fixed factor. The parameters in the omitted equations are derived using the symmetry and homogeneity constraints as suggested by McKay, Lawrence, and Vlastuin (1983).

separability was not tested for in the restricted profit model, it is likely to be rejected given the significant coefficients on the relative input prices in the output shares equations. Thus, the time coefficients in the above equations reflect the bias effects and scale effects. Assuming that the scale effects do not offset the bias effects, the estimates of ϕ_{it} suggest that technological change was hired labor-saving and chemical-using. With respect to the fixed inputs, technological change was, in general, family labor-saving and capital-using.

COMPARISON OF THE EMPIRICAL RESULTS

In this section the results from the eight econometric models discussed in this chapter are compared and a conclusion is reached regarding which specifications provide better descriptions of the aggregate technology. This evaluation is based on compliance with the theoretical properties for a well-behaved production technology, as well as on the inferences regarding technological change and the structural measures of the aggregate technology. More specifically, the comparisons are with respect to the overall productivity measures and the trends in rate of growth of productivity; the directions and magnitudes of factor biases; estimated price elasticities; and measures of scale effects. Whenever possible both the point estimates and the standard errors are reported.

Model Performance: Theoretical Consistency

The criteria used to obtain the set of models that qualify as best are monotonicity and curvature conditions, that is, theoretical consistency at the point of approximation. For the primal and dual models, this translates into concavity of the production function; concavity of the cost function in variable input prices and convexity in output levels; and convexity of the profit function in input and output prices. The monotonicity and curvature requirements in terms of the parameters of the translog cost and profit functions at the point of approximation are provided in appendixes A and B to chapter 2. One would, of course, prefer a model that is well behaved at all data points rather than for only a subset of the data; with the translog approximations only local consistency can be achieved.

The theoretical consistency results at the point of approximation for models 1 through 8 are summarized in table 5-11. The curvature conditions pertain to the signs of the diagonal elements of the Hessian matrix of second-order partials. Compliance of the diagonal elements ensures

TABLE 5-11. THEORETICAL CONSISTENCY OF PRODUCTION MODELS AT THE POINT OF APPROXIMATION (1970)

	Monotonicity[a]	Curvature conditions[a,b]
Model 1	Satisfied	Violated for equipment, animal capital, and inventories
Model 2	Satisfied	Satisfied
Model 3	Satisfied	Satisfied
Model 4	Satisfied	Satisfied for all inputs; violated for both outputs
Model 5	Satisfied	Satisfied
Model 6	Satisfied for all variable inputs and for family labor; violated (increasing) in quantity of family labor	Satisfied for variable inputs; violated for family labor
Model 7	Satisfied	Satisfied
Model 8	Satisfied	Satisfied

[a] Only statistically significant violations are noted.
[b] Curvature conditions reflect slopes of factor demands and output supplies and not the determinants of the Hessian matrix of second-order partials.

only that the factor demands or output supply functions are of the correct slope; ideally, one should also check the principal minors of the Hessian. The estimates of the principal minors of the Hessian at the point of approximation are straightforward to calculate; the standard errors are more difficult to obtain. However, without the standard errors this information is of limited value. For this reason, only the diagonal elements of the Hessian are discussed.

Among the full static equilibrium production models, models 2, 3, and 5 were theoretically consistent with respect to the monotonicity requirements and the signs of the diagonal elements of the Hessian. For the noninstantaneous adjustment models, only the two restricted profit functions merit further consideration. The Brown and Christensen cost function model (model 6) satisfied the restrictions with respect to the set of variable inputs and aggregate output, but was not decreasing in both quasi-fixed factors.

Productivity Measures and Economies of Scale

The rate of growth of TFP is conventionally defined as the rate of growth of an aggregate output index minus the rate of growth of an aggregate input index, $\text{TFP} = \dot{Q} - \dot{X}$ (see chapter 2). In this section, we adopt the Solow (1957)-Jorgenson and Griliches (1967) approach, which interprets this residual as technical progress or a shift in the production function. As demonstrated in chapter 2, under constant returns to scale and perfect competition in the input and output markets, the rate of technical progress is equivalent to the rate of TFP growth.

The comparisons of the productivity measures from the primal and dual production models require adjustments for returns to scale. The primal measure of technological change is defined as the proportional shift in the production function, $\dot{A} \equiv \partial \ln f(x,t)/\partial t$. $\dot{A}$ is related to the dual measure of technical change $\dot{C} \equiv \partial \ln C(p,Q,t)/\partial t$:

$$\dot{A} \equiv -\dot{C}\epsilon_{CQ}^{-1} \qquad (5\text{-}1)$$

where $\epsilon_{CQ} = \partial \ln C/\partial \ln Q$ is the cost-output elasticity, and ϵ_{CQ}^{-1} is the dual measure of returns to scale. Equation (5-1) implies that intertemporal shifts in the cost function are not identical to intertemporal shifts in the production function, unless the production structure can be characterized by constant returns to scale. Furthermore, the relationships between TFP and proportional shifts in the cost function and the production function are given by

$$\text{T}\dot{\text{F}}\text{P} = -\dot{C} + (1 - \epsilon_{CQ})\dot{Q} \qquad (5\text{-}2)$$

$$\text{T}\dot{\text{F}}\text{P} = \dot{A} + (\epsilon_{CQ}^{-1} - 1)\dot{X} \qquad (5\text{-}3)$$

If constant returns to scale do not exist, the intertemporal shifts in the cost or production function are not equal to the conventional measure of TFP.

The inferences with respect to productivity for U.S. agriculture are compared using the econometrically estimated production and cost models. Productivity measures are not directly forthcoming from the profit specifications because technical change shifts both the factor demand and output supply functions. For the translog production function (model 1), the primal measure of technological change evaluated at the point of approximation was 1.4 percent per year; the 95 percent confidence interval was approximately 1.2 percent to 1.6 percent. For the single-output translog cost function (model 2), the dual measure of technological progress evaluated at the point of approximation was extremely small, 0.05 percent per year, and not statistically different from zero. Furthermore, the primal measure of TFP growth computed from the parameter estimates of model 2 also was not statistically significant.

Of the remaining translog cost models only the nonhomothetic, nonneutral specification (model 5) was well behaved and merits further analysis. The rate of cost diminution, returns to scale, and the equivalent primal rate of technological progress at the point of approximation are reported in the first part of table 5-12. After adjustments are made for the scale effects, the primal rate of technological change exceeded the intertemporal shift in the cost function. The 95 percent confidence interval associated with the point estimate of the equivalent primal rate of technological change and the corresponding confidence interval from the production function model overlapped from 1.4 percent to 1.6 percent.

Also in table 5-12, the parametric measures of productivity are compared to the growth accounting measure of TFP reported in chapter 3. As all measures are based on the same data set, it is not surprising to find that the nonparametric measures fall within the 95 percent confidence intervals of the parametric measures of technological change.

Factor Biases

Biased technological change occurs when the production function shifts nonneutrally and thus shifts the expansion path and the factor shares of total cost. With a nonhomothetic technology, changes in output also lead to changes in the cost shares. This output or scale effect must be taken into account in the measurement of the biases, as explained in chapter 2. If the technology is homothetic, the scale effect is zero and the Hicksian technological change biases can be easily identified.

TABLE 5-12. MEASURES OF TOTAL FACTOR PRODUCTIVITY FOR U.S. AGRICULTURE, 1948–1982

Dual measures of TFP (based on Model 5) [a]		
Cost elasticity $\left(\sum_i \epsilon_{CQ_i} \right)$	1.099	(0.050)
Dual returns to scale $\left(\sum_i \epsilon_{CQ_i} \right)^{-1}$	0.909	(0.047)
Rate of cost diminution $(\partial \ln C / \partial t)$	−0.022	(0.002)
Primal rate of technological change $-(\partial \ln C/\partial t) \cdot \left(\sum_i \epsilon_{CQ_i} \right)^{-1}$	0.020	(0.003)
Parametric and nonparametric measures of conventional TFP:		
Parametric measures:		
Model 1 (using equation 3-2)	0.014	(0.001)
Model 5 (using equation 3-1)	0.020	(0.003)
Nonparametric measure: TFP index		
(table 3-5, chapter 3)	0.0157	

Note: Approximate standard errors are in parentheses.
[a] Evaluated at the point of approximation of the sample (1970).

TABLE 5-13. ESTIMATED FACTOR BIASES FOR NONNEUTRAL TRANSLOG MODELS OF U.S. AGRICULTURE (AT POINT OF APPROXIMATION)

	Model					
	2		5		7	
Bias measure:	B_i^c (1)	B_i^{ce} (2)	MB_i^c (3)	MB_i^{ce} (4)	B_i^p (5)	B_i^{pe} (6)
Labor (hired only)	−0.064	−0.048	−0.038	−0.034	−0.031	−0.020
Land and structures	0.006	−0.007			—	—
Equipment and animal capital	0.017	0.011	0.006	0.009	—	—
Materials	0.012	0.024	0.013	0.006		
Chemicals					0.019	0.034
Other					−0.004	0.001

Note: The biases for the translog models are calculated according to the following equations (see discussions in chapter 2):

B_i^c denotes the dual cost measure of bias, including a *scale effect* and a *bias effect*, $B_i^c = \partial \ln S_i / \partial t$, where S_i is the optimal (cost-minimizing) factor cost share.

B_i^{ce} denotes the dual Hicks measure of factor bias (expansion path preserving): $B_i^{ce} = B_i^c - (\partial \ln S_i / \partial \ln Q)^{-1} (\partial \ln C / \partial t)$.

MB_i^c and MB_i^{ce} are the multiple output counterparts to B_i^c and B_i^{ce}.

B_i^p denotes the dual measure of bias defined in terms of the profit-maximizing factor proportions, $B_i^p = \partial \ln S_i / \partial t$, where S_i is the profit-maximizing cost share derived from the profit function.

B_i^{pe} denotes the measure of the bias effect due to shift in expansion path: $B_i^{pe} = B_i^p - (\partial \ln S_i / \partial \ln p)(\partial \ln \pi / \partial \ln p)^{-1} (\partial \ln \pi / \partial t)$, where p denotes output price and π denotes profits.

The estimated factor biases for the subset of well-behaved production models are presented in table 5-13.[5] As these models are specified for a nonhomothetic technology, the bias measures in columns (1), (3), and (5) represent two distinct effects—a scale effect due to movement along the expansion path, and a bias effect due to a shift in the expansion path. The measures of Hicksian biases, constructed by adjusting the dual cost share measures for the scale effect, are provided in columns (2), (4), and (6).

For the two cost function models (model 2 and model 5), adjustment for the scale effects did not, in the end result, change the qualitative indication of the factor biases. There were, however, differing quantitative changes. For labor, the scale effect reinforced the direction of the

---

[5] The factor biases associated with model 8 are not reported. By estimating only the set of shares equations, we are unable to calculate the measures of the bias effect due to shifts in the expansion path.

Hicksian biases in both models. The same is true for materials and for land, structures, and equipment in model 5. However, for model 2, the scale effect tended to weaken the measure of bias for materials. For the profit function specification (model 7), the scale effect also tended to augment the direction of the Hicksian bias for labor, but weakened the direction of the Hicksian bias for chemicals and other materials. Overall, adjusting for scale appeared to make the greatest difference in the reported biases for the material inputs.

However measured, the cost and profit function models tended to yield the same qualitative factor biases: labor-saving and capital- and material-using. Cross-comparisons of the absolute magnitudes of the biases are not meaningful.

Factor Price Elasticities

In comparing the factor price elasticities from the set of econometric models, it is useful to recall Le Chatelier's proposition—that in the long run, when all factors can freely adjust to changes in parameter values, the factor demand curves are more elastic than in the short run, when some factors are held fixed. Theoretically, the fewer the constraints placed on the system the greater the absolute response of a factor to a change in its price will be.

In general, therefore, one would expect to observe two types of relationships from the models we have estimated. First, the partial equilibrium own-factor price elasticities should be smaller in absolute terms than the corresponding full equilibrium elasticities. In order to make this comparison one would need to calculate the optimal levels of the quasi-fixed factors from the estimates of the restricted cost or profit function, and use these levels in computing the long-run, or full static, equilibrium elasticities (Brown and Christensen, 1981, or Hazilla and Kopp, 1986). Thus, the comparison of the estimates of the own-factor price elasticities from the full and partial static equilibrium models should be made cautiously.

Second, the factor price elasticities estimated from the cost and profit function models should differ due to the treatment of output levels. In the profit function models, the output level is allowed to vary, whereas in the cost function models, the output level is fixed. Using Lopez's (1984) terminology, the "compensated" factor demands derived from parameter estimates of the profit function should be less responsive than the "uncompensated" factor demands.

Unless similar calculations have been made, the comparisons presented in tables 5-10 and 5-14 for the partial and full equilibrium factor

TABLE 5-14. FULL EQUILIBRIUM FACTOR PRICE ELASTICITIES

Elasticity	Model			
	2	3	4[a]	5[b]
ϵ_{LL}	−0.207	−0.120	−0.492	−0.535
ϵ_{LE}	0.097	−0.083	0.232	0.074
ϵ_{LM}	0.086	0.253	0.201	0.534
ϵ_{LA}	0.033	−0.050	−0.186	
ϵ_{EL}	0.073	−0.072	0.201	0.027
ϵ_{EE}	−0.146	0.038 (N)	−0.018 (N)	−0.061
ϵ_{EM}	−0.089	0.036	−0.131	0.033
ϵ_{EA}	−0.003	−0.003	0.157	
ϵ_{ML}	0.113	0.169	0.823	0.311
ϵ_{ME}	−0.156	0.028	−0.617	0.052
ϵ_{MM}	−0.068 (N)	−0.193	−0.876	−0.405
ϵ_{MA}	0.089	−0.002	0.534	
ϵ_{AL}	0.023	−0.037	−0.143	
ϵ_{AE}	−0.003	−0.002	0.113	
ϵ_{AM}	0.046	−0.002	0.101	
ϵ_{AA}	−0.193	0.041 (N)	−0.249	

Note: (N) indicates that the estimated elasticity is not statistically significant at the 5 percent level. All elasticities are evaluated at the point of approximation (1970). L denotes labor; E denotes equipment, animal capital, and replacement inventories; M denotes materials; A denotes land.

[a] Materials includes pesticides and fertilizers only.
[b] Land and structures is aggregated with equipment and animal capital.

price elasticities should be used only to draw general inferences. In this regard, several points are worth noting with respect to the factor demand elasticities. First, among the partial equilibrium models, there was a remarkable degree of consistency with respect to the magnitude of the own price elasticities for labor and chemicals. Also, for these two inputs as well as for "other purchased materials," the compensated factor demands from model 6 were less elastic than the uncompensated factor demands from models 7 and 8. This is consistent with *a priori* expectations.

Second, with the exception of materials and hired labor in the multiple-output profit function (model 8), all variable inputs were either net or gross complements in the production process. Although not reported in table 5-10, the variable inputs were estimated to be net substitutes for family labor and net complements of land, structures, and equipment.

Third, full equilibrium own price elasticities were fairly inelastic and in general smaller (in absolute terms) than the partial equilibrium elasticities. Intuitively, one might expect just the opposite—that is, that the short-run, or partial, equilibrium factor demands would be more inelastic relative to the long-run factor demands.

CONCLUSIONS AND IMPLICATIONS

A series of primal and dual models of aggregate production behavior has been estimated using the same set of data for the U.S. agricultural sector. The specifications have permitted the production structure to vary with respect to input and output aggregations, returns to scale, homotheticity, technological change biases, and treatment of fixed versus variable factors. The purpose of this analysis was to assess the sensitivity of the empirical results to the choice among production, cost, and profit function models. Several sets of elasticity estimates and productivity-related measures are reported. These comparisons highlight the fact that the empirical results and theoretical consistency are sensitive to model specification. The conclusions drawn from the analysis are of limited generality, as they are based on only one data set and represent empirical observations regarding the performance of these specific models around the point of approximation (1970).

The translog functional form has been relied on almost exclusively to provide a second-order approximation to the true functional form. Although the translog has become a popular production model because of its flexibility, economic tractability, and ease of mathematical manipulations, it should not be hailed unreservedly. Many researchers have found the estimated translog to be ill-behaved over portions of the data set, that is, the monotonicity and curvature properties hold only locally (Caves and Christensen, 1980). This was also evident in many of the models presented in this chapter.

In general, violations of theoretical consistency at the point of approximation cast doubts on the economic consistency of the underlying data, or on the appropriateness of the functional specification. The researcher has two options: he can make improvements to the data base or use an alternative functional form. The latter might include adding more restrictions to the existing functional form, by, for example, imposing the curvature properties, or using an alternative flexible form. As noted earlier, not all of the econometric models satisfied locally the monotonicity conditions and the curvature conditions. The specifications that were ill behaved were noted and excluded from further comparisons.

At present, the series of production models is not linked by any explicit sequence of nested hypotheses. An evaluation of the set of best models is based primarily upon theoretical consistency. Generally, the models that permitted nonneutral technical change and nonhomotheticity lead to cost and profit functions that were better behaved relative to the models that maintained neutral technical change or constant returns to scale or both. Often the tests for homotheticity have been rejected in favor of the nonhomothetic modeling assumption.

In the models that disaggregated the output index into crops and livestock indexes, the technical change effects could be more successfully distinguished from the scale effects. For example, model 2 and model 5 differ primarily by the treatment of the aggregate output index. Model 5, with multiple outputs, is superior for purposes of productivity measurement since scale effects can be differentiated from technical change. Likewise model 4 and model 5 differ in terms of their treatment of technical change. Model 4, with Hicks neutrality, is ill behaved with respect to the slopes of the output supply functions. One might add, however, that the cost of relaxing the neutral technical change and homotheticity assumptions in model 5 is a reduction in the number of feasible input categories.

Among the partial equilibrium specifications, the translog profit models (models 7 and 8) are better behaved relative to the translog cost model (model 6). The performance and inferences from the profit models are generally comparable.

In summary, both dual and primal production models can be used to test for returns to scale, neutral technological change, homotheticity, and other structural characteristics. Not surprisingly, the theoretical consistency of the estimated model differs with the treatment of inputs and outputs, more so than among production, cost, or profit specifications. We found the differences from the set of well-behaved specifications to be insubstantial. With few exceptions, the alternative specifications yielded qualitatively similar information with regard to productivity growth, factor biases, and price responses, although the relative magnitudes differed.

In the final analysis, the choice of model may relate to the relative ease with which the empirical productivity measures can be deduced from the specification, and the relative ease of quantifying various components of technology structure.[6] Pope (1982) provides similar insights into choices among primal and dual models for applied research. For example, the effects of government policies that adjust commodity prices may be more easily analyzed using a profit function model, because output prices are direct arguments.

Expanding our research in several areas would improve the inferences on the aggregate structure of U.S. agriculture and the estimates of pro-

[6] Discussion of the methodological and empirical issues involved in selection of a mode of operation can be found in Burgess (1975); Fuss and McFadden (1978); Blackorby, Primont, and Russell (1978); Berndt and Khaled (1979); Chambers (1982); Lopez (1982); and Pope (1982).

ductivity, scale, and the substitution and price elasticities presented in this chapter. First would be to employ the Diewert-Parker nonparametric tests for theoretical consistency. Second, additional improvements to the land and structures data to reflect the true rental or user cost rather than the capitalized value of government programs and *ex post* capital gains also would be beneficial. Many of the theoretical inconsistencies in the estimates reported in this chapter as well as in our unreported "trials and errors" appear to be related to the land data.

A third recommendation pertains to imposing the curvature conditions. If the data are reliable and the curvature violations are minor, then one might consider imposing the curvature conditions. However, if the data are simply inconsistent with any flexible functional form, a better strategy would be to improve the quality of the data base rather than imposing more structure on the model.

Finally, more research in the area of dynamic modeling is warranted. The partial static equilibrium models are a first step in that direction, although no attempt is made to model the adjustment process. Future research should explore the development of what Berndt, Morrison, and Watkins (1981) refer to as third-generation dynamic models, which explicitly represent the dynamics of the production process.

REFERENCES

Berndt, E. 1984. "A Memorandum on the Estimation and Interpretation of the Translog Short-Run Variable Cost Function," memorandum for the U.S. Bureau of Census.

―――, and M. Khaled. 1979. "Parametric Productivity Measurement and Choice Among Flexible Functional Forms," *Journal of Political Economy* vol. 87, no. 6, pp. 1220–1243.

―――, C. J. Morrison, and G. C. Watkins. 1981. "Dynamic Models of Energy Demand: An Assessment and Comparison," in E. R. Berndt and B. C. Field, eds., *Modeling and Measuring Natural Resource Substitution* (Cambridge, Mass., MIT Press).

Binswanger, H. P. 1974a. "A Cost Function Approach to the Measurement of Factor Demand Elasticities and of Elasticities of Substitution," *American Journal of Agricultural Economics* vol. 56, no. 2, pp. 377–386.

―――. 1974b. "The Measurement of Technological Change Biases with Many Factors of Production," *American Economic Review* vol. 64, no. 6, pp. 964–976.

―――, and V. E. Ruttan. 1978. *Induced Innovation* (Baltimore, Md., Johns Hopkins University Press).

Blackorby, C., D. Primont, and R. Russell. 1978. *Duality, Separability and Functional Structure: Theory and Economic Applications* (New York, Elsevier/North-Holland).

Brown, R. S. 1978. "Productivity, Returns and the Structure of Production in

U.S. Agriculture, 1947–74" (Ph.D. dissertation, University of Wisconsin, Madison).

———, and L. R. Christensen. 1981. "Estimating Elasticities of Substitution in a Model of Partial Static Equilibrium: An Application to U.S. Agriculture, 1947–74," in E. R. Berndt and B. C. Field, eds., *Modeling and Measuring Natural Resource Substitution* (Cambridge, Mass., MIT Press).

Burgess, D. F. 1975. "Duality Theory and Pitfalls in the Specification of Technologies," *Journal of Econometrics* vol. 3, no. 1, pp. 105–122.

Capalbo, S. M., and M. Denny. 1986. "Testing Long-Run Productivity Models for the Canadian and U.S. Agricultural Sectors," *American Journal of Agricultural Economics* (forthcoming).

———, T. T. Vo, and J. C. Wade. 1985. "An Econometric Database for the U.S. Agricultural Sector," Discussion Paper No. RR85-01, National Center for Food and Agricultural Policy (Washington, D.C., Resources for the Future).

Caves, D. W., and L. R. Christensen. 1980. "Global Properties of Flexible Functional Forms," *American Economic Review* vol. 70, no. 3, pp. 422–432.

———, L. R. Christensen, and J. A. Swanson. 1981. "Productivity Growth, Scale Economies, and Capacity Utilization in U.S. Railroads, 1955–74," *American Economic Review* vol. 71, no. 6, pp. 994–1002.

Chambers, R. G. 1982. "Relevance of Duality Theory to the Practicing Agricultural Economist: Discussion," *Western Journal of Agricultural Economics* vol. 7, no. 2, pp. 373–378.

Christensen, L. R., and D. W. Jorgenson. 1970. "U.S. Real Product and Real Factor Input 1929–1967," *Review of Income and Wealth* vol. 16, no. 1, pp. 19–50.

———, D. W. Jorgenson, and L. J. Lau. 1973. "Transcendental Logarithmic Production Frontiers," *Review of Economics and Statistics* vol. 55, no. 1, pp. 28–45.

Denny, M., and M. Fuss. 1977. "The Use of Approximation Analysis to Test for Separability and the Existence of Consistent Aggregates," *American Economic Review* vol. 67, no. 3, pp. 404–418.

Diewert, W. E. 1974. "Applications of Duality Theory," in M. Intrilligator and D. A. Kendrick, eds., *Frontiers of Quantitative Economics*, vol. II (Amsterdam, North-Holland).

———. 1985. "The Measurement of the Economic Benefits of Infrastructure Service." Discussion Paper No. 85-11. Economics Department, University of British Columbia, Vancouver, Canada.

———, and C. Parkan. 1985. "Tests for Consistency of Consumer Data," *Journal of Econometrics* vol. 30, no. 12, pp. 127–147.

Fuss, M., and D. McFadden, eds. 1978. *Production Economics: A Dual Approach to Theory and Applications* (Amsterdam, North-Holland).

———, and L. Waverman. 1981. "Multi-Product Multi-Input Cost Functions for a Regulated Utility: The Case of Telecommunications in Canada," in G. Fromm, ed., *Studies in Public Regulation* (Cambridge, Mass., MIT Press).

Hazilla, M., and R. Kopp. 1984. "Industrial Energy Substitution: Econometric Analysis of U.S. Data, 1958–1974," project report to Electric Power Research Institute, prepared by Resources for the Future, Washington, D.C.

———, and R. Kopp. 1986. "Restricted Cost Function Models: Theoretical and

Econometric Considerations," Discussion Paper No. QE86-05 (Washington, D.C., Resources for the Future).

Houthaker, H. S. 1960. "Additive Preferences," *Econometrica* vol. 28, no. 2, pp. 244–257.

Jorgenson, D. W., and Z. Griliches. 1967. "The Explanation of Productivity Change," *Review of Economic Studies* vol. 34, no. 3, pp. 249–283.

Lau, L. 1976. "A Characterization of the Normalized Restricted Profit Function," *Journal of Economic Theory* vol. 12, no. 1, pp. 131–163.

———. 1978. "Applications of Profit Functions," in M. Fuss and D. McFadden, eds., *Production Economics: A Dual Approach to Theory and Applications* (Amsterdam, North-Holland).

Lopez, R. E. 1980. "The Structure of Production and the Derived Demand for Inputs in Canadian Agriculture," *American Journal of Agricultural Economics* vol. 62, no. 1, pp. 38–45.

———. 1982. "Applications of Duality Theory to Agriculture," *Western Journal of Agricultural Economics* vol. 7, no. 2, pp. 353–366.

———. 1984. "Estimating Substitution and Expansion Effects Using a Profit Function Framework," *American Journal of Agricultural Economics* vol. 66, no. 3, pp. 358–367.

McFadden, D. 1978. "Cost, Revenue, and Profit Functions," in M. Fuss and D. McFadden, eds., *Production Economics, A Dual Approach to Theory and Applications* (Amsterdam, North-Holland).

McKay, L., D. Lawrence, and C. Vlastuin. 1983. "Profit, Output Supply, and Input Demand Functions for Multiproduct Firms: The Case of Australian Agriculture," *International Economic Review* vol. 24, no. 2, pp. 323–339.

Ohta, M. 1974. "A Note on the Duality Between Production and Cost Functions: Rate of Return to Scale and Rate of Technological Progress," *Economic Studies Quarterly* vol. 25, no. 1, pp. 63–65.

Pesaran, M. H., and A. S. Deaton. 1978. "Testing Non-Nested Nonlinear Regression Models," *Econometrica* vol. 46, no. 3, pp. 677–694.

Pope, R. 1982. "To Dual or Not to Dual," *Western Journal of Agricultural Economics* vol. 7, no. 2, pp. 337–351.

Ray, S. C. 1982. "A Translog Cost Function Analysis of U.S. Agriculture, 1939–1977," *American Journal of Agricultural Economics* vol. 64, no. 3, pp. 490–498.

Samuelson, P. A. 1965. "A Theory of Induced Innovation Along Kennedy-Weisäcker Lines," *Review of Economics and Statistics* vol. 47, no. 4, pp. 343–356.

Silberberg, E. 1978. *The Structure of Economics: A Mathematical Analysis* (New York, McGraw-Hill).

Solow, R. M. 1957. "Technical Change and the Aggregate Production Function," *Review of Economics and Statistics* vol. 39, no. 3, pp. 312–320.

Varian, H. R. 1978. *Microeconomic Analysis* (New York, Norton).

6
PRODUCTIVITY MEASUREMENT AND THE DISTRIBUTION OF THE FRUITS OF TECHNOLOGICAL PROGRESS:
A MARKET EQUILIBRIUM APPROACH

RAMON E. LOPEZ

In 1971 Yujiro Hayami and Vernon Ruttan argued that relative factor prices in an economy determine the orientation of technological progress. Since then the measurement of technological change in agriculture and, in particular, the measurement of induced biases in technological change, has become a popular exercise, as evidenced by the overall focus of this book and the review by Capalbo and Vo in chapter 3. The rediscovery of duality theory and the development of the so-called flexible functional forms in the early 1970s gave additional impetus to this area of research by permitting a relatively less restrictive econometric modeling of production relations and technological change. Although they usually analyze aggregate sectoral technologies, these studies are characterized by their strong microeconomic orientation, that is, they typically use the behavioral implications of the competitive firm (cost minimization or profit maximization) but ignore market equilibrium relations that may be dependent on the extent and nature of technological progress.

Technological progress in agriculture is viewed as strongly affected by market conditions but this process is usually considered as a one-way phenomenon. The effects of technological change on the market equilibrium conditions are ignored despite the explicit recognition that technological progress induces biases in the structure of factor demand. The neglect of changes in market equilibrium conditions associated with technological progress is most undesirable in agriculture, where the demand for land—a factor largely specific to agriculture—has probably been greatly affected by technological change. In addition, the competitive nature of agricultural production suggests that agricultural product prices and output levels are sensitive to technological change. Studies using the cost function approach are mute with respect to the output price implications of technological progress; they do not consider the dependence of output on technological change, and implicitly assume that land prices do not capitalize any of the gains or losses of technological change. Studies using the dual profit function approach assume that firms are able to retain the full extent of the gains from technological change by enhancing their profits accordingly. These studies are not, in a sense, consistent because they use certain implications of competition (that is, firms are price takers) but tacitly negate the market equilibrium conditions implied by their microeconomic models.

If competitive equilibrium prevails, output prices and the prices of inputs not in perfectly elastic supply (mainly land, in the case of agriculture) would absorb the full extent of the gains from technological change at least in the long run.[1] A problem with using the microeconomic approach in the analysis of sectoral productivity is that it sheds no light on the important issue of the measurement of the distribution of the fruits of technological progress. For example, this approach is not useful in answering questions such as how the gains of productivity growth are distributed among lower output prices (benefiting intermediaries and consumers), land prices (affecting land owners), and farm profits; or which types of technological progress are likely to have the greatest effects on agricultural product prices rather than on land prices and vice versa.

In the remainder of this chapter some issues concerning the use of a competitive market equilibrium rather than the traditional microeconomic approach to the measurement of productivity growth are discussed. An empirically feasible non-*ad hoc* model that integrates the measurement of technological progress biases and the measurement of the distribution of the fruits of technological change between output prices and land prices is presented. In addition to being free of the problems of the microeconomic approach discussed above, the methodology is less data demanding, by not requiring data on input or output quantities (except land), relying exclusively on price data. This is important given some serious problems of the input quantity series currently available (Gardner and coauthors, 1980).[2] Although not problem-free, data on output and input prices seem to be more reliable than quantity

---

[1] Another feature of these models is that they are static, in the sense that they assume that firms are able to adjust their production decisions instantaneously. There is thus no reason why the competitive equilibrium conditions, including zero profits, would not immediately be reached following productivity gains.

[2] To illustrate the deficiencies of the input quantity data, it is worthwhile citing the following in connection with USDA data on labor:

The labor input data are not derived from surveys of actual hours of labor or of workers committed to agricultural production. Instead, the labor input is calculated

data. A third advantage of the competitive market equilibrium approach is that, under some conditions, it permits one to estimate productivity growth for different farm commodity groups without knowledge of the allocation of farm inputs among these groups. Because quantity data on inputs such as labor, fertilizers, and so forth are not needed, one can avoid the almost impossible task of allocating these inputs among the disaggregated commodity groups and still obtain productivity growth estimates for each of the groups. In the final section, an example illustrating an empirical application of the model to U.S. agriculture is provided.

COMPETITIVE MARKET EQUILIBRIUM

The model exploits the equilibrium implication of a competitive industry with endogenous output prices. It is well known that in competitive equilibrium firms behave as if they minimize average cost with respect to output and that the output price, in equilibrium, is equal to the minimum average cost (Silberberg, 1974). Silberberg considered the case of an industry small enough to be unable to affect factor prices, that is, an industry in which all input supply functions were infinitely elastic. In the case of the agricultural industry, however, we need to relax this assumption, because farmland is approximately fixed and, hence, its rental price cannot be considered constant. In Silberberg's case, the industry supply function is infinitely elastic because firms can expand production by opening new plants producing at the same minimum average cost level. In agriculture, the fixity of land—and consequently the flexibility of its rental price—may preclude this. We shall assume hereafter that all other input prices are exogenous to the sector.[3]

Consider first a conditional cost function defined by

$$C(w, L, y; T, s) = \min_{x} \{wx : F(x, L,; T, s) = y\} \tag{6-1}$$

---

on a "requirements" basis, using estimated quantities of labor required for various production activities. The requirement coefficients are obtained on an individual commodity basis by consultation with state agricultural experiment stations and extension experts. This was last done on a comprehensive basis in 1964, and before that in agricultural census years. Since 1964, the 1959–64 trends have been extrapolated, with some modifications based on subjective judgments of changes in yields and developments in mechanization (Gardner and coauthors, 1980, p. 14).

[3] Agriculture is not a major industry within the economy and, hence, it may not be able to affect prices of inputs that are not specific to agriculture. Furthermore, most industries supplying inputs specific to agriculture (for example, fertilizers) are themselves relatively small, typically facing infinitely elastic supplies of inputs.

where w is a vector of input prices except land, L is the level of agricultural land, y is output, T is an index of technological change, s is a weather index, and x is a vector of input levels, excluding land.

The properties of $C(\cdot)$ in w are well known (concavity, monotonicity, and linear homogeneity). It is assumed that $C(\cdot)$ is increasing in y and s, decreasing in L, and strictly convex in y and L. If firms are price takers, the profit maximization problem of the representative firm is

$$\max_{y,L} py - C(w, L, y; T, s) - qL \tag{6-2}$$

where p is output price and q is the land rental price. The first order conditions associated with problem (6-2) are

(i) $p - C_y(w, L, y; T, s) = 0$
(ii) $q + C_L(w, L, y; T, s) = 0$ \hfill (6-3)

In addition, competitive equilibrium implies that

$$py - C(w, L, y; T, s) - qL = 0 \tag{6-4}$$

As total land is fixed, the representative firm should be regarded as having a fixed endowment of land as well. Hence, condition (6-3ii) needs to be interpreted as an equilibrium condition in the land market rather than as a behavioral equation.

Combining conditions (6-3) and (6-4), we obtain that in equilibrium

(i) $p = \dfrac{C(\cdot) + qL}{y} = C_y(\cdot)$
(ii) $q = -C_L(\cdot)$ \hfill (6-5)

Condition (6-5i) requires that, for any given q, the output price be equal to the average cost; condition (6-5ii) establishes the land market equilibrium, that is, the land rental price is equal to minus the marginal effect of land on costs. The second equality in (6-5i) requires at the same time that the average cost be at its minimum level. Hence, defining

$$\text{MTAC}(w, q; s, T) \equiv \min_{y} \left\{ \frac{C(w, L, y : s, T) + qL}{y} \right\} \tag{6-6}$$

we obtain the following competitive equilibrium conditions:

(i) $p = \text{MTAC}(w, q; s, T)$
(ii) $q = -C_L(w, L, y^*; s, T)$ \hfill (6-7)

where y^* is the level of output solving equation (6-6).

By postulating suitable functional forms for MTAC(·) and C(·), one could jointly estimate equations (6-7i) and (6-7ii). Several problems arise in doing this, however. In the first place, the functions MTAC(·) and $C_L(\cdot)$ are not independent and, hence, one would need to obtain an explicit solution for MTAC(·) derived from the postulated C(·) function (or first postulate a MTAC(·) function and then explicitly derive the underlying cost function). This could be a cumbersome and difficult procedure to implement. A second problem of using equation (6-7) is that we need to consider output y^* and q as explanatory variables. These variables are endogenous and, moreover, as discussed earlier, capture a large proportion of technological change. Hence, this procedure would be no more appropriate than the traditional cost function approach. However, it can be shown that by imposing relatively weak restrictions on the structure of the conditional cost function, the above problems can be easily overcome. Consider the following structure for the conditional cost function:

$$C(w, L, y; T, s) = y \cdot g(w, L; T, s) + c(w, y; T, s) \qquad (6\text{-}8)$$

where $g(\cdot)$ and $c(\cdot)$ satisfy all the properties of a conditional cost function specified above. Structure (6-8) imposes no restrictions on the production technology with respect to the way in which inputs relate to each other (including L) or in the way in which output levels relate to inputs other than land. However, it does impose restrictions on the interdependences between output and land. Structure (6-8) implies that for given w, T, and s the effect of changes in land endowments on the marginal cost is independent of the level of output. That is, the marginal cost curve shifts in a parallel manner when land endowments change (that is, C_{yy} is independent of L). It also implies that C_{LL} is independent of output levels. This exhausts the prior restrictions on the production technology imposed by structure (6-8). Using (6-8) the minimum average total cost is defined by

$$\text{MTAC} = \min_y \left\{ \frac{y\, g(w, L; T, s) + C(w, y^*; T, s) + qL}{y} \right\}$$

$$= g(w, L; T, s) + \min_y \left\{ \frac{C(w, y; T, s) + qL}{y} \right\} \qquad (6\text{-}9)$$

Using the market equilibrium condition $q = -C_L$, we obtain that

$$\text{MTAC} = g(w, L; T, s) + \min_y \frac{C(w, y; T, s) - y\, g_L(w, L; T, s)L}{y}$$

$$= g(w, L; T, s) - g_L(w, L; T, s)L + \min_y \left\{ \frac{C(w, y; T, s)}{y} \right\} \qquad (6\text{-}10)$$

Defining

$$\tilde{\text{MAC}}(w; T, s) \equiv \min_y \left\{ \frac{C(w, y; T, s)}{y} \right\} \tag{6-11}$$

we obtain that

$$\text{MTAC} = g(w, L; T, s) - g_L(w, L; T, s)L + \tilde{\text{MAC}}(w; T, s) \tag{6-12}$$

Therefore, in equilibrium we have

(i) $p = g(w, L; T, s) - g_L(w, L; T, s)L + \tilde{\text{MAC}}(w; T, s)$
(ii) $q = y^* g_L(w, L; T, s).$ (6-13)

Another way of writing equilibrium conditions (6-13) is

(i) $p(1 - \mu_L) = g(w, L; T, s) + \tilde{\text{MAC}}(w; T, s)$
(ii) $p\mu_L = -g_L(w, L; T, s) \cdot L$ (6-14)

where $\mu_L \equiv qL/py^*$ is the share of the rental cost of land in the value of output.

In order to estimate system (6-14) one needs only to postulate appropriate functional forms for the functions $g(w, L; T, s)$ and $\tilde{\text{MAC}}(w; T, s)$. The $g(\cdot)$ function satisfies all the properties of a regular cost function; $\tilde{\text{MAC}}(\cdot)$ satisfies the properties of a minimum average cost function, that is, it is nondecreasing, concave, and linearly homogeneous in w. Moreover,

$$g(w, L; T, s) + \tilde{\text{MAC}}(w; T, s) = \min_y \left\{ \frac{y g(w, L; T, s) + c(w, y; T, s)}{y} \right\}$$

$$= \min_y \left\{ \frac{C(w, L, y; T, s)}{y} \right\} \tag{6-15}$$

That is, the right-hand side of equation (6-14i) coincides with the minimum average *conditional* cost, and hence, we can use the envelope theorem to show that

$$\frac{x_i}{y^*} = g_{w_i}(\cdot) + \tilde{\text{MAC}}_{w_i}(\cdot) \text{ for } i = 1, \ldots, N, \quad i \neq L \tag{6-16}$$

where x_i are the input demands conditional on the level of land L. In estimating the model one must use equation (6-14) and, depending on the quality of input quantity data, may or may not estimate equation (6-16) jointly with equation (6-14).

Measuring technological change and in general the production structure via equation system (6-14) offers several advantages: (a) it uses the competitive market equilibrium conditions in both the output and land markets, thus clearly showing the allocation of the fruits of technological progress between changing output prices and changing land rental value shares; (b) although it uses the output market equilibrium condition, it does not demand the use of output demand relations, which, as we shall see, are implicit in the land rental value shares of the left-hand side in equation (6-14). Only exogenous input prices, state of technological change, and other exogenous variables, such as weather, are required to explain p and μ_L and thus the production system (fully represented by the right-hand side of equation (6-14)) can be analyzed without interferences from the demand side; (c) data on input quantities other than land are not required nor are cost data required. That is, the model is much less data-demanding than the more conventional models; (d) as is shown next, the extent and nature of technological progress (that is, factor biases) can be easily derived once equations (6-14) have been estimated.

It is worth noting that this method can be applied to estimate an aggregate market equilibrium price and to disaggregate farm commodity prices provided that economies or diseconomies of joint production across outputs are negligible (that is, that the marginal cost of producing commodity i is independent of the level of production of commodity j). This condition might be satisfied at certain levels of commodity aggregation. Lopez (1984), for example, rejected the hypothesis of existence of economies or diseconomies of joint production between animal outputs and crops in Canadian agriculture. Therefore, at the very least, one could estimate the productivity and other parameters for animal outputs and crop outputs using the following regressions:

(i) $p^A(1 - \mu^A)_L = g^A(w^A, L^A; T) + \tilde{MAC}^A(w^A; T)$
(ii) $p^A \mu_L^A = -g_L^A(w^A, L^A; T) L^A$
(iii) $p^c(1 - \mu_L^c) = g^c(w^c, L^c; T, s) + \tilde{MAC}^c(w^c; T, s)$
(iv) $p^c \mu_L^c = -g_L^c(w^c, L^c; T, s) L^c$ $\qquad(6\text{-}14^1)$

where the superscripts A correspond to variables related to animal outputs and the superscripts c correspond to variables related to crop production. The econometric procedure should consider the possible endogeneity of the level of land devoted to animal production (L^A) and the one use for crop production (L^c), which are explanatory variables in system (6-14[1]). If little interchange of land between cropland and land devoted to animal production (that is, graze land) exists, then L^A and L^c can be regarded as exogenous and no simultaneity problems will arise. Estimation of system (6-14) may allow one to determine whether technological progress has had a greater effect on the crops or the animal production sector, that is, to verify whether there exists in agriculture—as it does in manufacturing—much diversity of productivity growth between the various subsectors and to determine whether the character of technological progress (that is, factor biases) has been different in the two major subsectors.

FACTOR BIASES OF TECHNOLOGICAL CHANGE

Studies based on the microeconomic approach estimate partial, rather than total, biases of technological change. Partial biases are those measures that account only for the direct effect of technological progress on factor shares and do not consider the indirect effects of technological progress via induced changes in land prices and output quantities or prices. In the case of a conditional cost function (that is, if land can be seen as fixed), we need not be concerned about land prices in obtaining measures of total biases of technological change from system (6-14). However, since output levels are not independent of technological change, any measure of total bias should account for the market equilibrium output changes due to technological progress. The factor shares in the context of a conditional cost function are

$$\mu_i = \frac{\partial \ln c(w, y, L; s, T)}{\partial \ln w_i} \quad \forall i \neq L \tag{6-17}$$

The total bias of technological change is

$$\frac{\partial \mu_i}{\partial T} = \frac{\partial^2 \ln c}{\partial \ln w_i \partial T} + \frac{\partial^2 \ln c}{\partial \ln w_i \partial \ln y} \frac{\partial \ln y}{\partial T} \tag{6-18}$$

Clearly, studies using the conventional microeconomic cost approach measure only the first right-hand side term and ignore the second right-hand side term. Most studies actually use an unconditional cost function $c(w, q, y; s, T)$, in which case the total technological change factor bias is

$$\frac{\partial \mu_i}{\partial T} = \frac{\partial^2 \ln c}{\partial \ln w_i \partial T} + \frac{\partial^2 \ln c}{\partial \ln w_i \partial \ln q} \frac{\partial \ln q}{\partial T} + \frac{\partial^2 \ln c}{\partial \ln w_i \partial \ln y} \frac{\partial \ln y}{\partial T} \tag{6-19}$$

Note that the indirect effects of technological change on μ_i may very well dominate the direct effects and, thus, total technological change factor bias may be the opposite of the partial bias measure. This can occur even if a conditional cost approach is used, if the technology is nonhomothetic. If a profit function approach is used, the technological change bias measure is still partial because only the direct effect of T on y is accounted for, not the indirect effects via changes in q and p induced by technological change. The total technological change biases can be obtained from our approach. For notational simplicity define $\text{MAC}(w, L; T, s) \equiv g(w, L; T, s) + \tilde{\text{MAC}}(w, T, s)$. Then using equation (6-16) we obtain

$$p\mu_i = w_i \text{MAC}_{w_i} \forall i \neq L \tag{6-20}$$

where μ_i is the share of factor x_i in total sales. Because in equilibrium total sales equal total cost, μ_i is also the cost share of x_i. Taking logs of equation (6-20) and differentiating with respect to T, we obtain

$$\frac{\partial \ln \mu_i}{\partial T} = \frac{\partial \ln \text{MAC}_{w_i}}{\partial T} - \frac{\partial \ln p}{\partial T} \tag{6-21}$$

From equations (6-14i) and (6-14ii) we obtain that

$$p = \text{MAC}(\cdot) - g_L(\cdot) L \tag{6-22}$$

and, therefore,

$$\frac{\partial \ln p}{\partial T} = \frac{\partial \ln (\text{MAC} - g_L L)}{\partial T} \tag{6-23}$$

Using equation (6-23) in equation (6-21) we obtain an expression for the total technological change bias for factor i:

$$\frac{\partial \ln \mu_i}{\partial T} = \frac{\partial \ln \text{MAC}_{w_i}}{\partial T} - \frac{\partial \ln (\text{MAC} - g_L L)}{\partial T} \quad \forall i \neq L \tag{6-24}$$

Since equation (6-14) permits the estimation of both the $\text{MAC}(\cdot)$ and $g(\cdot)$ functions, we can measure total technological change biases for each factor.

To measure total technological change bias on land, we differentiate equation (6-14ii) to obtain

$$\frac{\partial \ln \mu_L}{\partial T} = -\frac{\partial \ln g_L}{\partial T} - \frac{\partial \ln p}{\partial T} \tag{6-25}$$

and, hence,

$$\frac{\partial \ln \mu_L}{\partial T} = -\frac{\partial \ln g_L(\cdot)}{\partial T} - \frac{\partial \ln (\text{MAC} - g_L L)}{\partial T} \tag{6-26}$$

Thus, estimation of the market equilibrium equations (6-14) permits the measurement of the total, rather than the partial, factor biases induced by technological change. These total biases account for the direct partial or microeconomic biases plus the indirect biases derived from changes in land prices as well as in output level and price that are, in turn, caused by technological change.

An Example

To illustrate the potential divergence between partial and total factor intensity biases associated with technological change, consider a generalized Leontief (GL) cost function. Suppose one estimates a GL unconditional cost function in order to measure technological change biases. For simplicity, assume that the production technology exhibits constant returns to scale. In this case the cost function can be written as

$$\tilde{C} = y \left[a_{00} q + 2 \sum_i b_{0i} q^{1/2} w_i^{1/2} + \sum_{i \neq 0} \sum_{j \neq 0} b_{ij} w_i^{1/2} w_j^{1/2} + \gamma_0 q T + \sum_{i \neq 0} \gamma_i w_i T \right] \tag{6-27}$$

where we have maintained the same notation previously used. The estimating factor demand intensities are

(i) $\dfrac{L}{y} = a_{00} + \sum_{i \neq 0} b_{0i} \left(\dfrac{w_i}{q} \right)^{1/2} + \gamma_0 T$

(ii) $\dfrac{x_i}{y} = b_{00} + \sum_{j \neq 0} b_{ij} \left(\dfrac{w_j}{w_i} \right)^{1/2} + b_{i0} \left(\dfrac{q}{w_i} \right)^{1/2} + \gamma_i T \qquad \forall i \neq 0$ \qquad (6-28)

The parameter γ_0 measures the effect of technological change on land intensity and the γ_i parameters measure similar effects for all other inputs.

It can be easily verified that the conditional cost function (that is, conditional on a given level of land) associated with equation (6-27) is

$$C = y \left[\frac{1}{\frac{L}{y} - \gamma_0 T - a_{00}} \right] \left[\sum_{i \neq 0} b_{0i} w_i^{1/2} \right]^2 + y \left[\sum_{i \neq 0} \sum_{j \neq 0} b_{ij} w_i^{1/2} w_j^{1/2} + \sum_{i \neq 0} \gamma_i w_i \right] \tag{6-29}$$

By minimizing $C(y, L, w) + qL$ with respect to L we obtain exactly $\tilde{C}(\cdot)$, as in equation (6-27). If competitive equilibrium prevails, we have $q = -\partial C/\partial L$, that is,

$$q = \left[\frac{\sum_{i \neq 0} b_{0i} w_i^{1/2}}{\frac{L}{y} - \gamma_0 T - a_{00}} \right]^2 \tag{6-30}$$

and, hence, we can regard equation (6-30) as a market equilibrium relation. The effect of technological progress on q can be obtained from equation (6-30):

$$\frac{\partial q}{\partial T} = \frac{2 \left[\sum_{i \neq 0} b_{0i} w_i^{1/2} \right]^2}{\left[\frac{L}{y} - \gamma_0 T - a_{00} \right]^3} \cdot \gamma_0 \tag{6-31}$$

The second-order conditions for minimization of $C(\cdot) + qL$ require that

$$C_{LL} = \frac{\left[\sum_{i \neq 0} b_{0i} w_i^{1/2} \right]^2}{\left[\frac{L}{y} - \gamma_0 T - a_{00} \right]^3} \frac{1}{y} > 0 \tag{6-32}$$

and, hence, the sign of $\partial q/\partial T$ will be identical to the sign of γ_0. From equation (6-28) we obtain that the total effect of technological progress on L/y is

$$\frac{d(L/y)}{dT} = \gamma_0 - \sum_{i \neq 0} b_{0i} \left(\frac{w_i}{q} \right)^{1/2} \frac{1}{q} \frac{\partial q}{\partial T} \tag{6-33}$$

Using equations (6-30) and (6-31) in (6-33) we obtain

$$\frac{d(L/y)}{dT} = \gamma_0 - \sum_{i \neq 0} b_{0i} \left(\frac{w_i}{q} \right)^{1/2} \frac{2\gamma_0}{\frac{L}{y} - \gamma_0 T - a_{00}} \tag{6-34}$$

Using equation (6-28i) in (6-34), we obtain

$$\frac{d(L/y)}{dT} = \gamma_0 - 2\gamma_0 = -\gamma_0 \tag{6-35}$$

Thus, the total effect of technological change on the land intensity ratio is exactly opposite to the partial effect. By similar procedures one can show that

$$\frac{d(x_i/y)}{dT} = \gamma_i + b_{i0}\left(\frac{q}{w_i}\right)^{1/2} \frac{2\gamma_0}{\sum_{j\neq 0} b_{0j}\left(\frac{w_j}{q}\right)^{1/2}}, \forall i \neq 0 \qquad (6\text{-}36)$$

Note that by equation (6-32)

$$\sum_{j\neq 0} b_{0j}\left(\frac{w_j}{q}\right)^{1/2} > 0$$

and, hence, if $b_{i0} > 0$ and $\gamma_0 > 0$, then

$$\frac{d(x_i/y)}{dT} > \gamma_i$$

that is, the partial technological change bias underestimates the factor i using effect. Of course, depending on the signs of b_{i0} and γ_0, other outcomes are possible. The effect of technological change on the average cost is also subject to errors if the microeconomic approach is used

$$\frac{d(c/y)}{dT} = \gamma_0 q + \gamma_0 \frac{\partial q}{\partial T} + \sum_{i\neq 0} \gamma_i w_i$$

$$= \gamma_0 q \left[1 + \frac{2}{\sum_{i\neq 0} b_{0i}\left(\frac{w_i}{q}\right)^{1/2}}\right] + \sum_{i\neq 0} \gamma_i w_i \qquad (6\text{-}37)$$

If, for example, $\gamma_0 > 0$, then the partial microeconomic estimate of technological progress on average cost, that is, $\gamma_0 q + \sum_{i\neq 0} \gamma_i w_i$, underestimates the total effect of technological change in lowering average cost.

PRICE EFFECTS OF TECHNOLOGICAL CHANGE: FOOD PRICES VERSUS LAND PRICES

We have argued that technological progress in agriculture is largely absorbed by changes in output prices and land prices. In this section we try to isolate the key parameters affecting the distribution of the fruits of technological change between variations in land prices and output prices. The analysis using equations (6-14) allows one to measure the distribution of the technological progress gains between output prices and the share of land (see equations (6-23) and (6-25)). However, it is not possible to isolate the effect on land prices from the effects on land shares. In order to do so, we need to make explicit an output demand equation that was only implicit in equation (6-14). Moreover, in order to increase the level of generality of the analysis, we shall now assume that land supply is not completely inelastic. Thus, the basic model used here is the following

(i) $p = g(w, L^D; T, s) + \tilde{MAC}(w; T, s) + \dfrac{qL^D}{y^s}$

(ii) $q = -y^s g_L(w, L^D; T, s)$

(iii) $y^D = D(p); y^D = y^s$

(iv) $L^s = L^s(q); L^s = L^D$ (6-38)

where y^s is output supply, y^D is output demand, L^D and L^s stand for land demand and supply, respectively. It is assumed that $D_p < 0$ and $L_p^s \geq 0$.

Totally differentiating equation (6-38) with respect to p, q, and T and after some trivial algebraic manipulation we obtain

(i) $\dfrac{dp}{dT} = \dfrac{(g_T + \tilde{MAC}_T)(1 - \sigma\epsilon_{Lq}^s) - Lg_{LT}}{|D|}$

(ii) $\dfrac{dq}{dT} = \dfrac{-(1 - \mu\epsilon_{yp}^D)g_{LT}y + (g_T + MAC_T)\epsilon_{yp}^D \dfrac{q}{p}}{|D|}$ (6-39)

where ϵ_{yp}^D is the demand elasticity for output; ϵ_{Lq}^s is the land supply elasticity;

$$\sigma \equiv \dfrac{\partial \ln g_L}{\partial \ln L}$$

is a measure of curvature of the inverse land demand equation; and $|D|$ is the determinant of the following matrix

$$D = \begin{bmatrix} (1 - \mu_L \epsilon_{yp}^D) & -\dfrac{L}{y} \\ -\epsilon_{yp}^D \dfrac{q}{p} & -(1 - \sigma\epsilon_{Lq}^s) \end{bmatrix}$$

Note that given our assumptions $\epsilon_{yp}^D < 0$, $\epsilon_{Lq}^s \geq 0$, and $\sigma < 0$. Hence, $|D| > 0$. Also, note $g_T + \tilde{MAC}_T < 0$, that is, the effect of technological progress is to decrease minimum average cost. Therefore, the signs of

dp/dT and dq/dT depend mainly on g_{LT}. If $g_{LT} > 0$, then $dp/dT < 0$ and dq/dT is ambiguous. If $g_{LT} < 0$, then land prices increase after technological change but the effect on the output price is ambiguous. Only if $g_{LT} = 0$, can one be sure that the effect of technological change is normal for both output price and land price, that is, in this case $dp/dT < 0$ and $dq/dT > 0$. It can be easily shown that if the microeconomic bias of technological progress (that is, for given prices) is land-using, then $g_{LT} < 0$ and, hence, if this type of technological change occurs, we expect land prices to increase and an ambiguous effect on output prices. In general, from equation (6-39) it is clear that the greater is g_{LT} the more effective is technological change in lowering output prices. The implication of this simple result is important: land-using technological change will induce less of an impact on food prices than land-saving technological progress. This proposition in combination with the induced innovation hypothesis signifies that after periods of rapid increases in land prices, technological progress will induce stronger downward pressures on farm output prices than after periods during which land has been cheap relative to other factors of production. One may speculate that perhaps the drastic increase in land prices during the 1970s was, at least in part, responsible for the current deterioration of real farm prices.

APPLICATION TO U.S. AGRICULTURE

In this section we illustrate the applicability of the model estimating the market equilibrium effects of technological progress for U.S. agriculture for the period 1950–80. We consider three variable inputs assumed to be in elastic supply (that is, with exogenous prices), namely, labor, capital, and intermediate materials. Land and output prices are endogenous, satisfying the market equilibrium conditions. The postulated minimum average cost function is a generalized Leontief, yielding from equation (6-14i):

$$p(1 - \mu_L) = \sum_{i=1}^{3} \alpha_{0i} w_i L + \sum_{i=1}^{3} \gamma_{0i} w_i LT + \tfrac{1}{2} L^2 \sum_{i=1}^{3} a_{0i} w_i$$
$$+ \sum_{i=1}^{3} \sum_{j=1}^{3} b_{ij} w_i^{1/2} w_j^{1/2} + T \sum_{i=1}^{3} \beta_{0i} w_i \qquad (6\text{-}40)$$

and from equation (6-14ii):

$$p \mu_L = -L \left[\sum_{i=1}^{3} \alpha_{0i} w_i + \sum_{i=1}^{3} \gamma_{0i} w_i T + \sum_{i=1}^{3} a_{0i} w_i L \right] \qquad (6\text{-}41)$$

where subscripts 1 refer to labor, subscripts 2 to capital, and subscripts 3 to intermediate materials. Using equation (6-14) it is clear that the first three right-hand side terms in equation (6-40) comprise the function $g(\cdot)$; the last two terms in equation (6-40) constitute the function $\tilde{MAC}(\cdot)$. Factor demands could be jointly estimated with equations (6-40) and (6-41) but given the deficiencies in quantity data indicated in footnote 2 we prefer to omit their estimation.

The data used are annual data for aggregate U.S. agriculture observations for output prices, land shares, and the prices of labor, capital (rental), and intermediate inputs. The index of technological change used (T) is a time trend variable. In calculating land shares, the average rental rate was used, which, multiplied by total agricultural land, gives the land rental value. It was assumed that average land rents were a good proxy for the average actual land rental price. Index numbers for the rental price of capital were taken from Shoemaker (1984). A farm wage rate index was constructed based on the wage rate for hired labor published by USDA. These data consist of average hourly wage by workers receiving cash wages only. A price index for intermediate materials was constructed as an average for U.S. price indexes for seed, fertilizer, and miscellaneous inputs weighted by the expenditure shares for the three components. Price indexes for each one of these categories are published by USDA. Finally, an aggregate output price index was constructed from prices of seven output categories provided in Ball (1984). These included cash grains, field crops excluding cash grains, vegetables, fruits, livestock excluding dairy, poultry and eggs, and dairy products.

Table 6-1 shows the estimated coefficients of equations (6-40) and (6-41) obtained using a seemingly unrelated estimator. In general the goodness-of-fit and statistical significance of the estimated parameters were highly satisfactory as reflected by the high system R^2 and T-statistics. Moreover, the estimated minimum average cost function satisfied the required regularity conditions. It was increasing in factor prices at all sample points, is decreasing in land levels (L) also at all sample points, decreasing in T, and satisfying the concavity condition in factor prices at all observations. The only condition that was not satisfied at all sample points was the convexity in L, which was met at only 40 percent of the observations. We note that the symmetry conditions were *a priori* imposed in the model to economize degrees of freedom. Thus, the estimated model is consistent with the theoretical properties discussed in the first section and, moreover, the magnitude of the estimates is in general quite plausible. The estimates suggest that labor and capital are strong substitutes (note the positive and significant value of the coefficient b_{12}) and labor and other intermediate materials are also substitutes.

TABLE 6-1. ESTIMATES OF THE MARKET EQUILIBRIUM EQUATIONS

Coefficient	Value	T-statistics
b_{11}	−37.18	−3.58
b_{22}	16.71	1.55
b_{33}	1.99	0.26
b_{12}	7.05	7.67
b_{13}	6.66	4.85
b_{23}	−0.032	−0.05
α_{01}	57.86	2.98
α_{02}	−32.84	−1.73
α_{03}	−5.40	−0.41
β_{01}	2.51	2.48
β_{02}	−2.83	−2.75
β_{03}	−0.99	−1.39
γ_{01}	−1.50	−1.42
γ_{02}	2.23	2.18
γ_{03}	0.44	0.64
a_{01}	−59.93	−3.54
a_{02}	28.39	1.69
a_{03}	4.21	0.36

System $R^2 = 0.995$

Capital and intermediate materials appear to be neither substitutes nor complements, as reflected by the lack of significance of the b_{23} coefficient.

The estimates of the effect of technological progress indicate important quantitative effects on output price and factor shares. Table 6-2 shows the average annual effect of technological change on the various variables of importance. Thus, technological change has played an important role in depressing farm output prices and land values, whereas it has been land-saving and a user of labor, capital, and intermediate goods. On average, farm output prices have dropped by almost 2 percent per annum due to productivity growth, a sizable effect. The effects of technological change on factor shares include the direct quantity effects plus the indirect effects via land price and output price changes. This is in contrast to conventional analysis, which accounts only for the former, ignoring price effects. The result that technological progress has been labor intensive rather than labor saving is drastically different from those obtained using microeconomic models (see the review by Capalbo and Vo in chapter 3 of this book). It turns out that, although the direct effects of technological change (that is, the constant prices effects) lead to reducing the share of labor, the indirect effects are very strong in offsetting the former effects. The lower land prices and lower output prices both pointed towards increasing the share of labor in total costs. Most of

TABLE 6-2. AVERAGE MARKET EQUILIBRIUM ANNUAL EFFECTS OF TECHNOLOGICAL PROGRESS ON KEY VARIABLES, 1950–1980

Variable	Percentage change
Output price	−1.96
Land share	−1.66
Labor share	2.88
Capital share	1.41
Intermediate materials share	1.39

the induced decrease in the share of land is due to a large depressing effect of technological change on land prices.

CONCLUSION

The most important message of this chapter is the necessity of accounting for market equilibrium effects of technological change. The microeconomic approach, by ignoring these effects, leaves aside perhaps the most important part of technological change—that part that is manifested in output and input prices variation. In the absence of adjustment costs or other sources of slow factor adjustment, there is no reason that technological progress will not be entirely manifested in output and input price variability even in the short run. In a more realistic setting, where slow adjustments of some factors do exist, one would expect that part of technological change will not be expressed in price variation in the short run.

A framework for the measurement of technological progress relying on a competitive market equilibrium approach has been developed. This method possesses several advantages over the microeconomic conventional approach: it is able to capture the price effects of technological progress; it is substantially less data-demanding, by requiring only input and output price and not quantity data; and it allows one to measure productivity growth by major subsectors within agriculture. The model proposed is, however, long run in nature, because it assumes the existence of zero profit equilibrium. This may be a limitation of the model if there exist serious restrictions to the adjustments of some factors in agriculture. This suggests the necessity of measuring technological change within a dynamic temporary market equilibrium framework that explicitly accounts for quasi-fixed factors.

A third aspect of the chapter has concerned the factors affecting the distribution of the fruits of technological progress between output and land prices (that is, between consumers and land owners). The major result in this section was that land-using technological change is likely to induce a smaller effect on food prices than land-saving technological progress. If the induced innovation hypothesis is valid, this implies that periods of rapid increases in land rental prices will lead to periods of stronger downward pressures on farm output prices than will periods of relatively low or decreasing land prices. This is so because rapidly increasing land prices will stimulate the generation of land-saving technological progress, which has a stronger negative impact on farm output prices.

Finally, the working of the proposed market equilibrium model was illustrated by estimating it using U.S. agriculture aggregate data. The desirable statistical properties of the estimated model as well as its consistency with the theoretical restrictions discussed in the first section suggest that consideration of market equilibrium relations in measuring technological change is a necessary avenue for further analyses. Moreover, it appears that the quantitative results obtained from such a model lead to substantially different results about the consequences of technological change from those obtained with the conventional microeconomic models.

REFERENCES

Antle, J. 1984. "The Structure of U.S. Agricultural Technology, 1910–1978," *American Journal of Agricultural Economics* vol. 66, no. 4, pp. 414–421.

Ball, E. 1984. "Measuring Agricultural Productivity: A New Look," Staff Report No. 840330 (Washington, D.C., USDA).

Binswanger, H. 1974. "The Measurement of Technical Change Biases with Many Factors of Production," *American Economic Review* vol. 64, no. 6, pp. 964–976.

Gardner, B., and coauthors. 1980. "Measurement of U.S. Agricultural Productivity: A Review of Current Statistics and Proposals for Change," Technical Bulletin No. 1614 (Washington, D.C., USDA).

Hayami, Y., and V. Ruttan. 1971. *Agricultural Development: An International Perspective* (Baltimore, Md., Johns Hopkins University Press).

Kako, T. 1978. "Decomposition Analysis of Derived Demand for Factor Inputs: The Case of Rice Production in Japan," *American Journal of Agricultural Economics* vol. 60, no. 4, pp. 628–635.

Lopez, R. 1980. "The Structure of Production and the Derived Demand for Inputs in Canadian Agriculture," *American Journal of Agricultural Economics* vol. 62, no. 1, pp. 38–45.

———. 1984. "Estimating Substitution and Expansion Effects Using a Profit Function Framework," *American Journal of Agricultural Economics* vol. 66, no. 3, pp. 358–367.

Shoemaker, R. 1984. "Productivity and Technological Change in U.S. Agriculture: A Decomposition Analysis, 1950–80" (M.S. thesis, University of Maryland, College Park).

Silberberg, E. 1974. "The Theory of the Firm in Long-Run Equilibrium," *American Economic Review* vol. 64, no. 4, pp. 734–741.

7
INTERTEMPORAL AND INTERSPATIAL ESTIMATES OF AGRICULTURAL PRODUCTIVITY

MICHAEL HAZILLA AND RAYMOND J. KOPP

INTRODUCTION

Technological progress has long been recognized as one of the major determinants of U.S. dominance in world agricultural production. It is thus not surprising that historically much research has been devoted to the study of the nature of this process. Today interest in the measurement of technological change is particularly important because of an ever-increasing world population and the implicit recognition of a U.S. role in alleviating world food shortages. The accurate measurement of technical change is vital to an understanding of its nature and causes.

We note at the outset that we are equating the measurement of technological change with the measurement of intertemporal total factor productivity. Intertemporal total factor productivity is usually interpreted in primal space as the rate of change over time of an index of outputs divided by an index of inputs (growth accounting), or by a rate of shift in a production function (structural analysis). The mechanism causing these rates of change is assumed to be technological progress and the measurement of technological progress is thus equivalent to the measurement of a change in intertemporal total factor productivity. Furthermore, under the maintained assumptions of producer cost minimization and competitive factor markets, technological change can be equivalently measured in the dual space by (a) the rate of change of production cost minus the rate of change of an index of outputs minus the rate of change of an index of all input prices, or (b) a rate of shift in a cost function (the dual interpretation of the production function).

Interspatial total factor productivity has a somewhat different interpretation. Interspatial productivity can be defined in the primal as the logarithmic difference in an index of outputs between two production regions divided by the log difference of an index of inputs. Equivalently, the dual interpretation would be the exponentiation of the log difference in production cost minus the log difference in an index of outputs minus the log difference in an index of input prices. Interspatial productivity differences arise not from the dynamic process of technological change but from static technological differences across producing regions.

Past economic studies of technological change have usually employed a growth accounting framework.[1] The primary motivation for pursuing this approach was the ease with which various index numbers could be computed. These index numbers depend on no unknown parameters and are simple algebraic aggregates based on price and quantity data. Despite methodological problems related to construction of the indexes, as well as problems associated with the appropriate measure of particular inputs, growth accounting estimates generally provide a great deal of information regarding productivity.[2]

Solow's 1957 paper served to highlight an important nexus between growth accounting approaches to measuring productivity and those approaches analyzing production structures characterized by neoclassical models. Diewert (1976, 1978, 1979), in a series of papers, developed the relationship between the functional forms used to compute index numbers and the functional forms used to characterize underlying production or cost (aggregator) functions. Exploiting Diewert's findings, Denny and Fuss (1983a) published a general approach for measuring intertemporal and interspatial total factor productivity. In applying this theoretical model to intertemporal and interspatial comparisons, Denny and Fuss provided evidence that suggests caution when using the growth accounting framework (based on index numbers) to quantify productivity changes, since such procedures ignore second-order price effects.

The purpose of this paper is to illustrate how the Denny-Fuss approach can be adapted for agricultural productivity measurement. We use a unique data set derived from the Firm Enterprise Data System (FEDS) (see Krenz, 1975a,b) that, in principle, could be used to study all major crops in the continental United States. Our results summarize a productivity analysis of corn and soybean production in the five-state area that

---

[1] A still classic but somewhat dated review of past studies is Nadiri (1970).

[2] A well-known and publicized controversy relates to the appropriate measurement of capital in the explanation of productivity change. See Denison's (1969) comments on the Jorgenson-Griliches (1967) method for deriving the appropriate measure of capital services that allows for varying levels of utilization. More recently, see the work of Berndt and Fuss (1982) for an innovative approach that adjusts the value of the capital service flow using Tobin's-q rather than the adjustment for variations in capital services and capacity utilization.

includes Illinois, Indiana, Iowa, Missouri, and Ohio. We address the following specific questions:

1. What has been the intertemporal change in productivity for the Midwest states between 1974 and 1978?
2. Using bilateral comparisons relative to Illinois, what are the interspatial differences in productivity for Indiana, Iowa, Missouri, and Ohio during 1974 and 1978?

An unfortunate complication in our analysis is the inability to dichotomize the intertemporal change in productivity between 1974 and 1978 into a portion attributable solely to technological progress and a portion attributable to random fluctuations in weather conditions. Because of a lack of sufficient degrees of freedom, severe collinearity, and a focus on two time periods, we are prohibited from including both a time trend to capture technological progress and a weather control variable in our econometric model of agricultural production. Furthermore, even if sufficient degrees of freedom were available insufficient independent variation exists between weather and time in our two period analysis statistically to distinguish one effect from another. As a consequence—and in the belief that weather variation was more pronounced than technological progress between 1974 and 1978—we have included in the econometric model only a control variable for weather. Accordingly, in this model, intertemporal changes in productivity are attributed to differences in weather.

In the second section we develop the theoretical basis for intertemporal and interspatial productivity measurement. The section is divided into three subsections. In the first subsection we describe a growth accounting framework utilizing Fisher-Tornqvist superlative index numbers. In the second subsection, we develop an alternative growth accounting approach utilizing indexes that employ the share derivatives (second order effects) from an econometric cost function model that we describe in the third subsection. The microeconomic database employed in the analysis is reviewed in the third section. In the fourth section we present our major findings and conclude with some final remarks in the fifth section.

THEORETICAL MODELS OF INTERTEMPORAL AND INTERSPATIAL PRODUCTIVITY

In this section we develop theoretical models that underlie our measure of intertemporal and interspatial productivity. The development of this section follows closely Denny and Fuss (1983a).

Growth Accounting Models: First-Order Approximations

We assume that the agricultural production process in region i at time t can be represented by the dual cost function

$$C_{it} = \bar{F}_{it}(\bar{P}_{it}, Q_{it}, T_t, D_i) \tag{7-1}$$

where t = time period
 i = spatial region
 Q_{it} = agricultural output region i, time t
 T_t = state of technology at time t (defined as a weather index)
 D_i = spatial indicator for region i
 $\bar{P}_{it}$ = a vector of k-input prices faced in the ith region at time t

Applying Diewert's (1976) quadratic lemma to a logarithmic approximation of $F_{it}(\ )$ defined as[3]

$$\log C_{it} = f(\log \bar{P}_{it}, \log Q_{it}, T_t, D_i) \tag{7-2}$$

where $f(\ )$ is quadratic, yields

$$\Delta \log C = \log C_{hb} - \log C_{oa}$$

$$= \tfrac{1}{2}\left(\left.\frac{\partial f}{\partial D}\right|_{D=D_h} + \left.\frac{\partial f}{\partial D}\right|_{D=D_o}\right)[D_h - D_o]$$

$$+ \tfrac{1}{2}\Sigma_k\left(\left.\frac{\partial f}{\partial \log P_k}\right|_{P_k=P_{khb}} + \left.\frac{\partial f}{\partial \log P_k}\right|_{P_k=P_{koa}}\right)$$

$$\cdot [\log P_{khb} - \log P_{koa}]$$

$$+ \tfrac{1}{2}\left(\left.\frac{\partial f}{\partial \log Q}\right|_{Q=Q_{hb}} + \left.\frac{\partial f}{\partial \log Q}\right|_{Q=Q_{oa}}\right)$$

$$\cdot [\log Q_{hb} - \log Q_{oa}]$$

$$+ \tfrac{1}{2}\left(\left.\frac{\partial f}{\partial T}\right|_{T=T_b} + \left.\frac{\partial f}{\partial T}\right|_{T=T_a}\right)[T_b - T_a] \tag{7-3}$$

---

[3] The quadratic lemma states that the difference between the values of a quadratic function evaluated at two points is equal to the average of the gradient evaluated at both points multiplied by the difference between the points. Whereas the Taylor series requires knowledge of the quadratic Hessian evaluated at the expansion point, the quadratic lemma requires only first-order (gradient) information at both the point of expansion and point of evaluation.

where h and o represent two distinct regions and a and b represent two distinct time periods. The partial derivatives $\partial f/\partial D$ and $\partial f/\partial T$ are defined even though D and T are spatial and temporal binary (discontinuous) indicator variables.[4] Equation (7-3) states that the change in costs is equal to a linear function of interspatial, input price, output, and intertemporal differences.

Let θ_{ho}^1 and μ_{ba}^1 denote the Fisher-Tornqvist approximation to the continuous Divisia dual rate of interspatial and intertemporal productivity. In other words we define

$$\theta_{ho}^1 = \tfrac{1}{2}\left(\left.\frac{\partial f}{\partial D}\right|_{D=D_h} + \left.\frac{\partial f}{\partial D}\right|_{D=D_o} \right)[D_h - D_o] \tag{7-4}$$

and that

$$\mu_{ba}^1 = \tfrac{1}{2}\left(\left.\frac{\partial f}{\partial T}\right|_{T=T_b} + \left.\frac{\partial f}{\partial T}\right|_{T=T_a} \right)[T_b - T_a] \tag{7-5}$$

Using these definitions and assuming constant returns to scale and perfectly competitive factor markets, we can rewrite equation (7-3) as[5]

$$\Delta \log C = \tfrac{1}{2}\Sigma_k[S_{khb} + S_{koa}] \cdot [\log P_{khb} - \log P_{koa}] \\ + [\log Q_{hb} - \log Q_{oa}] + \theta_{ho} + \mu_{ba} \tag{7-6}$$

Following Denny and Fuss, suppose interspatial comparisons are of interest and the technology indicator is equal across time periods ($T_a = T_b$). Then

$$\theta_{ho}^1 = [\log C_h - \log C_o] - [\log Q_h - \log Q_o] \\ - \tfrac{1}{2}\Sigma_k[S_{kh} + S_{ko}][\log P_{kh} - \log P_{ko}] \tag{7-7}$$

defines our first-order (denoted by superscript 1) dual measure of interspatial productivity between region h and reference region o at a particular time.

---

[4] Denny and Fuss (1983b, p. 420) demonstrate that Diewert's quadratic lemma is still applicable for binary (noncontinuous) variables so that "one can ignore the discreteness of the variables."

[5] In this expression S_{khb} denotes the kth input cost share in region h during time period b; S_{koa} denotes the kth input cost share in region o, time period a. Other variables are interpreted similarly.

Similarly the first order intertemporal productivity measure (denoted $\mu_{ba}{}^1$) for two time periods a and b when $D_h = D_o$, is defined

$$\mu_{ba}{}^1 = [\log C_b - \log C_a] - [\log Q_b - \log Q_a] \\ - \tfrac{1}{2}\Sigma_k[S_{kb} + S_{ka}] \cdot [\log P_{kb} - \log P_{ka}] \tag{7-8}$$

Equations (7-7) and (7-8) specify the Denny-Fuss first-order accounting equations. These measures are easily computed using price, quantity, and output data and do not require any econometric estimation. We shall demonstrate how these indexes are computed for an intertemporal and interspatial productivity analysis.

Growth Accounting Models: Second-Order Approximations

To derive the Denny-Fuss general second-order accounting equation, consider a second-order Taylor Series expansion of equation (7-2) around the rth data point, evaluated at data point s [6]

$$\ln C^s = \ln C^r + \Sigma_k \left(\frac{\partial f}{\partial \ln P_k}\bigg|_{P_k = P_k{}^r}\right)(\ln P_k^s - \ln P_k^r)$$

$$+ \left(\frac{\partial f}{\partial T}\bigg|_{T=T^r}\right)(T^s - T^r) + \left(\frac{\partial f}{\partial D}\bigg|_{D=D^r}\right)(D^s - D^r)$$

$$+ \left(\frac{\partial f}{\partial \ln Q}\bigg|_{Q=Q^r}\right)(\ln Q^s - \ln Q^r)$$

$$+ \tfrac{1}{2}\left[\left(\Sigma_k \Sigma_l \frac{\partial^2 f}{\partial \ln P_k \partial \ln P_l}\bigg|_{\substack{P_k = P_k{}^r \\ P_l = P_l{}^r}}\right)\right.$$

$$\left. \cdot (\ln P_k^s - \ln P_k^r)(\ln P_l^s - \ln P_l^r)\right]$$

$$+ \left(\frac{\partial^2 f}{\partial D \partial D}\bigg|_{D=D^r}\right)(D^s - D^r)^2 + \left(\frac{\partial^2 f}{\partial T \partial T}\bigg|_{T=T^r}\right)(T^s - T^r)^2$$

---

[6] These data points are assumed to represent particular intertemporal-interspatial combinations, so that we economize on the number of superscripts needed to denote a particular data point.

$$+ \Sigma_k \left(\frac{\partial^2 f}{\partial \ln P_k \partial D} \bigg|_{\substack{P_k = P_k^r \\ T = T^r}} \right) (\ln P_k^s - \ln P_k^r)(D^s - D^r)$$

$$+ \Sigma_k \left(\frac{\partial^2 f}{\partial \ln P_k \partial T} \bigg|_{\substack{P_k = P_k^r \\ T = T^r}} \right) (\ln P_k^s - \ln P_k^r)(T^s - T^r)$$

$$+ \left(\frac{\partial^2 f}{\partial T \partial D} \bigg|_{\substack{T = T^r \\ D = D^r}} \right) (T^s - T^r)(D^s - D^r)$$

$$+ \left(\frac{\partial^2 f}{\partial Q \partial Q} \bigg|_{Q = Q^r} \right) (\ln Q^s - \ln Q^r) \tag{7-9}$$

where subscripts r, s denote data points. In a similar manner we can also expand equation (7-2) about the point s, and evaluate the cost function gradient and Hessian elements at s to yield a function identical to equation (7-9) except that the s and r indexes have been interchanged. The difference between this approximating function and equation (7-9) yields the following Denny-Fuss general second-order accounting equation that includes intertemporal and interspatial effects as well as second-order and interaction effects

$$\ln C^s - \ln C^r = \tfrac{1}{2} \Sigma_k \left(\frac{\partial f}{\partial \ln P_k} \bigg|_{P_k = P_k^r} + \frac{\partial f}{\partial \ln P_k} \bigg|_{P_k = P_k^s} \right)$$

$$\cdot (\ln P_k^s - \ln P_k^r)$$

$$+ \left(\frac{\partial f}{\partial T} \bigg|_{T = T^r} + \frac{\partial f}{\partial T} \bigg|_{T = T^s} \right)(T^s - T^r)$$

$$+ \left(\frac{\partial f}{\partial D} \bigg|_{D = D^r} + \frac{\partial f}{\partial D} \bigg|_{D = D^s} \right)(D^s - D^r)$$

$$+ \tfrac{1}{4} \left[\Sigma_k \Sigma_l \left(\frac{\partial^2 f}{\partial \ln P_k \partial \ln P_l} \bigg|_{\substack{P_k = P_k^r \\ P_l = P_l^r}} - \frac{\partial f}{\partial \ln P_k \partial \ln P_l} \bigg|_{\substack{P_k = P_k^s \\ P_l = P_l^s}} \right) \right.$$

$$\cdot (P_k^s - P_k^r)(P_l^s - P_l^r)$$

$$+ \left(\frac{\partial^2 f}{\partial T \partial T} \bigg|_{T = T^r} - \frac{\partial^2 f}{\partial T \partial T} \bigg|_{T = T^s} \right)(T^s - T^r)(T^s - T^r)$$

$$+ \Sigma_k \left(\frac{\partial^2 f}{\partial \ln P_k \, \partial T} \bigg|_{\substack{P_k = P_k^r \\ T = T^r}} - \frac{\partial^2 f}{\partial \ln P_k \, \partial T} \bigg|_{\substack{P_k = P_k^s \\ T = T^s}} \right)$$

$$\cdot (P_k^s - P_k^r)(T^s - T^r)$$

$$+ \left(\frac{\partial^2 f}{\partial D \, \partial D} \bigg|_{D = D^r} + \frac{\partial^2 f}{\partial D \, \partial D} \bigg|_{D = D^s} \right)(D^s - D^r)(D^s - D^r)$$

$$+ \Sigma_k \left(\frac{\partial^2 f}{\partial \ln P_k \, \partial D} \bigg|_{\substack{P_k = P_k^r \\ D = D^r}} - \frac{\partial^2 f}{\partial \ln P_k \, \partial D} \bigg|_{\substack{P_k = P_k^s \\ D = D^s}} \right)$$

$$\cdot (P_k^r - P_k^s)(D^r - D^s)$$

$$\left. + \left(\frac{\partial^2 f}{\partial T \, \partial D} \bigg|_{\substack{T = T^r \\ D = D^r}} - \frac{\partial^2 f}{\partial T \, \partial D} \bigg|_{\substack{T = T^s \\ D = D^s}} \right)(T^s - T^r)(D^s - D^r) \right] \quad (7\text{-}10)$$

The second-order interspatial logarithmic productivity is derived by collecting terms that contain $(D^s - D^r)$ in equation (7-10) to yield

$$\theta_{sr}^2 = \theta_{sr}^1 - \tfrac{1}{4} \left(\Sigma_k \Sigma_l \frac{\partial^2 f}{\partial \ln P_k \, \partial \ln P_l} \bigg|_{\substack{P_k = P_k^s \\ P_l = P_l^s}} \right.$$

$$\left. - \frac{\partial^2 f}{\partial \ln P_k \, \partial \ln P_l} \bigg|_{\substack{P_k = P_k^r \\ P_l = P_l^r}} \right)(\ln P_k^s - \ln P_k^r)(\ln P_l^s - \ln P_l^r) \quad (7\text{-}11)$$

where we assume technological commonality $(T_t^s = T_t^r)$ but permit spatial differences $D_i^s \neq D_i^r$. Similarly, the Denny-Fuss general second-order intertemporal logarithmic productivity index is derived by collecting $(T^s - T^r)$ terms in equation (7-10) to yield

$$\mu_{sr}^2 = \mu_{sr}^1 - \tfrac{1}{4} \left(\Sigma_k \Sigma_l \frac{\partial^2 f}{\partial \ln P_k \, \partial \ln P_l} \bigg|_{\substack{P_k = P_k^s \\ P_l = P_l^s}} \right.$$

$$\left. - \frac{\partial^2 f}{\partial \ln P_k \, \partial \ln P_l} \bigg|_{\substack{P_k = P_k^r \\ P_l = P_l^r}} \right)(\ln P_k^s - \ln P_k^r)(\ln P_l^s - \ln P_l^r) \quad (7\text{-}12)$$

where in this case we assume that $D_i^s = D_i^r$ but $T_t^s \neq T_t^r$.

Unlike the first-order equations (7-7 and 7-8), the second-order indexes require Hessian derivatives that in turn require a complete structural model of agricultural production. If the Denny-Fuss second order accounting equation (7-10) is appropriate, the bias that results from using the first order Fisher-Tornqvist intertemporal index can be measured by

$$\text{BIAS}(\mu_{sr}^l) = \frac{1}{4}\left(\Sigma_k \Sigma_l \frac{\partial^2 f}{\partial \ln P_k \, \partial \ln P_l}\bigg|_{\substack{P_k = P_k^s \\ P_l = P_l^s}}\right.$$

$$\left. - \frac{\partial^2 f}{\partial \ln P_k \, \partial \ln P_l}\bigg|_{\substack{P_k = P_k^r \\ P_l = P_l^r}}\right)(\ln P_k^s - \ln P_k^r)(\ln P_l^s - \ln P_l^r) \quad (7\text{-}13)$$

whereas the interspatial index bias is given by

$$\text{BIAS}(\theta_{sr}^l) = \frac{1}{4}\left(\Sigma_k \Sigma_l \frac{\partial^2 f}{\partial \ln P_k \, \partial \ln P_l}\bigg|_{\substack{P_k = P_k^s \\ P_l = P_l^s}}\right.$$

$$\left. - \frac{\partial^2 f}{\partial \ln P_k \, \partial \ln P_l}\bigg|_{\substack{P_k = P_k^r \\ P_l = P_l^r}}\right)(\ln P_k^s - \ln P_k^r)(\ln P_l^s - \ln P_l^r) \quad (7\text{-}14)$$

Intertemporal and Interspatial Translog Cost Function

The second-order biases in equations (7-13) and (7-14) will vanish if the second derivatives of cost with respect to input prices are the same for all regions and time. To examine the size of the second-order effects, we must construct an econometric model that permits the estimation of time- and space-dependent second derivatives. Such a model is described as

$$\ln C = A + Q + \sum_i^{m-1} \alpha_i^0 D_i + \sum_i^{m-1} \theta_i^0 \ln T$$

$$+ \sum_k^n \left(\beta_k + \sum_i^{m-1} \alpha_{ki}^1 D_i + \sum_i^m \theta_{ki}^1 D_i \ln T\right) \ln P_k$$

$$+ \frac{1}{2}\sum_k^n \sum_l^n \left(\gamma_{kl} + \sum_i^{m-1} \alpha_{kli}^2 D_i + \sum_i^m \theta_{kli}^2 D_i \ln T\right) \ln P_k \ln P_l \quad (7\text{-}15)$$

where n = counts inputs (k, l = capital, labor, energy, materials, and acreage)

m = counts spatial areas (i = Indiana, Iowa, Missouri, and Ohio, with Illinois serving as the base region)

A = intercept

Q = gross output

D_i = interspatial binary variables

α_i^0 = zero-order interspatial difference coefficients

T = intertemporal index (weather control variable)

θ^0 = zero-order intertemporal coefficient

β_k = first-order price coefficients

α_{ki}^1 = first-order interspatial difference coefficients

θ_k^1 = first-order intertemporal coefficient

γ_{kl} = second-order price coefficients

α_{kli}^2 = second-order interspatial difference coefficients

θ_{kli}^2 = second-order intertemporal coefficients

The translog model (7-15) provides for zero-, first-, and second-order differences due to interspatial and intertemporal technology differences.

Symmetry and linear homogeneity in prices are imposed on equation (7-15) with the following parametric restrictions

$\gamma_{kl} = \gamma_{lk}$
$\alpha_{kli}^2 = \alpha_{lki}^2$ for each i
$\theta_{kli}^2 = \theta_{lki}^2$ for each i
$\Sigma_k \beta_k = 1$
$\Sigma_k \alpha_{ki}^1 = \Sigma_k \theta_{ki}^1 = 0$ for each i
$\Sigma_k \gamma_{kl} = \Sigma_l \gamma_{kl} = \Sigma_k \Sigma_l \gamma_{kl} = 0$
$\Sigma_k \alpha_{kli}^2 = \Sigma_l \alpha_{kli}^2 = \Sigma_k \Sigma_l \alpha_{kli}^2 = 0$ for each i
$\Sigma_k \theta_{kli}^2 = \Sigma_l \theta_{kli}^2 = \Sigma_k \Sigma_l \theta_{kli}^2 = 0$ for all i \hfill (7-16)

In estimation we make use of the following cost share equations implied by Shephard's lemma (1953, 1970) and estimate a system of cost function and share equations

$S_k = \beta_k + \Sigma_i \alpha_{ki}^1 D_i + \Sigma_i \theta_{ki}^1 D_i \ln W$
$\quad + \Sigma_l (\gamma_{kl} + \Sigma_i \alpha_{kli}^2 D_i + \Sigma_i \theta_{kli}^2 D_i \ln W) \ln P_l$ \hfill (7-17)

DATA CONSTRUCTION USING FIRM ENTERPRISE DATA SYSTEM BUDGETS

Firm Enterprise Data System

To the best of our knowledge, the data set we employ in this paper has never been used in the analysis of agricultural productivity. Because of its short time dimension, the Firm Enterprise Data System (FEDS) is not suited to the measure of intertemporal productivity; however, its detailed input and output list and fine regional resolution suit FEDS well to interspatial comparisons. The origin of the FEDS can be traced to the Agriculture and Consumer Protection Act of 1973. This act established target prices for wheat, feed grains, and cotton and permitted the target prices to be adjusted on the basis of an index of prices paid by producers for production inputs, wages, interest, and taxes. The same act directs the secretary of agriculture to carry out studies of cost and production for wheat, feed grains, cotton, and dairy products; these studies thus formed the basis for assembling FEDS.

The Firm Enterprise Data System divides the United States into ten major regions, then into states and finally into FEDS areas. Normally, budgets exist at the most disaggregate level, which is the FEDS area. This disaggregation permits econometric models to be constructed at the FEDS region level, the state level, or in some cases even the substate level. The ability to construct region-specific models is likely to be important, because the technology of agricultural production for a crop can differ substantially even among states. Moreover, many government policies, economic variables, and weather conditions are region-specific, and thus dictate finely detailed models.

It is important to emphasize, and to note by way of contrast, that the level of input detail contained in a single production budget far exceeds the detail found in most industrial sector models of the U.S. economy. Disaggregation of industrial model capital normally proceeds no further than structures and equipment, but the FEDS capital inventory often includes fifty to sixty items. Furthermore, quantities are measured in the ideal economic unit of service hours of particular capital types.

Aggregation of FEDS Budget Data to the KLEMA Level

The first step in creating a five-factor data set using the individual budgets is to extract and classify budget data. To begin, we classify the various budget items into one of five categories corresponding to capital (K), labor (L), energy (E), purchased material inputs (M), and acreage (A). This is relatively straightforward and proceeds as follows.

Capital input expenditure, given on a machine-hour basis, is the aggregate of depreciation, interest, insurance, and tax expenses. We exclude machine repair and lubricant expenditures, which we treat as purchased inputs, and classify fuel cost as an energy expenditure. Total capital expenditures and total capital hours (quantity) are computed as the sum of individual capital components. The implicit capital service price is computed as the ratio of total capital expenditures to total hours.

Labor and energy price and quantity data are also available directly as line items in a budget. Labor quantity (hours) and wage rate (dollars per hour) are taken from the annual labor requirements section of the budget. Annual gasoline, LP gas, and diesel fuel requirements, measured in gallons, provide the quantity measure for disaggregate energy components. To construct an aggregate of individual energy components, we share weight the price index for gasoline, LP gas, and diesel. Total energy expenditure is derived as the sum of expenditures on individual energy components and the imputed energy quantity is computed as the ratio of total energy expenditures to the share-weighted aggregate energy price index.

Purchased inputs (materials) include herbicides, insecticides, seeds, nitrogen, phosphate, lime, and drying costs, as well as miscellaneous inputs, such as machinery repair and lubricants. To create a price aggregate of purchased inputs, we share weight individual components in the same manner as the construction of the energy price aggregate. Aggregate material expenditure is determined as the sum of individual material expenditures; the implicit quantity index is computed as the ratio of aggregate material expenditures and the share-weighted aggregate material price index.

Aggregate expenditure on land equals the sum of the budget land charge and general farm overhead. This expenditure is calculated on a per acre basis and serves as our per acre land service price variable.

The above quantity data represent factor inputs utilized per acre. Multiplying by the number of acres planted for any given FEDS area generates the total expenditures on each category of inputs. These data comprise our econometric FEDS data set.

Finally, we must construct a measure of weather variation. An examination of weather station data supplied by the National Oceanic and Atmospheric Administration reveals significantly different interspatial weather patterns for the months of July 1974 and July 1978. Specifically, during July 1974 the Midwest experienced very high temperatures and very low rainfall. During the other months of the growing season weather patterns between the two years were quite similar. Given the differences, we have formed an aridity variable using the ratio of July precipitation to July temperature for each of our FEDS areas for each year.

RESULTS

In this section we present results based on FEDS data and the models developed above. In particular, we evaluate the first-order growth accounting equation to compute the Fisher-Tornqvist intertemporal and interspatial productivity indexes. Secondly, we present our results using the general second-order accounting equation. Because this latter index requires structural estimation of a corn belt cost function model, we also present in this section the econometric findings that underlie the second-order index. Using equations (7-13) and (7-14) and employing the second-order effects obtained from the econometric model, we compute the bias associated with the first-order growth accounting equations.

Intertemporal and Interspatial Productivity: Fisher-Tornqvist Comparisons

Before considering the intertemporal and interspatial productivity indexes, it is useful to examine an example of index computation for corn. The basic components needed in the index computation are displayed in tables 7-1 through 7-4. Tables 7-1 and 7-2 present input quantity and price data used to compute the intertemporal index for Illinois and the interspatial index for Indiana. Table 7-3 presents the cost and output data and table 7-4 presents the values of the indexes.

We begin by computing the primal growth accounting measure of productivity. To do so, we derive the logarithmic change in output: Illinois produced 830,759 and 1,191,056 (thousand) bushels of corn in 1974 and 1978 so that the logarithmic output change is 0.3603. As we are working with logarithms, exponentiation yields the percentage change (43 percent) in Illinois corn output, from 1974 to 1978.

The next step is to determine the growth in aggregate inputs. To do this, we turn to table 7-1 and compute the logarithmic quantity changes for each of the five inputs and present these results in row 3. An example will clarify. In 1974 Illinois used 36,449 thousand manhours of labor, whereas in 1978 labor manhours decreased to 29,729. The logarithmic difference is equal to -0.2038, an 18 percent decrease in manhours.

In a similar fashion we compute the remaining logarithmic input changes as -0.1593, -0.1382, -0.3194, and 0.0856 for capital, energy, materials, and land, respectively. To derive the change in the total quantity index, we share weight individual component changes, using the

TABLE 7-1. NUMERICAL EXAMPLE FOR COMPUTING THE
INTERTEMPORAL PRODUCTIVITY INDEX

Row	Data description	Capital	Labor	Energy	Materials	Acreage
	Intertemporal Productivity (Primal) Comparison Index Data					
1	1978 quantity, Illinois	40,129	29,729	219,145	326,865	10,723
2	1974 quantity, Illinois	47,059	36,449	251,636	449,871	9,843
3	Logarithmic quantity change	−0.1593 (0.8527)[a]	−0.2038 (0.8156)	−0.1382 (0.8709)	−0.3194 (0.7266)	0.0856 (1.0894)
4	Average cost share (1974 and 1978)	0.1208	0.0418	0.0420	0.3277	0.4677
5	Share weighted quantity change	−0.0192	−0.0085	−0.0058	−0.1047	0.0400
	Intertemporal Productivity (Dual) Comparison Index Data					
6	1978 price, Illinois	7.89	3.36	0.47	2.60	102.70
7	1974 price, Illinois	4.83	2.37	0.33	1.39	99.75
8	Logarithmic price change	0.4907 (1.6335)	0.3491 (1.4178)	0.3536 (1.4242)	0.6262 (1.8705)	0.0291 (1.0295)
9	Average cost share (1974 and 1978)	.1208	.0418	.0420	.3277	.4677
10	Share weighted price change	0.0593	0.0146	0.0149	0.2052	0.0136

[a] Numbers in parentheses denote the percent of 1978 input quantity or price relative to 1974. Illinois farmers were using 15 percent less capital in 1978, but 8 percent more acreage, whereas the input prices increased by 63, 41, 42, 87, and 2 percent for capital, labor, energy, material, and acreage, respectively.

average of the 1974 and 1978 input cost shares specified in row 4. This yields −.0982. When exponentiated this number yields .9065, which simply states that the aggregate share-weighted quantity index in 1978 was approximately 90 percent of the 1974 levels.

The differences between the logarithmic rate of change in output and input (.4585 = .3603 − (−.0982)) is the logarithmic rate of productivity change. Exponentiating yields 1.58, which may be interpreted as a 58 percent increase in the output-input ratio between 1974 and 1978. In other words, Illinois farmers could produce 1.58 times as much corn in 1978 as in 1974 using the 1974 input bundle.

To compute the intertemporal productivity index dual to the primal productivity measure above, we consider the difference between the logarithmic change in costs and the sum of the logarithmic change in output and input prices. From table 7-3 Illinois cost of production was $2,003,639,000 and $2,472,471,000 in 1974 and 1978, respectively—equal to a 23 percent cost increase—while output increased by 43 percent.

TABLE 7-2. NUMERICAL EXAMPLE FOR COMPUTING THE INTERSPATIAL PRODUCTIVITY INDEX

Row	Data description	Capital	Labor	Energy	Materials	Acreage
	Interspatial Productivity (Primal) Comparison Index Data					
1	1978 quantity, Indiana	20,999	17,554	124,055	212,903	5,917
2	1978 quantity, Illinois	40,129	29,729	219,145	326,865	10,723
3	Logarithmic quantity differential	−0.6476 (0.5233)	−0.5268 (0.5905)	−0.5690 (0.6661)	−0.4287 (0.6514)	−0.5946 (0.5518)
4	Average cost share (Indiana and Illinois)	.1319	.0419	.0415	.3383	.4464
5	Share weighted quantity differential	−0.0854	−0.0221	−0.0236	−0.1450	−0.2654
	Interspatial Productivity (Dual) Comparison Index Data					
6	1978 price, Indiana	8.48	3.24	.43	2.05	99.28
7	1978 price, Illinois	7.89	3.36	.47	2.60	102.70
8	Logarithmic price differential	0.0721 (1.0748)	−0.0364 (.9643)	−0.0889 (.9149)	−0.2377 (.7884)	−0.0339 (.9667)
9	Average cost share (Indiana and Illinois)	.1319	.0419	.0415	.3383	.4464
10	Share weighted price differential	0.0095	−0.0015	−0.0037	−0.0804	−0.0151

Note: Numbers in parentheses denote the percent difference in Indiana input quantity or price relative to Illinois. Indiana farmers were using 52 percent less capital than Illinois farmers. Similarly Indiana farmers paid approximately $.04 less per gallon of fuel, reflecting their energy price index, which is 91 percent of Illinois'.

TABLE 7-3. COST AND OUTPUT DATA FOR COMPUTING COMPARISON INDEXES

Row	Data description	Cost (in thousands of dollars)	Output (in thousands of bushels)
1	Illinois 1978	2,472,471	1,191,056
2	Illinois 1974	2,003,639	830,759
3	Indiana 1978	1,312,641	637,193
4	Logarithmic change, Illinois (1974–1978)	0.2103	0.3603
5	Logarithmic differential (Indiana, Illinois)	−0.6332	−0.6255
6	Percentage increase, Illinois (1974–78)	23[a]	43
7	Percentage difference (Indiana, Illinois)	53	53[b]

[a] Computed as $(C_{1978}/C_{1974}) \cdot 100$

[b] Computed as $(Q_{\text{Indiana}}/Q_{\text{Illinois}}) \cdot 100$

To compute the change in the aggregate price index, we compute a weighted average of rates of growth of its component (KLEMA) prices. In row 10 of table 7-1 we show, for example, that labor's share-weighted growth is 0.0146. To derive this number we take the logarithmic change in the labor wage rate, which increased from 2.37 to 3.36 per hour from 1974 to 1978. This change equals 0.3491, or a 41 percent increase in labor cost weighted by labor cost share (.0418). Other component price changes are similarly derived. Adding up weighted average rates of growth yields 0.3076, which upon exponentiation yields 1.36. In other words, the aggregate price index increased by about 36 percent from 1974 to 1978.

We compute the intertemporal dual productivity measure as -0.4576 (.2103 $-$ .3603 $-$ .3076), which corresponds to the log change in cost less the log change in output and log change in the input price index. Exponentiating this value yields .6328, implying that Illinois farmers could, with constant input prices, produce the 1974 output in 1978 at only 63 percent of the 1974 production cost.

To compute the interspatial comparison index, we follow the same procedure used to compute the intertemporal index, except that we now compare the efficiency of Indiana corn production relative to Illinois. We begin by considering levels of corn production. In 1978 Illinois produced 1,191,056 (thousand) bushels of corn, whereas Indiana produced 637,193 (thousand) bushels. Thus Indiana produced about 54 percent of Illinois output in corn. Given Indiana's smaller level of output, it is not surprising that it uses less inputs, as shown in table 7-2. To determine Indiana's productivity relative to Illinois' we compute the relative difference in the aggregate input index of the two states. This index is approximately 58.19, so that, while Indiana's smaller scale farming is able to produce 53 percent of Illinois' corn output, it takes an Indiana farmer 58.19 percent of the levels of Illinois' inputs to do so. Indiana is thus relatively less productive.

How much more productive would Indiana farmers have to become to remove these interspatial productivity differences? Assuming that the same relative inputs are used, it is easy to show that Indiana corn output would have to increase to approximately 693,045 thousand bushels, an increase of about 8 percent. In this case the exponentiated version of the interspatial index is 1, denoting productivity equality between Indiana and Illinois farmers.

The interspatial dual productivity measure is computed as the logarithmic difference between Indiana's and Illinois' costs less the sum of the logarithmic difference in output and aggregate input prices. From table 7-4 both Indiana's cost and its output levels are approximately 53

TABLE 7-4. COMPUTATION OF COMPARISON INDEXES

	Share weight		Logarithmic change		
Index	Input quantity	Input price	Output	Cost	Computed value
Intertemporal productivity (primal)	−0.0982 (0.9065)		0.3603 (1.4338)		0.4585 (1.5817)
Intertemporal relative cost efficiency (dual)		0.3076 (1.3602)	0.3603 (1.4338)	0.2103 (1.2340)	−0.4576 (0.6328)
Interspatial productivity (primal)	−0.5415 (0.5819)		−0.6255 (0.5350)		−0.0840 (0.9194)
Interspatial relative cost efficiency (dual)		−0.0912 (0.9128)	−0.6255 (0.5350)	−0.6332 (0.5309)	0.0835 (1.0871)

Note: Numbers in parentheses denote percentage change. For example, using table 7-5, the quantity of Illinois corn output is 830,759 and 1,191,056 thousand bushels in 1974 and 1978, respectively, so that the 1978-to-1974 output ratio equals 1.4338. Note that this is equivalent to Exp(0.3603).

percent of Illinois'. The share-weighted aggregate price index for Indiana is about 9 percent less than for Illinois. Computing the overall efficiency index yields 0.0835 ($\log \Delta$ cost − $\log \Delta$ output − $\log \Delta$ aggregate input price) and exponentiating yields 1.0871. Since this index is greater than 1, Indiana is about 8 percent less efficient than Illinois.

From the primal measure of interspatial productivity given above, we know that if Indiana farmers increased 1978 output to 693,045 thousand bushels, holding inputs constant, they would be as efficient as farmers in Illinois. Using the dual interspatial index, it is also clear that this same output increase is sufficient to remove the interspatial productivity difference. That is, the interspatial measure ($\log \Delta$ cost − $\log \Delta$ output − $\log \Delta$ input price) yields a zero value (−.6332 − (−.5415) − (−.0912)) when Indiana output rises to 693,045.

Turning now to the computed first-order productivity indexes, we present in table 7-5 the intertemporal and interspatial indexes for both corn and soybeans in the corn belt. Between 1972 and 1978, Indiana experienced the greatest increase in productivity for both corn and soybeans, with Missouri, Ohio, Iowa, and Illinois all enjoying substantial weather-induced productivity gains.

Considering interspatial differences and using Illinois as the reference base, we see that Iowa was the most productive in both 1974 and 1978, while Ohio and Missouri had the poorest relative performance in 1974 and 1978, respectively. It is interesting to note that from 1974 to 1978 disparity in productivity levels declined somewhat, but that the ordinal rankings of the states remained largely unchanged.

TABLE 7-5. PRODUCTIVITY DIFFERENCES USING
FIRST-ORDER APPROXIMATION

State	Intertemporal — $\text{EXP}(\mu_{t,t+1}^1)$ 1974–1978	Interspatial — $\text{Exp}(\theta_{io}^1)$	
		1974	1978
Corn			
Illinois	0.6323	1.0000	1.0000
Indiana	0.5367	1.2734	1.0876
Iowa	0.5936	0.9311	0.8673
Missouri	0.5694	1.2994	1.1749
Ohio	0.5893	1.2676	1.1979
Soybeans			
Illinois	0.6381	1.0000	1.0000
Indiana	0.6261	1.0880	1.0608
Iowa	0.6588	0.8393	0.8583
Missouri	0.6804	1.1171	1.1991
Ohio	0.6206	1.0465	1.0263

Interspatial Productivity: Second-order Approximation

In this analysis we limit our consideration to second-order biases due to spatial differences in the belief that such differences are a result of systematic economic and agronomic influences, whereas second-order intertemporal differences are a result primarily of random weather fluctuations. The Denny-Fuss second-order growth accounting equation (7-11) is calculated using second-order share elasticity estimates ($\partial \ln C / \partial \ln P_i \partial \ln P_j$) that are permitted to vary by region. These share elasticities are estimated using equation (7-17), where $\gamma_{kl} + \theta_{kli}^2 \ln T$ is the elasticity estimated for the base region and $\gamma_{kl} + \alpha_{kli}^2 + \theta_{ki}^2 \ln T$ is the estimate for the ith region. Unfortunately, degrees of freedom limitations prohibit share elasticities to vary by state. These limitations force us to group the five states of the corn belt into two regions (assuming common second-order effects in each).

We consider two groupings of the states: (a) Iowa, Missouri and Illinois, Indiana, Ohio and (b) Ohio, Indiana and Illinois, Missouri, Iowa. These groupings assume that contiguous states have common second-order effects. Equation (7-15) is estimated twice for each of the two groupings, once for corn production and once for soybeans. The first- and second-order estimates of interspatial productivity and the resulting bias for each crop and state grouping are displayed in table 7-6. The results presented in table 7-6 clearly demonstrate the second-order biases to be small and thus, for this particular data set, support the use of simple index number-based productivity measurement techniques.

TABLE 7-6. ESTIMATES OF FIRST- AND SECOND-ORDER INTERSPATIAL PRODUCTIVITY FOR CORN AND SOYBEAN PRODUCTION

Region	1974			1978		
	First-order	Second-order	Bias	First-order	Second-order	Bias
			Corn			
Iowa, Missouri	1.000	1.000		1.000	1.000	
Illinois, Indiana, Ohio	1.146	1.143	0.0020	1.170	1.169	0.0008
			Soybeans			
Iowa, Missouri	1.000	1.000		1.000	1.000	
Illinois, Indiana, Ohio	1.137	1.139	−0.0010	1.063	1.063	0.0001
			Corn			
Ohio, Indiana	1.000	1.000		1.000	1.000	
Illinois, Missouri, Iowa	0.762	0.762	0.0002	0.839	0.839	0.0000
			Soybeans			
Ohio, Indiana	1.000	1.000		1.000	1.000	
Illinois, Missouri, Iowa	0.881	0.880	0.0004	0.935	0.935	0.0000

CONCLUSION

The importance of improving the methods for measuring technological change is likely to increase in the future, as higher evaluatory standards are imposed on agricultural policy. Because agricultural policy in the past has not been oriented towards highly aggregate issues, it is unlikely that studies using macrodata will yield important insights into an intrinsically micronature of U.S. agriculture. Agricultural policy has long emphasized particular regions of the country and targeted particular crops. Consequently, we conclude that an aggregate approach largely ignores individual agents' behavior in a highly diversified agricultural economy and that a microorientation is more appropriate.

Most researchers would agree with our position advocating the analysis of agricultural productivity at the microlevel. Nevertheless, the paucity of microdata forces many researchers to employ aggregated data sets. In some instances, detailed surveys have been conducted and can form the basis for microanalysis, but all too often these surveys are region- (most probably state-specific) or crop-specific, thus preventing a large-scale microanalysis across the entire United States covering all major field crops. Our survey of potential databases has led us to believe that FEDS can be a valuable tool not only for the analysis of agricultural productivity but for the analysis of many other agricultural issues as well. We have found no alternate data set that rivals FEDS in terms of region, crop, and input disaggregation.

The results of our study are meant to illustrate data, technique, and interspatial productivity differences rather than to highlight the weather-induced productivity gains from 1974 to 1978 in the U.S. corn belt. Controlling for weather variation, we find substantial interspatial productivity differences in this relatively homogeneous region, and that the simple Fisher-Tornqvist growth accounting equation is not seriously biased by its neglect of second-order effects. The small second-order bias holds the promise that the measurement of agricultural productivity could be conducted using microdata without resorting to sophisticated econometric techniques.

REFERENCES

Berndt, E. R., and M. A. Fuss. 1982. "Productivity Measurement Using Capital Asset Valuation to Adjust for Variations in Utilization," Working Paper No. 895 (Cambridge, Mass., National Bureau of Economic Research).

Denison, E. F. 1969. "Some Major Issues in Productivity Analysis: An Examination of Estimates by Jorgenson and Griliches," *Survey of Current Business* vol. 49, no. 5, pp. 1–28.

Denny, M., and M. Fuss. 1983a. "A General Approach to Intertemporal and Interspatial Productivity Comparisons," *Journal of Econometrics* vol. 23, no. 3, pp. 315–330.

———, and M. Fuss. 1983b. "The Use of Discrete Variables in Superlative Index Number Comparisons," *International Economic Review* vol. 24, no. 2, pp. 419–421.

Diewert, W. E. 1976. "Exact and Superlative Index Numbers," *Journal of Econometrics* vol. 4, no. 2, pp. 115–145.

———. 1978. "Superlative Index Numbers and Consistency in Aggregation," *Econometrica* vol. 46, no. 4, pp. 883–900.

———. 1979. "The Economic Theory of Index Numbers: A Survey," Discussion Paper No. 79-09 (Vancouver, British Columbia, Department of Economics, University of British Columbia).

Jorgenson, D. W., and Z. Griliches. 1967. "The Explanation of Productivity Change," *The Review of Economic Studies* vol. 34, no. 99, pp. 249–282.

Krenz, R. D. 1975a. "The USDA Firm Enterprise Data System: Capabilities and Applications," *Southern Journal of Agricultural Economics* vol. 7, no. 1, pp. 33–38.
———. 1975b. "Current Efforts at Estimation of Costs of Production in ERS," *American Journal of Agricultural Economics* vol. 57, no. 5, pp. 929–933.

Nadiri, M. I. 1970. "Some Approaches to the Theory and Measurement of Total Factor Productivity: A Survey," *Journal of Economic Literature* vol. 8, no. 4, pp. 1137–1177.

Shephard, R. W. 1953. *Cost and Production Functions* (Princeton, N.J., Princeton University Press).
———. 1970. *Theory of Cost and Production Functions* (Princeton, N.J., Princeton University Press).
Solow, R. 1957. "Technical Change and the Aggregate Production Function," *Review of Economics and Statistics* vol. 39, no. 3, pp. 312–320.

8
AN ECONOMETRIC METHODOLOGY FOR MULTIPLE-OUTPUT AGRICULTURAL TECHNOLOGY:
AN APPLICATION OF ENDOGENOUS SWITCHING MODELS*

WALLACE E. HUFFMAN

As noted by Capalbo and Vo in chapter 3, econometric productivity (or production function) studies have historically been of a single-output, several-input type. The early studies employed the primal approach, fitting (aggregate) output to several variable inputs. Later studies employed the dual approach by fitting the profit function or input demand functions derived from underlying cost or profit functions.[1]

Many of these earlier studies have serious deficiencies as frameworks within which to learn about agricultural productivity. First, the single-output methods assume that technology is separable between inputs and outputs. This implies that decisions on optimal relative output (input) quantities are unaffected by input (output) prices (Lau, 1972; Christensen, 1975). For example, the relative quantity of corn to soybeans supplied is unaffected by the price of fertilizer. More importantly, production is assumed to be either nonjoint in inputs or joint in a special way. When applied to agriculture, nonjointness in inputs ignores technical interdependencies among outputs and interdependencies induced by allocable fixed factors (Christensen, 1975; Lau, 1972; Shum-

---

*Arne Hallam, Richard Shumway, and other participants in the conference made useful suggestions on an earlier draft of this paper. Financial assistance was obtained from the USDA-CSRS under a cooperative agreement and from the Iowa Agriculture and Home Economics Experiment Station, Journal Paper No. J-12309 of the Iowa Agriculture and Home Economics Experiment Station, Ames, Iowa, Project No. 2516.

[1] Studies employing the primal methodology include Cobb and Douglas (1928); Arrow, Chenery, Minhas, and Solow (1961); Griliches (1963a, 1963b, 1964); Cline (1975); Evenson (1968, 1980); Huffman (1976); and Akino and Hayami (1974). Studies employing the dual methodology include Berndt and Christensen (1973, 1974); Berndt and Wood (1975); Evenson and Binswanger (1980); Lau and Yotopoulos (1971, 1972); Yotopoulos and Lau (1973); Lopez (1980); Antle and Aitah (1983); Jamison and Lau (1982); and Woodland (1977).

way, Pope, and Nash, 1984). If both crop and livestock outputs are supplied, production is almost certainly joint in inputs. A potentially fertile area of agricultural productivity research seems to be the disaggregation of outputs so that the relative and absolute supply response to exogenous variables—for example, prices, allocable fixed factors, and public sector agricultural research—can be quantified.

Attempts to employ multiple-output, multiple-input production technology in econometric studies have met with only mixed results. Just, Zilberman, and Hockman (1983) were successful when they used the primal approach to estimate production functions for seven Israeli vegetables. Muller (1980), Shumway (1983a,b), Saez and Shumway (1983), Weaver (1983), and Weaver and Lass (1983) used the dual methodology and a multiple-output profit function to derive estimates of output supply and input demand functions. Muller, using data for paper manufacturing, and Shumway (1983a), using data for agriculture, obtained estimates for some supply elasticities that were negative, which is inconsistent with theory. In general, researchers have obtained disappointing results when they have attempted to disaggregate output.

My interest in methodology for multiple-output models was stimulated by an attempt to fit a set of output supply and input demand functions to state aggregate pooled time series/cross-sectional data for U.S. cash grain farms. Major outputs of these farms are feed grains, wheat, soybeans, cotton, and livestock, but some states do not produce one or more of these outputs, for example, soybeans and cotton. When these outputs are not produced, the decision functions on outputs and inputs are different. These corner and interior solution combinations are the source of multiple economic and econometric structures for agricultural supply and demand functions. When least-squares econometric methods are applied to estimate a traditional set of supply and demand or profit (cost) share equations under these conditions, the estimator is statistically inconsistent (Maddala, 1983; Heckman, 1979; Hanemann, 1984).

The purpose of this paper is to present an econometric model that accounts for the different economic structures of farmers' choice functions on outputs and inputs when some outputs are not produced. It employs the methodology of endogenous switching in a multivariate econometric model (Maddala, 1983, pp. 283–289; Amemiya, 1974). Multiple sets of conditional supply and demand (share) equations can be combined into one set of unconditional supply and demand (share) equations when each set is weighted by its probability of occurrence. These unconditional supply and demand equations can then be estimated inexpensively by multiple-stage application of least squares (Maddala, 1981, 1983; Fishe, Trost, and Lurie, 1981), yielding inexpensive and consistent tobit-type estimates of the coefficients. Estimation of one system of unconditional choice functions has advantages over estimation of several systems of conditional choice functions when the observation units are distributed unequally among the different economic structures. This advantage will in some cases be outweighed by additional problems with near multicollinearity.

The paper has the following organization. The first section presents a brief discussion of the economic model. The econometric model for dealing with multiple corner solutions in a system of equations is presented in the second section. The third section contains a discussion of procedures applicable for estimating the model inexpensively. Conclusions are presented in the final section.

THE ECONOMIC MODEL

Agriculture is a competitive multiproduct industry. Farmers may be represented as profit maximizers who produce several outputs. Outputs require many of the same fixed and variable inputs, and production may be joint. Some examples of jointness are caused by technical interdependencies: land, machinery, human capital (Shumway, Pope, and Nash, 1984). The farmer may have in the short or intermediate horizon a fixed quantity of these inputs that can be allocated to several different outputs. Thus, production technologies that are technically independent become interdependent through a resource constraint. Furthermore, an increment to public agricultural research expenditures—for example, biotechnology—may increase the quantity supplied of several plant and livestock outputs.

Standard regularity conditions for agricultural technology are maintained. Firms are assumed to make m current choices on inputs and outputs, denoted Y_i, $i = 1, \ldots, m$. Following the netput concept (Varian, 1978), $Y_i > 0$, $i = 1, \ldots, j$ denotes outputs and $Y_i < 0$, $i = j + 1, \ldots, m$ denotes inputs. In making choices on Y, firms are assumed to treat a set of factors Z_k, $k = 1, \ldots, q$ as fixed.

The relationship among choices (Y) and fixed factors is represented in primal form by a well-behaved transformation function, $F(Y, Z) = 0$. In addition, if (expected) prices of inputs and outputs ($P^e > 0$) are exogenous to firm-level decisions and firms maximize expected profit,[2] then by

---

[2] In this model, farmers are assumed to face uncertain prices for outputs when input decisions are made. The transformation function is deterministic. Farmers are assumed to be risk neutral about their preferences on price uncertainty. Incorporating production uncertainty and a risk averse attitude toward uncertainty would greatly complicate duality theory.

duality theory (Diewert, 1973; Lau, 1972) a well-behaved profit function exists that relates maximized expected profit to the prices of current choices and the fixed factors:

$$\pi^e = \pi^e(P^e, Z) \tag{8-1}$$

The expected profit function is twice continuously differentiable, convex, linear homogeneous, and monotonic in P^e (Diewert, 1973; Lau, 1972).

Applying Hotelling's lemma, a system of (profit) maximizing output supply and input demand functions is obtained by differentiating π^e with respect to P^e:

$$\frac{\partial \pi^e}{\partial P_i^e} = Y_i^*(P^e, Z), \qquad i = 1, \ldots, m \tag{8-2}$$

Thus, choices are determined by expected prices (P^e) and the Zs. These choice functions are also homogeneous of degree zero in all prices.[3]

Comparative static results follow directly from the properties of π^e

$$\frac{\partial^2(\pi^e)}{\partial P_i^e \partial P_j^e} = \frac{\partial Y_i^*}{\partial P^e} \begin{Bmatrix} > 0 \text{ for } i = j \\ \gtreqless 0 \text{ for } i \neq j \end{Bmatrix}, i,j = 1, \ldots, m$$

and

$$\frac{\partial^2 \pi^e}{\partial P_i^e \partial Z_k} = \frac{\partial Y_i^*}{\partial Z_k} \gtreqless 0, \qquad i = 1 \ldots m, k = 1 \ldots q \tag{8-3}$$

Convexity of π^e or concavity of $F(\cdot)$ implies that supply curves have positive slopes and demand curves have negative slopes. Because cross-partial derivatives of π^e are independent of the order of differentiation, cross-price effects on current choices are symmetric across choice functions, that is, $\partial Y_i^* / \partial P_j^e = \partial Y_j^* / \partial P_i^e$, $i \neq j$; $i,j = 1, \ldots, m$. Within-equation cross-price effects are, however, indeterminant in sign.

THE ECONOMETRIC MODEL

In the econometric model, m decisions on inputs and outputs are assumed to be made jointly. Some (two or more) of these decisions are assumed to result in extreme values (zeros) or corner solutions for optimal choices on outputs and (or) inputs. To make the model less abstract,

---

[3] When some optimal values of choice variables are zero, not all parameters of the transformation function can be uniquely identified from the profit function.

I limit the number of choices to a maximim of nine; I limit the number of choices that have extreme values to two outputs; and I designate the profit function to be normalized quadratic.

The normalized quadratic profit function is one of several functional forms providing a second-order approximation to any profit function (Diewert, 1973). It has the advantage of yielding simple supply and demand functions that have real quantities (or quantity indexes) as dependent variables and that are linear functions of relative prices. Linear aggregation of real quantities across farms is appropriate for obtaining state or regional aggregates of inputs and outputs. Elasticities of output supply and input demand are easily evaluated at sample mean values of prices and quantities. The function has the disadvantage that it is not globally convex in prices, and homogeneity of degree one in prices is imposed rather than tested.

The quadratic normalized expected profit function is

$$\pi = \alpha_0 + \sum_{i=1}^{8} b_{i0} P_i \sum_{R=1}^{Q} g_k Z_k$$
$$+ \frac{1}{2} \left\{ \sum_{i=1}^{8} \sum_{j=1}^{8} b_{ij} P_i P_j + \sum_{k=1}^{q} \sum_{l=1}^{q} d_{kl} Z_k Z_l \right\} + \sum_{i=1}^{8} \sum_{k=1}^{q} c_{ik} P_i Z_k \qquad (8\text{-}4)$$

where $\pi = \pi^e/P_0^e$; $P_i = P_i^e/P_0^e$; and α, b_{ij}, c_{ik}, and d_{kl}, for all $\forall i, j, k, l$ are unknown parameters. Applying equation (8-2), the output supply and input demand functions are obtained. Eight of these equations are assumed to be stochastic and to contain a random disturbance term V_i, $i = 1, \ldots, 8$. This basic system of eight stochastic supply and demand equations is

$$Y_i = b_{i0} + \sum_{j=1}^{8} b_{ij} P_j + Z' \underline{c} + V_i, \qquad i = 1, \ldots, 8 \qquad (8\text{-}5)$$

where Z' is a $1 \times q$ vector of fixed factors and c is a $q \times 1$ vector of unknown parameters. The disturbance vector V_i is assumed to have a population mean of a zero vector. The equation for the ninth choice variable is obtained residually

$$Y_0 = \alpha_0 + \sum_{R=1}^{Q} g_k Z_R - \frac{1}{2} \sum_{i=1}^{8} \sum_{j=1}^{8} b_{ij} P_i P_j + \frac{1}{2} \sum_{R=1}^{Q} \sum_{l=1}^{Q} h_{kl} Z_R Z_l \qquad (8\text{-}6)$$

The symmetry conditions are simply the cross-equation restrictions $b_{ij} = b_{ji}$, $i \neq j$; $i, j = 1, \ldots, 8$.

In a set of N observations, some of the observed Y_is may contain a significant share of zeros as extreme values or corner solutions. For example, in pooled time series/cross-sectional agricultural data using states as cross-sectional observations, farmers do not produce cotton and soybeans in all states and years. Farmers in some states do not produce soybeans in early years, although they are a significant output in later years. Other commodities may be produced in all states and years.

The structure of the production technology is different when one or both of the commodities are not produced. Let Y_7 and Y_8 denote the outputs possessing zeros or corner solutions, for example, cotton and soybeans, respectively. It is useful to think of partitioning N observations on production units into four mutually exclusive subsamples based upon the outcomes of the discrete choices on Y_7 and Y_8: (a) N_1 observations, where all nine Y_is have nonzero values; (b) N_2 observations, where $Y_8 = 0$ and $Y_i \neq 0$, $i = 0, \ldots, 7$; (c) N_3 observations, where $Y_7 = 0$ and $Y_i \neq 0$, $i = 0, \ldots, 6, 8$; and (d) N_4 observations, where $Y_7 = Y_8 = 0$ and $Y_i \neq 0$, $i = 1, \ldots, 6$. The system of structural equations for each subsample can now be obtained. For subsamples S_1 through S_4, the four sets of regimes for the eight conditional supply and demand functions are, respectively

$$Y_i = b_{i01} + \sum_{j=1}^{8} b_{ij1} P_j + Z' \underline{c}_1 + V_{i1}, \qquad i = 1, \ldots, 8 \tag{8-7}$$

$$Y_i = b_{i02} + \sum_{j=1}^{7} b_{ij2} P_j + Z' \underline{c}_2 + V_{i2}, \qquad i = 1, \ldots, 7, Y_8 = 0 \tag{8-8}$$

$$Y_i = b_{i03} + \sum_{j=1}^{6} b_{ij3} P_j + b_{i83} P_8 + Z' \underline{c}_3 + V_{i3}, \qquad i = 1, \ldots, 6, 8, Y_7 = 0 \tag{8-9}$$

$$Y_i = b_{i04} + \sum_{j=1}^{6} b_{ij4} P_j + Z' \underline{c}_4 + V_{i4}, \qquad i = 1, \ldots, 6, Y_7 = Y_8 = 0 \tag{8-10}$$

where an additional subscript is added to coefficients and disturbance terms so that different regimes (r) can be distinguished.

Because farmers' output and input decisions are the result of optimizing behavior, there is endogenous switching among the four multi-equation regimes. Stated another way, observations are not randomly assigned to each of the four regimes or subsamples. Thus, the disturbance terms of the conditional demand and supply functions do not have a zero mean, that is, $E(V_{ir}/S_r) \neq 0$, $i = 1, \ldots, 8$; $r = 1, \ldots, 4$. If these sets of equations were fitted directly to the observations in each of the respective subsamples by least squares, the estimated coefficients would be biased (Maddala, 1983; Heckman, 1979).

All observations have a nonzero probability of being assigned to each of the four subsamples or regimes. The condition that $Y_i > 0$ implies that

ENDOGENOUS SWITCHING MODELS   235

$V_i > -b_{i0} - \sum_{j=1}^{8} b_{ij} P_j - Z'\underline{c}$. Thus, the probability of any observation being included in each of the four regimes (subsamples) is determined by evaluating the following bivariate probabilities

$M_{11} \equiv P(S_1) = P(Y_1, \ldots, Y_8 \neq 0)$

$= P\left[V_7 > -b_{70} - \sum_{j=1}^{8} b_{7j} P_j - Z'\underline{c}_7, V_8 > -b_{80} - \sum_{j=1}^{8} b_{8j} P_j - Z'\underline{c}_8\right]$ (8-11)

$M_{10} \equiv P(S_2) = P(Y_1, \ldots, Y_7 \neq 0, Y_8 = 0)$

$= P\left[V_7 > -b_{70} - \sum_{j=1}^{8} b_{7j} P_j - Z'\underline{c}_7, V_8 \leq -b_{80} - \sum_{j=1}^{8} b_{8j} P_j - Z'\underline{c}_8\right]$ (8-12)

$M_{01} \equiv P(S_3) = P[Y_1, \ldots, Y_6 \neq 0, Y_7 = 0, Y_8 > 0]$

$= P\left[V_7 \leq -b_{70} - \sum_{j=1}^{8} b_{7j} P_j - Z'\underline{c}_7, V_8 > -b_{80} - \sum_{j=1}^{8} b_{8j} P_j - Z'\underline{c}_8\right]$ (8-13)

$M_{00} \equiv P(S_4) = P[Y_1, \ldots, Y_6 \neq 0, Y_7 = Y_8 = 0]$

$= P\left[V_7 \leq -b_{70} - \sum_{j=1}^{8} b_{7j} P_j - Z'\underline{c}_7, V_8 \leq -b_{80} - \sum_{j=1}^{8} b_{8j} P_j - Z'\underline{c}_8\right]$ (8-14)

We assume each of the triple sets of random disturbances has a trivariate normal distribution. The marginal density function is designated by $p = P_{V_i V_7 V_8}(V_i, V_7, V_8)$, $i = 1, \ldots, 6$, and the conditional mean of V_i is $E(V_i/V_7, V_8) = \gamma_{7i8} E(V_7/V_7, V_8) + \gamma_{8i7} E(V_8/V_7, V_8)$, where γ_{7i8} and γ_{8i7} are unknown regression coefficients, $i \neq 7, 8$ (Johnson and Kotz, 1972). The variances of V_7 and V_8 are represented by σ_7^2 and σ_8^2, respectively, and σ_{78} is the covariance between V_7 and V_8.

By applying the rule for evaluating conditional mean of a random variable and other rules presented in Johnson and Kotz (or Maddala, 1983), the nonzero mean values of the disturbances of the conditional demand and supply equations for each selection rule or regime become clear. The nonzero values occur because the disturbances $V_1 - V_6$ (disturbances where $Y_i \neq 0$) are generally correlated with V_7 and V_8 (disturbances where commodities sometimes have extreme values of zero) and because V_7 and V_8 are generally correlated with each other. Compactly the means of the disturbances are

$E[V_{i1}/Y_1, \ldots, Y_8 \neq 0] = \gamma_{7i8} \dfrac{A_{11}}{M_{11}} + \gamma_{8i7} \dfrac{A_{21}}{M_{11}}, \quad i = 1, \ldots, 6,$ or

$\qquad = \dfrac{A_{11}}{M_{11}}, \quad i = 7;$ or

$\qquad = \dfrac{A_{21}}{M_{11}}, \quad i = 8$ (8-15)

$$E[V_{i2}/Y_1, \ldots, Y_7 \neq 0, Y_8 = 0] = \gamma_{7i8} \frac{A_{12}}{M_{10}} + \gamma_{8i7} \frac{A_{22}}{M_{10}}, \quad i = 1, \ldots, 6, \text{ or}$$

$$= \frac{A_{12}}{M_{10}}, \quad i = 7 \quad (8\text{-}16)$$

$$E[V_{i3}/Y_1, \ldots, Y_6 \neq 0, Y_7 = 0, Y_8 > 0] = \gamma_{7i8} \frac{A_{13}}{M_{01}}$$

$$+ \gamma_{8i7} \frac{A_{23}}{M_{01}}, \quad i = 1, \ldots, 6, \text{ or}$$

$$= A_{23}/M_{01}, \quad i = 8 \quad (8\text{-}17)$$

$$E[V_{i4}/Y_1, \ldots, Y_6 \neq 0, Y_7 = Y_8 = 0] = \gamma_{7i8} \frac{A_{14}}{M_{00}} + \gamma_{8i7} \frac{A_{24}}{M_{00}}, \quad i = 1, \ldots, 6 \quad (8\text{-}18)$$

where $A_{11} = \sigma_7^2 f_7 (1 - F_8) + \sigma_{78} f_8 (1 - F_7)$
$A_{21} = \sigma_8^2 f_8 (1 - F_7) + \sigma_{78} f_7 (1 - F_8)$
$A_{12} = \sigma_7^2 f_7 (1 - F_8^*) - \sigma_{78} f_8 (1 - F_7)$
$A_{22} = \sigma_8^2 f_8 (1 - F_7) + \sigma_{78} (1 - F_8^{**})$
$A_{13} = \sigma_7^2 f_7 (1 - F_8^{**}) - \sigma_{78} f_8 (1 - F_7^*)$
$A_{23} = \sigma_8^2 f_8 (1 - F_7^{**}) + \sigma_{78} f_7 (1 - F_8^*)$
$A_{14} = \sigma_7^2 f_7 F_8 - \sigma_{78} f_8 F_7$
$A_{24} = \sigma_8^2 f_8 F_7 - \sigma_{78} f_7 F_8$
f_7 = density of $V_7 \sim N(0, \sigma_7^2)$,

evaluated at $f\left(-b_{70} - \sum_{j=1}^{8} b_{7j} P_j - Z' \underline{c}_7\right)$

f_8 = density of $V_8 \sim N(0, \sigma_8^2)$,

evaluated at $f\left(-b_{80} - \sum_{j=1}^{8} b_{8j} P_j - Z' \underline{c}_8\right)$

F_7 = distribution function of $N(0, \sigma_7^{*2})$

where $\sigma_7^{*2} = \sigma_7^2 - (\sigma_{78}^2/\sigma_8^2)$, evaluated at

$$F\left[-b_{70} - \sum_{j=1}^{8} b_{7j} P_j - Z' \underline{c}_7 - \left(\frac{\sigma_{78}}{\sigma_8^2}\right)\left(-b_{80} - \sum_{j=1}^{8} b_{8j} P_j - Z' \underline{c}_8\right)\right]$$

$$F_7^* = F\left[b_{70} + \sum_{j=1}^{8} b_{7j} P_j + Z' \underline{c}_7 - \left(\frac{\sigma_{78}}{\sigma_8^2}\right)\left(-b_{80} - \sum_{j=1}^{8} b_{8j} P_j - Z' \underline{c}_8\right)\right]$$

$$F_7^{**} = F\left[-b_{70} - \sum_{j=1}^{8} b_{7j} P_j - Z' \underline{c}_7 + \left(\frac{\sigma_{78}}{\sigma_8^2}\right)\left(-b_{80} - \sum_{j=1}^{8} b_{8j} P_j - Z' \underline{c}_8\right)\right]$$

F_8 = distribution function of $N(0, \sigma_8^{*2})$ where

$\sigma_8^{*2} = \sigma_8^2 - (\sigma_{78}^2/\sigma_7^2)$ evaluated at

$$F\left[-b_{80} - \sum_{j=1}^{8} b_{8j} P_j - Z'\underline{c}_8 - \left(\frac{\sigma_{78}}{\sigma_7^2}\right)\left(-b_{70} - \sum_{j=1}^{8} b_{7j} P_j - Z'\underline{c}_7\right)\right]$$

$$F_8^* = F\left[b_{80} + \sum_{j=1}^{8} b_{8j} P_j + Z'\underline{c}_8 - \left(\frac{\sigma_{78}}{\sigma_7^2}\right)\left(-b_{70} - \sum_{j=1}^{8} b_{7j} P_j - Z'\underline{c}_7\right)\right]$$

$$F_8^{**} = F\left[-b_{80} - \sum_{j=1}^{8} b_{8j} P_j - Z'\underline{c}_8 + \left(\frac{\sigma_{78}}{\sigma_7^2} - b_{70} - \sum_{j=1}^{8} b_{7j} P_j - Z'\underline{c}_7\right)\right]$$

Two options for estimating the coefficients of equations (7-7) through (7-10) are presented. First, the structural equations for each subsample or regime r can be corrected for the nonzero mean value of its disturbances, then the system of equations can be fitted to the N_r observations (see Amemiya, 1974; Wales and Woodland, 1983; Weaver, 1983; Hanemann, 1982). Second, after each set of equations has been corrected for nonzero mean disturbance terms, the conditional structural equations for each commodity can be combined to obtain one unconditional structural supply or demand equation. These equations can be fitted by least squares to the $N = N_1 + N_2 + N_3 + N_4$ observations.

If the N observations are evenly divided among the subsamples, then the first and second procedures are equivalent. The first one, however, will be computationally cheaper. If the N observations are not evenly divided among the subsamples, then the second procedure may be superior (Maddala, 1981). Frequently, one of the subsamples (regimes), say the rth, contains too few observations. Additional information is, however, available on these parameters from the other $N - N_r$ observations in other regimes because each of them has a nonzero probability of being included in the rth regime or subsample. Thus, by applying the second procedure, the total N degrees of freedom can be more efficiently allocated among the four economic structures for the purpose of estimating all structural coefficients with precision.

The unconditional mean value of the endogenous variables (the quantities of outputs supplied and of inputs demanded) is derived as the weighted average of the structural equations from each of the four subsamples (or regimes)

$$EY_i = \sum_{r=1}^{4} E(Y_i/S_r) P_r(S_r), \quad i = 1, \ldots, 8 \tag{8-19}$$

Furthermore, $E(Y_8/S_2) = E(Y_7/S_3) = E(Y_7/S_4) = E(Y_8/S_4) = 0$ by construction. These commodities are not produced in regimes 3 and (or) 4.

Equations for the unconditional values of each of the endogenous variables are obtained by adding a zero mean random disturbance term (Ψ_i) to each of the eight equations (8-19). These unconditional supply and demand functions are

$$Y_i = b_{i01} M_{11} + \sum_{j=1}^{8} b_{ij1} P_j M_{11} + Z' M_{11} \underline{c}_{i1}$$

$$+ b_{i02} M_{10} + \sum_{j=1}^{7} b_{ij2} P_j M_{10} + Z' M_{10} \underline{c}_{i2}$$

$$+ b_{i03} M_{01} + \sum_{j=1}^{6} b_{ij3} P_j M_{01} + b_{i83} P_8 M_{01} + Z' M_{01} \underline{c}_{i3}$$

$$+ b_{i04} M_{00} + \sum_{j=1}^{6} b_{ij4} P_j M_{00} + Z' M_{00} \underline{c}_{i4}$$

$$+ A \underline{\gamma}_i + \psi_i, \quad i = 1, \ldots, 6 \tag{8-20}$$

$$Y_7 - A_{11} - A_{12} = b_{701} M_{11} + \sum_{j=1}^{8} b_{7j1} P_j M_{11} + Z' M_{11} \underline{c}_{71}$$

$$+ b_{702} M_{10} + \sum_{j=1}^{7} b_{7j2} P_j M_{10} + Z_1 M_{10} \underline{c}_{72} + \psi_7 \tag{8-21}$$

$$Y_8 - A_{21} - A_{23} = b_{801} M_{11} + \sum_{j=1}^{8} b_{8j1} P_j M_{11} + A' M_{11} \underline{c}_{81} + b_{803} M_{01}$$

$$+ \sum_{j=1}^{6} b_{8j3} P_j M_{01} + b_{883} P_j M_{01} + Z_1 M_{01} \underline{c}_{91} + \psi_8 \tag{8-22}$$

where $A' = (A_{11} A_{12} A_{21} A_{22} A_{13} A_{23} A_{14} A_{24})'$, γ_i is a parameter vector conformable with A, and $E\psi_i = 0$, $i = 1, \ldots, 8$. Thus, in these unconditional supply (demand) equations, each of the four econometric structures or regimes is weighted by the probability that an observation is included in that structure. This methodology effectively increases the degrees of freedom for estimating the structure associated with a relatively small subsample size.

Furthermore, economic theory suggests that the following cross-equation symmetry equations should hold

$$b_{121} = b_{211}, b_{131} = b_{311}, \ldots, b_{181} = b_{811}$$
$$b_{122} = b_{212}, b_{132} = b_{312}, \ldots, b_{172} = b_{712}$$
$$b_{123} = b_{213}, b_{133} = b_{313}, \ldots, b_{163} = b_{613}, b_{183} = b_{813}$$
$$b_{124} = b_{214}, b_{134} = b_{314}, \ldots, b_{184} = b_{814} \tag{8-23}$$

ENDOGENOUS SWITCHING MODELS 239

Imposing these conditions reduces significantly the number of different unknown parameters to be estimated. Tests of the null hypothesis of nonjoint choices can also be performed by imposing restrictions, $b_{12r} = b_{21r} = 0, r = 1, \ldots, 4$.

Because $E\psi_i = 0$, estimation of equations (8-20) through (8-22) subject to the restrictions of equations (8-23) by multivariate ordinary least squares results in tobit-type estimates of the supply and demand functions. Thus, the marginal effect of P_j on EY_i, the unconditional mean, is

$$\frac{\partial EY_i}{\partial P_j} = \sum_{r=1}^{4} b_{ijr} P(S_r), \qquad i,j = 1, \ldots, 8 \tag{8-24}$$

One set of linear unconditional choice functions has been derived that is corrected for multidimensional sample selectivity that arises when there are multiple corner solutions (or zeros) in sets of joint choices. These equations are linear in unknown parameters, provided outside estimates of sample-selection correction terms are available. Thus, inexpensive estimation procedures can be applied.

A PROCEDURE FOR ESTIMATING THE ECONOMETRIC MODEL

Equations (8-20) through (8-22) can be estimated inexpensively by ordinary least squares (OLS) provided that estimates of M_{ij} and A_{ij} are obtained (Maddala, 1981) and that near multicollinearity is not serious. The estimates of M_{ij} and A_{ij} can be derived from first-round consistent estimates of underlying parameters. Then estimates $\hat{\sigma}_7^2$, $\hat{\sigma}_8^2$, $\hat{\sigma}_{78}$, $\hat{M}_{ij}$, $\hat{f}_7$, $\hat{f}_8$, $\hat{F}_7$, and $\hat{F}_8$ will replace their respective counterparts in equations (8-20) through (8-22).

First, estimates of σ_7^2, σ_8^2, and σ_{78} are required before the above procedure can be implemented. This is so because σ_7^{*2} and σ_8^{*2}, which are parameters of F_7 and F_8, are derived from σ_7^2, σ_8^2, and σ_{78}. First-round estimates of these variances and covariances can be obtained by first fitting the following equations

$$Y_7 = b_{70} + \sum_{j=1}^{8} b_{7j} P_j + Z'\underline{c}_7 + V_7$$

$$Y_8 = b_{80} + \sum_{j=1}^{8} b_{8j} P_j + Z'\underline{c}_8 + V_8$$

to the subsample S_1, where only interior values of endogenous variables occur. Residuals ($\hat{V}_i$) from these equations can be employed to obtain

$$\hat{\sigma}_7^2 = \hat{V}_7' \hat{V}_7 /N_1, \hat{\sigma}_8^2 = \hat{V}_8' \hat{V}_8 /N_1, \hat{\sigma}_{78} = \hat{V}_7' \hat{V}_8 /N_1$$

where $\hat{\sigma}_7^{*2} = \hat{\sigma}_7^2 - (\hat{\sigma}_{78}^2/\hat{\sigma}_8^2)$ and $\hat{\sigma}_8^{*2} = \hat{\sigma}_8^2 - (\hat{\sigma}_{78}^2/\hat{\sigma}_7^2)$

Second, univariate and bivariate probit equations for the discrete outcomes on Y_7 and Y_8 must be fitted. The estimated parameters in these equations can be employed to obtain instruments for f, F, and M. Third, instruments for the conditional mean values for all of the disturbance terms, equations (8-15) through (8-18), can be obtained by evaluating these means using the output from steps one and two. Fourth, after replacing M_{ij}, f_7, f_8, F_7, and F_8 with the instruments derived in step three and after imposing the cross-equation symmetric restrictions, equation (8-23), OLS estimation can be applied to obtain estimates of the coefficients of the structural supply and demand equations from the N observations.[4] This fourth stage yields tobit-type estimates of the structural parameters. Maddala (1983) and Lee, Maddala, and Trost (1980) present procedures for obtaining good asymptotic standard errors for these coefficients.[5]

Near multicollinearity becomes a serious potential problem in fitting equations (8-20) through (8-22) by OLS because each of the conditional supply (demand) equations contains many of the same variables. Thus, in the unconditional supply (demand) equations, a variable may appear as many as four times, although it is multiplied by a probability. All variables that enter a given conditional supply (or demand) equation are weighted by the same probability in the unconditional supply (demand) equations. Thus, these probabilities also increase the possible problems of near multicollinearity.

Ridge regression is an alternative estimator to OLS when near multicollinearity is a problem. Lin and Kmenta (1984) have shown that ridge regression outperforms OLS on the basis of mean square error, mean absolute error, or maximum absolute error when medium to high collinearity of regressors occurs. Some econometricians, however, believe that the ridge regression requires mechanistic and unreasonable modification of the moment matrix for a regression equation.

---

[4] Wayne Fuller has suggested to me that trying to estimate all the parameters of the supply and demand equations in the four different structures from a modest sized set of obervations is asking too much of the data. He suggested that some cross-price effects on choices may reasonably be set equal to zero.

[5] Nelson (1984) discusses the efficiency of two-step estimators.

CONCLUSION

This paper has considered the problem of obtaining estimates of the parameters of output supply and input demand (or profit share) equations when some of the outputs are not supplied by some of the observations. The combinations of corner and interior solutions for endogenous variables are the source of multiple economic and econometric structures for agricultural supply and demand (share) equations. I have proposed that the methodology of endogenous switching in a multivariate econometric model be applied to derive a low cost and consistent estimator. In principle, this methodology should help in obtaining estimates of supply elasticities that have positive signs. The primary disadvantage of this procedure is that near multicollinearity may become a serious problem when it is applied.

REFERENCES

Akino, M., and Y. Hayami. 1974. "Sources of Agricultural Growth in Japan, 1860–1965," *Quarterly Journal of Economics* vol. 88, no. 3, pp. 454–479.

Amemiya, T. 1974. "Multivariate Regression and Simultaneous Equation Models When the Dependent Variables Are Truncated Normal," *Econometrica* vol. 42, no. 6, pp. 999–1012.

Antle, J. M., and A. S. Aitah. 1983. "Rice Technology, Farmer Rationality, and Agricultural Policy in Egypt," *American Journal of Agricultural Economics* vol. 65, no. 4, pp. 667–674.

Arrow, K. J., H. B. Chenery, B. Minhas, and R. M. Solow. 1961. "Capital-Labor Substitution and Economic Efficiency," *Review of Economics and Statistics* vol. 43, no. 3, pp. 225–250.

Berndt, E. R., and L. R. Christensen. 1973. "The Translog Function and the Substitution of Equipment, Structures and Labor in U.S. Manufacturing 1929–68," *Journal of Econometrics* vol. 1, no. 1, pp. 81–114.

———, and L. R. Christensen. 1974. "Testing for the Existence of a Consistent Aggregate Index of Labor Inputs," *American Economic Review* vol. 64, no. 3, pp. 391–404.

———, and D. O. Wood. 1975. "Technology and the Derived Demand for Energy," *Review of Economics and Statistics* vol. 57, no. 4, pp. 259–268.

Binswanger, Hans P. 1974. "The Measurement of Technical Change Biases with Many Factors of Production," *American Economic Review* vol. 64, no. 6, pp. 964–976.

Christensen, Laurits R. 1975. "Concepts and Measurement of Agricultural Productivity," *American Journal of Agricultural Economics* vol. 57, no. 5, pp. 910–915.

Cline, P. L. 1975. "Sources of Productivity Change in United States Agriculture" (Ph.D. dissertation, Oklahoma State University, Stillwater).

Cobb, C. W., and P. H. Douglas. 1928. "A Theory of Production," *American Economic Review* vol. 18, no. 1, supplement pp. 139–165.

Diewert, W. E. 1973. "Functional Forms for Probit and Transformation Functions," *Journal of Economic Theory* vol. 6, no. 2, pp. 284–316.

Evenson, R. E. 1968. "The Contribution of Agricultural Research and Extension to Production" (Ph.D. dissertation, University of Chicago).

———. 1980. "A Century of Agricultural Research and Productivity Change Research, Invention, Extension and Productivity Change in U.S. Agriculture: An Historical Decomposition Analysis," in A. Arait, ed., *Research and Extension Productivity in Agriculture* (Moscow, Idaho).

———, and H. P. Binswanger. 1980. "Estimates of Labor Demand Functions: Indian Agriculture," Working Paper No. 356 (New Haven, Conn., Yale Economic Growth Center).

———, and D. Jha. 1973. "The Contribution of Agricultural Research System to Agricultural Production in India," *Indian Journal of Agricultural Economics* vol. 28, no. 4, pp. 212–230.

Fishe, R. P. H., R. P. Trost, and P. M. Lurie. 1981. "Labor Force Earnings and College Choice of Young Women: An Examination of Selectivity Bias and Comparative Advantage," *Economics of Education Review* vol. 1, no. 2, pp. 169–191.

Griliches, Zvi. 1963a. "The Sources of Measured Productivity Growth, U.S. Agriculture, 1940–1960," *Journal of Political Economy* vol. 71, no. 4, pp. 331–346.

———. 1963b. "Estimates of the Aggregate Agricultural Production Function from Cross-Sectional Data," *Journal of Farm Economics* vol. 45, no. 2, pp. 419–432.

———. 1964. "Research Expenditures, Education, and the Aggregate Agricultural Production Function," *American Economic Review* vol. 54, no. 6, pp. 961–974.

———, and V. Ringstad. 1971. *Economies of Scale and the Form of the Production Function* (Amsterdam, North-Holland).

Hall, R. E. 1973. "The Specification of Technology with Several Kinds of Output," *Journal of Political Economy* vol. 81, no. 4, pp. 878–892.

Hanemann, W. M. 1982. "Quality and Demand Analysis," in G. C. Rausser, ed., *New Directions in Econometrics Modeling and Forecasting in U.S. Agriculture* (Amsterdam, North-Holland).

———. 1984. "Discrete/Continuous Models of Consumer Demand," *Econometrica* vol. 52, no. 3, pp. 541–562.

Heckman, J. 1979. "Sample Selection Bias as a Specification Error," *Econometrica* vol. 47, no. 1, pp. 153–161.

Huffman, Wallace E. 1976. "The Productive Value of Human Time in U.S. Agriculture," *American Journal of Agricultural Economics* vol. 58, no. 4, pp. 672–683.

Jamison, D. T., and L. J. Lau. 1982. *Farmer Education and Farm Efficiency* (Baltimore, Md., Johns Hopkins University Press).

Johnson, N. L., and S. Kotz. 1972. *Distributions in Statistics: Continuous Multivariate Distributions* (New York, N.Y., Wiley).

Just, R. E., D. Zilberman, and E. Hockman. 1983. "Estimation of Multicrop Production Functions," *American Journal of Agricultural Economics* vol. 65, no. 4, pp. 770–780.

Lau, L. J. 1972. "Profit Functions of Technologies with Multiple Inputs and Outputs," *Review of Economic Statistics* vol. 54, no. 3, pp. 281–289.

———, and P. Yotopoulos. 1971. "A Test for Relative Efficiency and Application to Indian Agriculture," *American Economic Review* vol. 61, no. 1, pp. 94–109.

———, and P. Yotopoulos. 1972. "Profit, Supply and Factor Demand Functions," *American Journal of Agricultural Economics* vol. 54, no. 1, pp. 11–18.

Lee, L. F., G. S. Maddala, and R. P. Trost. 1980. "Asymptotic Covariance Matrices of Two-Stage Probit and Two-Stage Tobit Methods for Simultaneous Equations Models with Selectivity," *Econometrica* vol. 48, no. 2, pp. 491–503.

Lin, K., and J. Kmenta. 1984. "Ridge Regression under Alternative Loss Criteria," *Review of Economics and Statistics* vol. 64, no. 3, pp. 488–494.

Lopez, R. 1980. "The Structure of Production and the Derived Demand for Inputs in Canadian Agriculture," *American Journal of Agricultural Economics* vol. 62, no. 1, pp. 38–45.

Maddala, G. S. 1981. "Identification and Estimation Problems in Limited Dependent Variable Models," in A. S. Blinder and P. Freedman, eds., *Natural Resources, Uncertainty and General Equilibrium Systems: Essays in Memory of Rafael Lusky* (New York, N.Y., Academic Press).

———. 1983. *Limited Dependent and Qualitative Variables in Econometrics* (New York, N.Y., Cambridge University Press).

Muller, R. A. 1980. "Modelling Production with the Normalized Quadratic Restricted Profit Function," Working Paper No. 80-12 (Hamilton, Ontario, McMaster University, Department of Economics).

Nelson, F. P. 1984. "Efficiency of Two Step Estimators for Models with Endogenous Sample Selection," *Journal of Econometrics* vol. 24, no. 1, pp. 181–196.

Seaz, R. R., and C. R. Shumway. 1983. "U.S. Agricultural Product Supply and Input Demand Relationships Derived from Regional Data." Paper presented at annual meeting of the American Agricultural Economics Association, Purdue University, West Lafayette, Ind., August 1.

Shumway, C. Richard. 1983a. "Production Interrelationships in Texas Crop Agriculture." Paper presented at annual meeting of the Southern Agricultural Economics Association, Atlanta, February.

———. 1983b. "Supply, Demand, and Technology in a Multiproduct Industry: Texas Field Crops," *American Journal of Agricultural Economics* vol. 65, no. 4, pp. 748–760.

———, R. D. Pope, and E. K. Nash. 1984. "Allocatable Fixed Inputs and Jointness in Agricultural Production: Implications for Economic Modeling," *American Journal of Agricultural Economics* vol. 66, no. 1, pp. 72–78.

Varian, H. R. 1978. *Microeconomic Analysis* (New York, N.Y., Norton).

Wales, T. J., and A. D. Woodland. 1983. "Estimation of Consumer Demand Systems with Binding Non-Negativity Constraints," *Journal of Econometrics* vol. 21, no. 3, pp. 263–285.

Weaver, Robert D. 1983. "Multiple Input, Multiple Output Production Choices and Technology in the U.S. Wheat Region," *American Journal of Agricultural Economics* vol. 65, no. 1, pp. 45–56.

———, and D. A. Lass. 1983. "Corner Solutions in Duality Models: A Cross-Section Analysis of Dairy Production Decisions," Staff Paper No. 52 (University Park, Penn., Department of Agricultural Economics, Pennsylvania State University).

Woodland, A. D. 1977. "Estimation of a Variable Profit and of Planning Price Functions for Canadian Manufacturing, 1947–70," *Canadian Journal of Economics* vol. 10, no. 3, pp. 355–377.

Yotopoulos, Pan, and L. Lau. 1973. "A Test for Relative Economic Efficiency: Some Further Results," *American Economic Review* vol. 63, no. 1, pp. 214–223.

PART III
TOWARD EXPLAINING AGRICULTURAL PRODUCTIVITY

9

INDUCED TECHNICAL CHANGE IN AGRICULTURE*

VERNON W. RUTTAN AND YUJIRO HAYAMI

As late as the last century, almost all of the increase in food production was obtained by bringing new land into production (exceptions occurred in limited areas of East Asia, in the Middle East, and in Western Europe). By the end of this century, almost all of the increase in world food production will have to come from higher yields. In most of the world the transition from a resource-based to a science-based system of agriculture is occurring within a single century. In a few countries this transition began in the nineteenth century; in most of the developed countries it did not begin until the first half of this century. Most of the countries of the developing world have been caught up in the transition only since mid-century.

In our earlier research, we found it useful to classify the traditional literature on agricultural development under five headings: (a) resource exploitation, (b) conservation, (c) location, (d) diffusion, and (e) high-payoff input models.[1]

The limitations of these models led to our developing a model of agricultural development in which technological change is treated as an exogenous factor operating independently of other development processes. The induced innovation perspective was stimulated by historical evidence that different countries had followed alternative paths of technological change in the process of agricultural development. In the induced innovation model, the productivity levels achieved by farmers in the most advanced countries can be viewed as arranged along a productivity frontier. This frontier reflects the level of technical progress achieved by the most advanced countries in each resource endowment classification. These productivity levels are not immediately available to farmers in most low-productivity countries, as they can be made avail-

---

*This paper draws on material from Hayami and Ruttan (1985).
[1] See Hayami and Ruttan (1985), chapter 3.

able only by investing in the agricultural research capacity needed to develop technologies appropriate to the countries' natural and institutional environments and in the physical and institutional infrastructure needed to realize the new production potential opened up by technological advances.

There is clear historical evidence that technology has been developed to facilitate the substitution of relatively abundant (hence cheap) factors for relatively scarce (hence expensive) factors of production. The constraints imposed on agricultural development by an inelastic supply of land have, in countries such as Japan and Taiwan, been offset by the development of high-yielding crop varieties designed to facilitate the substitution of fertilizer for land. The constraints imposed by an inelastic supply of labor in countries such as the United States, Canada, and Australia have been offset by technical advances leading to the substitution of animal and mechanical power for manpower. In some cases the new technology—embodied in new crop varieties, new equipment, or new production practices—may not always be substitutes *per se* for land or labor. Rather, they are catalysts that facilitate the substitution of relatively abundant factors (such as fertilizer or mineral fuels) for relatively scarce factors.

Advances in output per unit of land area have been closely associated with advances in biological technology; advances in output per worker and advances in mechanical technology have also been closely related. These historical differences have given rise to cross-sectional differences in productivity and factor use. Advances in biological technology may result in increases in output per worker, as well as increases in output per unit area, if the rate of growth of output per hectare exceeds the rate of growth of the agricultural labor force. In the Philippines, for example, growth in output per worker prior to the mid-1950s was closely related to expansion in the area cultivated per worker. Since the early 1960s, growth in output per worker has been more closely associated with increases in output per unit of land area.

INDUCED INNOVATION IN THE THEORY OF THE FIRM

The theory of induced innovation has been developed mainly within the framework of the theory of the firm. Two schools have attempted to incorporate the innovative behavior of profit-maximizing firms into economic theory. One is the Hicks tradition, which focused on the factor-saving bias induced by changes in relative factor prices resulting from changes in relative resource scarcities (Hicks, 1932).[2] The other is that of Schmookler-Griliches, who focused on the influence of the growth of product demand on the rate of technical change (Griliches, 1957; Schmookler, 1962, 1966).[3]

A fully developed general equilibrium theory of induced innovation that is capable of explaining the dynamic process of agricultural development should incorporate the mechanisms by which changes in both product demand and factor endowments interact with each other to influence the rate and direction of technological change.

The Hicks theory of induced innovation implies that a rise in the price of one factor relative to other factor prices induces a sequence of technical changes that reduces the use of that factor relative to the use of other factor inputs. As a result, the constraints imposed by resource scarcity on economic growth are released by technological advances that facilitate the substitution of relatively abundant factors for relatively scarce factors.

The Hicks theory has been criticized by W. E. G. Salter (1960) and others for its lack of a microeconomic foundation based on the optimizing behavior of the innovating firm.[4] Salter defined the production function to embrace all possible designs conceivable by existing scientific knowledge and called the choice among these designs "factor substitution" instead of "technical change." Salter admits, however, that "relative factor prices are in the nature of signposts representing broad influences that determine the way technological knowledge is applied to production."

---

[2] Interest by economists in the issue opened up by Hicks lagged until the 1960s. Two papers by Fellner (1961, 1962) were particularly important in directing attention to the issue of induced technical change. These were followed by an extended dialogue over the theoretical foundations and the macroeconomic implications of induced technical change, beginning with Kennedy (1964) and continued in Samuelson (1965, 1966), Kennedy (1966, 1967), Ahmad (1966, 1967a, 1967b), and Fellner (1967). For other contributions see Drandakis and Phelps (1966), Conlisk (1969), and van de Klundert and de Groof (1977). For reviews of this and related literature see Wan (1971), Nordhaus (1973), and Binswanger (1978a). For a comprehensive review, see Thirtle and Ruttan (1987).

[3] Mowery and Rosenberg (1979) have argued that in a number of recent studies "the notion that market demand forces 'govern' the innovation process is simply not demonstrated by the empirical analyses which have claimed to support that conclusion" (p. 104). Scherer (1982), using a more complete data base, concludes that "the relationship between demand pull and the flow of innovations was much weaker, though still significant."

[4] See also Griliches (1957); Schmookler (1962, 1966); Mowery and Rosenberg (1979); and Scherer (1982).

Although we do not deny the case for Salter's definition, it is clearly not very useful in helping to understand the process by which new technical alternatives become available. We regard technical change as any change in production coefficients resulting from the purposeful resource-using activity directed to the development of new knowledge embodied in designs, materials, or organizations. In terms of this definition, it is entirely rational for competitive firms to allocate funds to develop a technology that facilitates the substitution of increasingly more expensive factors for less expensive factors. Ahmad (1966) clearly showed that if one factor becomes more expensive relative to another over time, the innovative efforts of entrepreneurs will be directed toward saving the factor that becomes more expensive, as long as entrepreneurs conceive alternative new technical possibilities that can be developed by the same amount of research costs.[5] Similarly, in a country in which the ratio of factor costs is higher than it is in a second country, innovative efforts will be directed toward saving the relatively more expensive factors.

More recently, Binswanger (1974a, 1978b) has developed an induced innovation model incorporating a research production function. By assuming decreasing marginal productivity of research resources in applied research and development, he was able to construct a model of induced factor-saving bias in technical change based on the profit-maximizing behavior of the firm without resorting to the restrictive assumption of a fixed research budget.[6] Binswanger also incorporated into the model the effect of product demand on research resource allocation. In his model, the growth in product demand increases the marginal value product of resources devoted to research, thereby increasing the optimum level of research expenditure for the profit-maximizing firm. The larger research budget implies a shift of the innovation possibility curve (IPC), defined as an envelope of unit isoquants corresponding to the alternative technologies that potentially can be developed for a given research budget at a given state of art, toward the origin. In the Binswanger model, technical change is guided along the IPC by changes in relative factor prices, whereas the IPC itself is induced to shift inward toward the origin by the growth in product demand. Thus, he was able to incorporate both the Hicks approach, which focused on the effect of relative factor prices on factor-saving bias, and the Schmookler-Griliches

[5] See also Fellner (1967), Ahmad (1967a,b), and Kennedy (1967).
[6] Two particularly bothersome assumptions in the Kennedy growth theory tradition of induced innovation theory were the assumption of (a) an exogenously given budget for research and development and (b) a stable "fundamental" tradeoff or transformation function (the IPF).

approach, which focused on the effect of product demand on the rate of technological change, into a single model of induced technical change.

There is no presumption in the theory of induced innovation that technical change is wholly of an induced character. There is a supply (exogenous) dimension to the process as well as a demand (endogenous) dimension. In addition to the effects of resource endowments and growth in demand, technical change reflects the progress of general science and technology. Progress in general science (or scientific innovation) that lowers the cost of technical and entrepreneurial innovations may influence technical change unrelated to changes in factor proportions and product demand (Nelson, 1959; Schmookler, 1966). Even in these cases, the rate of adoption and the impact on productivity of autonomous or exogenous changes in technology will be strongly influenced by the conditions of resource supply and product demand, as these forces are reflected through factor and product markets.

A model of induced technical innovation is illustrated in figure 9-1. The model incorporates the characteristics of both factor substitution and complementarity associated with advances of biological and mechanical technologies.

The process of advance in mechanical technology is shown in the left-hand panel of figure 9-1. I_0^* represents the innovation possibility curve (IPC) at time zero; it is the envelope of less elastic unit isoquants that correspond, for example, to different types of harvesting machinery. A certain technology, a reaper, for example, represented by I_0, is invented when the price ratio BB prevails for some time. Correspondingly, the minimum-cost equilibrium point is determined at P with a certain optimal combination of land, labor, and nonhuman power to operate the reaper. In general, the technology that enables cultivation of a larger area per worker requires a higher animal or mechanical power per worker. This implies the complementary relationship between land and power, which may be drawn as a line representing a certain combination of land and power $[A, M]$. In this simplified presentation, land-cum-power is assumed to be substituted for labor in response to a change in wage relative to an index of land and power prices, though, of course, in actual practice land and power are substitutable to some extent.

I_1^* represents the IPC of period 1. Let us assume, for example, that from period 0 to period 1, labor becomes more scarce relative to land due to the transfer of labor to industry in the course of economic development, resulting in the decline in land rent relative to wage rates. Let us also assume that the price of power declines relative to the wage rate for labor due to the supply of a cheaper power source from industry. The change in the price ratio from BB to CC induces the invention of another

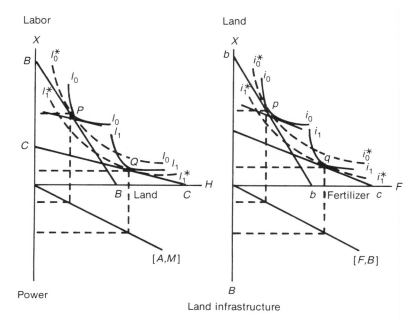

FIGURE 9-1. A model of induced technical change in agriculture. *Source:* Y. Hayami and V. W. Ruttan, *Agricultural Development: An International Perspective,* rev. ed. (Baltimore, Md., Johns Hopkins University Press, 1985).

technology, such as the combine, represented by I_1, that enables a farm worker to cultivate a larger land area using a larger amount of power.

The process of advance in biological technology is illustrated in the right-hand panel of figure 9-1. Here, i_0^* represents an IPC embracing less elastic land-fertilizer isoquants, such as i_0, corresponding to different crop varieties and cultural practices. When the fertilizer-land price ratio declines from bb to zz from period 0 to period 1, a new, more fertilizer-responsive technology, represented by i_1, is developed along i_1^*, the IPC of period 1. In general, the technology that facilitates substitution of fertilizer for land, such as fertilizer-responsive high-yielding crop varieties, requires better control of water and better land management. This suggests a complementary relationship between fertilizer and land infrastructure in the form of irrigation and drainage systems, as implied by the linear relationship $[F, B]$.[7]

---

[7] In the model of induced innovation in figure 9-2, we have treated, for pedagogical purposes, the impact of advances in mechanical and biological technology on factor ratios as if they were completely separable, even though they are interrelated. Furthermore, some biological innovations are labor-saving and some mechanical innovations are land-saving. Also, we do not deny the possibility of autonomous or innate bias in technical change

It would be misleading to leave the impression that induced innovation proceeds as a smooth adjustment along the IPC in response to changes in relative factor prices. In the dynamic process of development, the emergence of imbalance or disequilibrium is a critical element in inducing technical change and economic growth. Disequilibrium among the several elements in the system creates the bottlenecks that focus the attention of scientists, inventors, entrepreneurs, and public administrators on the solution of problems for attaining more efficient resource allocation.[8]

A requisite for agricultural productivity growth is the capacity of the agricultural sector to respond to changes in factor and product prices. This adaptation involves not only the movement along a fixed production surface but also the inducements of innovations leading to a shift in the production surface. For example, even if the price of fertilizer declines relative to the prices of land and farm products, increases in the use of fertilizer may be limited unless new crop varieties are developed that are more responsive than traditional varieties to high levels of biological and chemical inputs.

For illustrative purposes, the relationship between fertilizer use and yield may be drawn, as in figure 9-2, letting u_0 and u_1 represent the fertilizer-response curves of indigenous and improved varieties, respectively. For farmers facing u_0, a decline in the fertilizer price relative to the product price from p_0 to p_1 would not be expected to result in much increase in fertilizer application or in yield. The full effect of a decline in the fertilizer price on fertilizer use and output can be fully realized only if u_1 is made available to farmers through the development of more responsive varieties.

Conceptually, it is possible to draw a curve, such as U in figure 9-2, that is the envelope of many individual response curves, each representing a rice variety characterized by a different degree of fertilizer responsiveness. We shall identify this curve as a *metaproduction function*,

---

unrelated to changes in factor prices. If, for example, the rate of advance in mechanical technology exceeds the rate of advance in biological technology, because of autonomous bias in technological potential, it may force a rise in the land-labor ratio, even if there is no change in the land-labor price ratio. See, for example, Thirtle (1982, 1984).

[8] Rosenberg (1969) has suggested a theory of induced technical change based on "obvious and compelling need" to overcome the constraints on growth instead of relative factor scarcity and relative factor prices. The Rosenberg model is consistent with the model suggested here, since his "obvious and compelling need" is reflected in the market through relative factor prices. Timmer has pointed out (in an October 9, 1970 letter) that in a linear programming sense the constraints that give rise to the "obvious and compelling need" for technical innovation in the Rosenberg model represent the dual of the factor prices used in our model. For further discussion of the relationships between Rosenberg's approach and that outlined in this section see Hayami and Ruttan (1973).

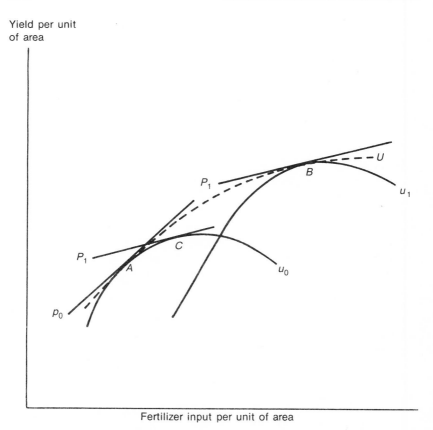

FIGURE 9-2. Shift in fertilizer response curve along the metaresponse curve. *Source:* Y. Hayami and V. W. Ruttan, *Agricultural Development: An International Perspective,* rev. ed. (Baltimore, Md., Johns Hopkins University Press, 1985).

or a *potential production function*. The metaproduction function can be regarded as the envelope of commonly conceived neoclassical production functions. In the short run, in which substitution among inputs is circumscribed by the rigidity of existing capital and equipment, production relationships can best be described by an activity with relatively fixed factor-factor and factor-product ratios. In the long run, in which the constraints exercised by existing capital disappear and are replaced by the fund of available technical knowledge, including all alternative feasible factor-factor and factor-product combinations, production relationships can be adequately described by the neoclassical production function. In the secular period of production, in which the constraints given by the available fund of technical knowledge are further relaxed to admit all potentially discoverable possibilities, production relationships can be described by a metaproduction function that describes all conceivable technical alternatives that might be discovered.[9]

The low efficiency in agricultural production in the LDCs is represented by point C in figure 9-2. This suboptimal equilibrium is the result of the lag in developing and adopting the fertilizer-responsive variety (u_1) in response to a decline in the relative price of fertilizer from p_0 to p_1. In this example, we have used the development of fertilizer-responsive rice varieties as a pedagogical device to illustrate how changes in factor prices induce the development of new short-run production functions along the long-run metaproduction function. Our more general hypothesis is that the relatively low production efficiency of LDC agriculture is explained mainly by the limited capacity of LDC agricultural research systems to develop new technology in response to changes in relative factor prices.

We do not consider that the metaproduction function is inherent in nature or that it remains completely stable over time. The metaproduction function will shift in response to the accumulation of general scientific knowledge. We do consider, however, that it is operationally feasible to assume a reasonable degree of stability for a technical "epoch," the time range relevant for many empirical analyses. Shifts in the metaproduction function are much slower than adjustments along the surface, or to the surface from below, of the metaproduction function, especially in LDCs.

PRODUCTION AND PRODUCTIVITY GROWTH

Japan and the United States are characterized by extreme differences in relative endowments of land and labor. In 1880, total agricultural land area per male worker in the United States was more than sixty times as great as in Japan; arable land area per worker in the United States was about twenty times as great as in Japan. The differences have widened over time. By 1980 total agricultural land area per male worker was

---

[9] See Brown (1966) for a discussion of short-run, long-run, and secular production processes. The relationship between U and the u_1s in figure 9-3 is somewhat analogous to the interfirm envelope of a series of intrafirm production functions (see Bronfenbrenner, 1944). For an alternative perspective that views technical change as a process of widening the elasticity of substitution possibilities among factors rather than in terms of a sequence of response functions, see Färe and Jansson (1974). For an empirical application see Grabowski and Sivan (1983).

more than 100 times as large and arable land area per male worker about fifty times as large in the United States as in Japan.

The relative prices of land and labor also differed sharply in the two countries. In 1880, in order to buy a hectare of arable land, it would have been necessary for a Japanese hired farm worker to work eight times as many days as a U.S. farm worker. In the United States, the price of labor rose relative to the price of land, particularly between 1880 and 1920. In Japan the price of land rose sharply relative to the price of labor, particularly between 1880 and 1900. By 1960 a Japanese farm worker would have had to work thirty times as many days as a U.S. farm worker in order to buy one hectare of arable land. This gap was reduced after 1960 partly due to extremely rapid increases in the wage rate in Japan during the two decades of "rapid" economic growth. In the United States, land prices rose sharply in the postwar period primarily because of the rising demand for land for nonagricultural use and the anticipation of continued inflation. Yet, in 1980 a Japanese farm worker still would have had to work eleven times as many days as a U.S. worker to buy one hectare of land.

In spite of these substantial differences in land area per worker and in the relative prices of land and labor, both the United States and Japan experienced relatively rapid rates of growth in production and productivity in agriculture (table 9-1). Overall agricultural growth performance for the entire 100-year period was similar in the two countries. In both countries total agricultural output increased at an annual compound rate of 1.6 percent and total inputs (aggregate of conventional inputs) increased at a rate of 0.7 percent. Total factor productivity (total output divided by total input) increased at an annual rate of 0.9 percent in both countries. Meanwhile, labor productivity measured by agricultural output per male worker increased at rates of 3.1 percent per year in the United States and 2.7 percent in Japan. It is remarkable that the overall growth rate in output and productivity were so similar despite the extremely different factor proportions that characterize the two countries.

Although there is a resemblance in the overall rates of growth in production and productivity, the time sequences of the relatively fast growing phases and the relatively stagnant phases differ between the two countries. In the United States, agricultural output grew rapidly up to 1900, after which time the growth rate decelerated. From the 1900s to the 1930s there was little gain in total productivity. This stagnation phase was succeeded by a dramatic rise in production and productivity in the 1940s and 1950s. Japan experienced rapid increases in agricultural production and productivity from 1880 to the 1910s, then entered

TABLE 9-1. ANNUAL COMPOUND RATES OF GROWTH IN OUTPUT, INPUT, PRODUCTIVITY, AND FACTOR PROPORTIONS IN U.S. AND JAPANESE AGRICULTURE, 1880–1980, SELECTED PERIODS

(percentage)

	Subperiods					Entire period 1880–1980
	1880–1900	1900–1920	1920–1940	1940–1960	1960–1980	
United States						
Output (net of seeds and feed)	2.2	0.8	1.3	1.9	1.9	1.6
Total inputs	1.6	1.4	0.2	0.1	0.3	0.7
Total productivity (output/total inputs)	0.6	−0.7	1.1	1.9	1.6	0.9
Number of male workers	1.1	0.2	−0.9	−3.7	−3.8	−1.5
Output per male worker	1.1	0.6	2.2	5.9	6.1	3.1
Agricultural land area	1.8	−0.1	−0.4	0.2	−0.2	0.3
Arable land area	2.7	1.1	−0.1	−0.1	0.1	0.7
Output per ha. of agricultural land	0.4	0.8	1.7	1.7	2.1	1.3
Output per ha. of arable land	−0.4	−0.3	1.4	2.0	1.8	0.9
Agricultural land area per male worker	0.7	−0.3	0.5	4.1	3.9	1.8
Arable land area per male worker	1.5	0.9	0.8	3.8	4.2	2.2
Japan						
Output (net of seeds and feed)	1.6	2.0	0.7	1.8	1.9	1.6
Total input	0.4	0.5	0.3	1.6	1.0	0.7
Total productivity	1.2	1.5	0.4	0.2	0.9	0.9
Number of male workers	0.1	−0.6	−0.9	−0.1	−4.2	−1.1
Output per male worker	1.5	2.6	1.6	1.9	6.3	2.7
Arable land area (= agricultural land area)	0.4	0.7	0.1	−0.04	−0.5	0.1
Output per ha. of arable land	1.2	1.3	0.6	1.8	2.4	1.5
Arable land area per male worker	0.4	1.3	1.0	0.1	3.8	1.2

Source: Data from Y. Hayami and V. W. Ruttan, *Agricultural Development: An International Perspective*, rev. ed. (Baltimore, Md., Johns Hopkins University Press, 1985), tables C-2 and C-3.

into a stagnation phase which lasted until the mid-1930s. Another rapid expansion phase commenced during the period of recovery from the devastation of World War II. Roughly speaking, the United States experienced a stagnation phase two decades earlier than Japan and also shifted to the second development phase two decades earlier.

In the course of U.S. and Japanese agricultural development, there have been substantial differences in the growth of labor productivity in the two countries. In the United States, land area per worker (A/L) rose much more rapidly than in Japan. In Japan, land productivity (Y/A) rose much more rapidly than in the United States. For the 1880–1980 period, increase in land area per worker explains about 70 percent of the labor productivity growth in the United States, whereas it explains less than 40 percent of that in Japan.

Both the United States and Japan experienced successive stages of rapid growth and relative stagnation followed by rapid growth in labor productivity. In the United States, the stagnation phase was associated with a reduction in the rate of growth in land area per worker (A/L). In Japanese agriculture it is clearly the movements in land productivity (Y/A) that are most closely associated with the sequence of the development and stagnation phases before World War II. Rapid increases in labor productivity (Y/L) in the United States and Japan after World War II were associated with increases in both land area per worker (A/L) and land productivity (Y/A). This similarity suggests that U.S. and Japanese agricultural growth patterns began to converge in the postwar period as the scarcity of labor relative to land increased in Japan and the scarcity of land relative to labor increased in the United States.

In the United States it was primarily the progress of mechanization that facilitated the expansion of agricultural production and productivity by increasing the area operated per worker. In Japan it was primarily the progress of biological technology, represented by seed improvements, which increased the yield response to higher levels of fertilizer application, that permitted the rapid growth in agricultural output in spite of severe constraints on the supply of land. U.S. agriculture has experienced significant biological innovations since the 1930s, and farm mechanization has progressed at an accelerating pace since the 1950s. In Japan, biological technology contributed significantly to the growth of production and productivity before World War II. In the postwar period, Japanese agriculture has made dramatic progress in mechanization corresponding to sharp increases in farm wage rates induced by the rapid transfer of labor to the industrial and service sectors.

The contrasting patterns of productivity growth and factor use in U.S. and Japanese agriculture can best be understood in terms of a process of dynamic (in the sense that production isoquants change in response to the changes in relative factor prices) adjustment to changing relative factor prices along a metaproduction function.

In the United States, the long-term decline in the prices of land and machinery relative to wages before 1960 could be expected to encourage the substitution of land and power for labor. This substitution generally involved the application of mechanical technology to agricultural production. Dramatic increases in land area and power per worker of the magnitude that occurred in the United States indicate a response to mechanical innovations that raised the marginal rate of substitution in favor of both land and power for labor.[10] This has been a continual process. The introduction of the tractor greatly raised the marginal rate of substitution of power for labor by making it much easier to command more power per worker. Substitution of higher-powered tractors for low-powered tractors had a similar effect.

In Japan the supply of land was inelastic and the price of land rose relative to wages. It was not profitable, therefore, to substitute land and power for labor. Instead, the new opportunities arising from continuous declines in the price of fertilizer relative to the price of land were exploited through advances in biological technology. Seed improvements were directed to the selection of more fertilizer-responsive varieties. The traditional varieties had equal or higher yields than the improved varieties at the lower level of fertilization, but did not respond to higher applications of fertilizer. With fixed biological technology represented by a certain variety of seed, the elasticity of substitution of fertilizer for land was low.

In Japan, because of the secularly rising trend of wages and drastically falling fertilizer prices relative to land prices, there was strong inducement for farmers and experiment station workers to develop biological innovations, such as the high-yielding fertilizer-responsive crop varieties. It is significant that in the United States the biological innovations represented by hybrid corn began about ten years after the rate of increase in arable land area per worker decelerated (around 1920), and that biological innovations and fertilizer application were accelerated after acreage restrictions were imposed by the government. It seems that the changes in land supply conditions, coupled with a dramatic decline in fertilizer prices, induced a more rapid rate of biological innovation in the United States after the 1930s. It may be that when the increase in fertilizer input per hectare resulting from this relative price decline

---

[10] This is consistent with the emphasis on the importance of the effect of mechanical innovations on the substitution between new and old machinery in terms of the relative price changes, as analyzed by David (1966). In fact, the decline in the price of new machines (relative to old machines) in efficiency terms represents a measure of the contribution of the farm machinery industry to technical changes in agriculture.

exceeded the amount of natural fertility depleted from the soil, the demand for biological innovations became a pressing need that, coupled with the change in the supply condition of arable land, induced the dramatic advantages in biological technology in the United States since the 1930s.

In terms of the induced innovation hypothesis, such adjustments in factor proportions in response to changes in relative factor prices represent movements along the isoproduct surface of a metaproduction function. These movements may be inferred from figures 9-3 and 9-4, in

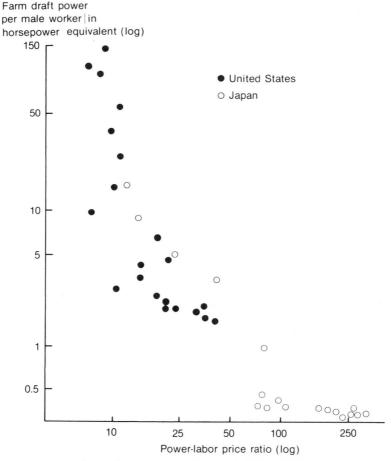

FIGURE 9-3. Relationship between farm draft power per male worker and power-labor price ratio, the United States and Japan, quinquennial observations for 1880–1980

INDUCED TECHNICAL CHANGE 261

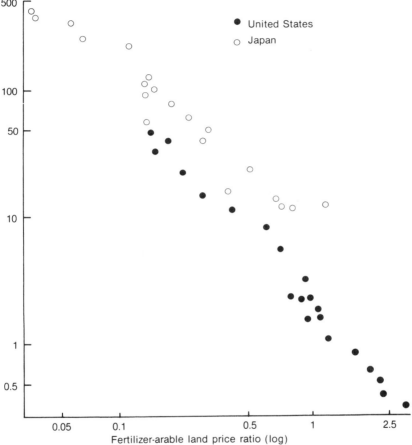

FIGURE 9-4. Relationship between fertilizer input per hectare of arable land and fertilizer-arable land price ratio, the United States and Japan, quinquennial observations for 1880–1980

which U.S. and Japanese data on the relationship between farm draft power (from both tractors and draft animals) per male worker and the machinery-labor price ratio, and between fertilizer input per hectare of arable land and the fertilizer-land price ratio, and between fertilizer input per hectare of arable land and the fertilizer-land price ratio, are plotted. Despite the enormous differences in climate and other environmental conditions, the relation between these variables is almost identical in both countries.

FACTOR SUBSTITUTION ALONG THE METAPRODUCTION FUNCTION

The hypothesis developed in the previous section can be summarized as follows: agricultural growth in the United States and Japan during the 1880–1980 period can best be understood when viewed as a dynamic factor substitution process. Factors have been substituted for each other along a metaproduction function in response to long-run trends in relative factor prices. Each point on the metaproduction surface is characterized by a technology that can be described in terms of specific sources of power, types of machinery, crop varieties, and animal breeds. Movements along this metaproduction surface involve technical changes. These technical changes have been induced to a significant extent by the long-term trends in relative factor prices.

As a test of this hypothesis, we have tried to determine the extent to which the variations in factor proportions, as measured by the land-labor, power-labor, and fertilizer-land ratios, can be explained by changes in factor price ratios. In a situation characterized by a fixed technology, it seems reasonable to presume that the elasticities of substitution among factors are small. This permits us to infer that innovations were induced, if the variations in these factor proportions are explained consistently by the changes in price ratios. The historically observed changes in those factor proportions in the United States and Japan are so large that it is hardly conceivable that these changes represent substitution along a given production surface describing a constant technology (figures 9-3 and 9-4).[11]

In order to specify adequately the regression form, we have to be able to infer the shape of the underlying metaproduction function and the functional form of the relationship between changes in the production function and in factor price ratios. Because of a lack of adequate a priori information, we have simply specified the regression in log-linear form with little claim for theoretical justification. If we can assume that the production function is linear and homogeneous, the factor proportions can be expressed in terms of factor price ratios alone and are independent of product prices.

---

[11] Griliches (1958) has shown, using a distributed lag model, that increases in fertilizer input by United States farmers can be explained solely in terms of the decline in fertilizer prices. The relation he estimated can be identified as the movement along the metaproduction function. A very high elasticity of substitution between labor and capital in the order of 1.7 was estimated for U.S. agriculture by Kislev and Peterson (1982). We consider their estimates to represent the elasticity of substitution along the metaproduction function rather than the elasticity along a given technology production function.

Considering the crudeness of the data and the purpose of this analysis, we used quinquennial observations (stock variables measured at five-year intervals and flow variables averaged for five years) instead of annual observations for the regression analysis.[12] A crude form of adjustment is built into our model, as our data are quinquennial observations and prices are generally measured as the averages of the five years preceding the year when the quantities are measured (for example, the number of workers in 1910 is associated with the 1906–1910 average wage).

The results of regression analyses are summarized in tables 9-2a and 9-2b and 9-3a and 9-3b. Tables 9-2a and 9-2b present the regressions for land-labor and power-labor proportions. In those regressions we originally included other variables, such as the fertilizer-labor price ratio and the exponential time trend. But, probably because of high intercorrelations, either the coefficients for those variables were nonsignificant or they resulted in implausible results for the other coefficients. Those variables were dropped in the subsequent analysis.

Table 9-2a shows the results for the United States. A major anomaly in our initial U.S. regressions on land-labor and power-land ratios (not shown) was that very poor results were obtained in terms of signs and significance levels of estimated coefficients, coefficients of determination, and the Durbin-Watson statistics if the regressions were estimated for 1880–1980, whereas good results were obtained for 1880–1960. This anomaly seems to be explained by the deficiency of land price data. Our land price was measured by the average unit of land in farms. This is a measure of the price of land as a stock but not the price of the service of land for agricultural production. As is well known, agricultural land prices in the United States diverged rapidly from agricultural land rents during the 1960–1980 period because of the rising demand for nonagricultural uses and the anticipation of continued inflation.[13] As a result, the stock price of land rose relative to the farm wage rate after 1960. But land rents seem to have declined relative to the farm wage rate although national aggregate time-series data on agricultural land rents are not yet available.

To adjust for the divergence between the stock and the service prices of agricultural land, the regressions in table 9-2a include a time dummy variable that is specified as zero for 1880–1960 and one after 1960.

---

[12] For earlier periods annual observations were either unavailable, or if available, were very crude and often inconsistent in terms of definition and measurement (for example, measured on the crop-year basis for some years and on the calendar-year basis for others).

[13] For the divergence between agricultural land prices and land rents in the United States since 1960, see Doll and Widdows (1982).

TABLE 9-2a. REGRESSIONS OF LAND-LABOR RATIOS
AND POWER-LABOR RATIOS ON RELATIVE FACTOR PRICES:
UNITED STATES, 1880–1980 QUINQUENNIAL OBSERVATIONS

		Coefficients of price of					
Regression number	Dependent variables	Land relative to farm wage	Machinery relative to farm wage	Time dummy	Coeff. of det. ($\bar{R}^2$)	S.E.	Durbin-Watson statistics
	Land-labor ratios:						
(1)	Agricultural land per male worker	−0.248 (0.191)	−0.313 (0.107)	0.984 (0.118)	0.922	0.155	1.82
(2)	Arable land per male worker	−0.042 (0.182)	−0.592 (0.102)	0.902 (0.112)	0.945	0.148	1.92
(3)	Agricultural land per work hour	−0.182 (0.206)	−0.267 (0.115)	0.971 (0.127)	0.898	0.167	1.55
(4)	Arable land per work hour	0.024 (0.195)	−0.545 (0.109)	0.889 (0.120)	0.929	0.158	1.67
	Power-labor ratios:						
(5)	Horsepower per male worker	−1.040 (0.466)	−1.060 (0.261)	1.839 (0.287)	0.928	0.378	1.73
(6)	Horsepower per work hour	−0.974 (0.480)	−1.013 (0.269)	1.826 (0.295)	0.919	0.389	1.65

Note: Equations are linear in logarithm. Standard errors of the estimated coefficients are in parentheses. The time dummy variable is zero for 1880–1960 and one for 1965–1980.

Source: Data from Y. Hayami and V. W. Ruttan, *Agricultural Development: An International Perspective*, rev. ed. (Baltimore, Md., Johns Hopkins University Press, 1985) appendix C.

About 90 percent of the variation in the land-labor ratio and in the power-labor ratio is explained by the changes in their price ratios together with the dummy variables. The coefficients are all negative, except the land-price coefficient in regression (4). Such results indicate that the marked increases in land and power per worker in U.S. agriculture over the past 100 years have been closely associated with declines in the price of land, and of power and machinery relative to the farm wage rate. The hypothesis that land and power should be treated as complementary factors is confirmed by the negative coefficients. This seems to indicate that, in addition to the complementarity along a fixed production surface, mechanical innovations that raise the marginal rate of substitution of power for labor tend also to raise the marginal rate of substitution of land for labor.

The results from using the same regressions for Japan (table 9-2b) are greatly inferior in terms of statistical criteria. This is probably because the ranges of observed variation in the land-labor and power-labor ratios are too small in Japan to detect meaningful relationships between the factor proportions and price ratios. It may also reflect the fact that the mechani-

TABLE 9-2b. REGRESSIONS OF LAND-LABOR RATIOS
AND POWER-LABOR RATIOS ON RELATIVE FACTOR PRICES:
JAPAN, 1880–1980 QUINQUENNIAL OBSERVATIONS

Regression number	Dependent variables	Coefficients of price of		Coeff. of det. ($\bar{R}^2$)	S.E.	Durbin-Watson statistics
		Land relative to farm wage	Machinery relative to farm wage			
	Land-labor ratios:					
(7)	Arable land per male worker	−0.147 (−0.068)	−0.408 (0.034)	0.893	0.123	1.00
(8)	Arable land per work hour	0.069 (0.067)	−0.354 (0.060)	0.680	0.215	0.48
	Power-labor ratios:					
(9)	Horsepower per male worker	0.221 (0.375)	−1.146 (0.188)	0.695	0.675	0.37
(10)	Horsepower per work hour	0.143 (0.430)	−1.091 (0.216)	0.615	0.773	1.74

Note: Equations are linear in logarithm. Standard errors of the estimated coefficients are in parentheses.

Source: Data from Y. Hayami and V. W. Ruttan, *Agricultural Development: An International Perspective,* rev. ed. (Baltimore, Md., Johns Hopkins University Press, 1985) appendix C.

cal innovations in Japan were developed and adopted primarily to increase yield, rather than as a substitute for labor, for the period before World War II.

For the United States, the results of the regression analyses of the determinants of fertilizer input per hectare of arable land are presented in table 9-3a. They indicate that variations in the fertilizer-land price ratio alone explain more than 90 percent of the variation in fertilizer use. Over a certain range, fertilizer can be substituted for human care of plants (for example, weeding). A more important factor in Japanese history would be the effects of substitution of commercial fertilizer for the labor allocated to the production of self-supplied fertilizers, such as animal and green manure.[14]

---

[14] Biological innovations represented by improvements in crop varieties, characterized by greater response to fertilizer, tend to be land-saving and labor-using. The yield potential of the improved varieties is typically achieved only when high levels of fertilization are combined with high levels of crop husbandry and water management. On this score, the introduction of high-yielding varieties enhances the substitution of fertilizer and labor for land. On the other hand, commercial fertilizers have significant labor-saving effects as they substitute for self-supplied fertilizers. In Japan, the production of such self-supplied fertilizers as manure, green manure, compost, and night soil traditionally has occupied a significant portion of a farmer's work hours. With the increased supply of commercial fertilizers, farmers can divert their labor to the improvements in cultural practices in such forms as better seedbed preparation and weed control.

TABLE 9-3a. REGRESSIONS OF FERTILIZER INPUT PER HECTARE
OF ARABLE LAND ON RELATIVE FACTOR PRICES:
UNITED STATES, 1880–1980

Regression number	Coefficients of price of			Coeff. of det. ($\bar{R}^2$)	S.E.	Durbin-Watson statistics
	Fertilizer relative to land	Labor relative to land	Machinery relative to land			
(11)	−1.512 (9.119)	0.850 (0.212)	−0.025 (0.233)	0.983	0.177	2.02
(12)	−1.521 (0.053)	0.843 (0.216)		0.984	0.189	2.02
(13)	−1.641 (0.063)			0.972	0.250	0.88
(14)	−1.295 (0.092)	1.118 (0.129)	−0.066 (0.176)	0.991	0.129	2.01
(15)	−1.328 (0.038)	1.076 (0.114)		0.992	0.134	2.04
(16)	−1.524 (0.075)			0.954	0.318	1.04

Note: Based on quinquennial observations. Equations are linear in logarithm. Standard errors of the estimated coefficients are in parentheses.

Source: Data from Y. Hayami and V. W. Ruttan, *Agricultural Development: An International Perspective*, rev. ed. (Baltimore, Md., Johns Hopkins University Press, 1985) appendix C.

A comparison of table 9-3a with table 9-3b indicates a striking similarity in the structure of demand for fertilizer in the United States and Japan. Despite enormous differences between the United States and Japan in climate, initial factor endowments, social and economic institutions, and organization, the results in these two tables seem to suggest that the inducement mechanism of innovations and the response of farmers to economic opportunities have been essentially the same.

Overall, the results of the statistical analyses are consistent with the hypothesis stated at the beginning of this section. In both Japan and the United States, factors have been substituted for each other along a metaproduction function, primarily in response to long-run trends in factor prices.

A TEST OF THE INDUCED TECHNICAL CHANGE HYPOTHESIS

The discussion of the induced technical change hypothesis presented in the previous section has established clearly the plausibility of the induced technical change hypothesis. But in order to test more rigorously the induced innovation hypothesis, it is necessary to decompose changes in factor proportions into (a) the effect of factor substitution along a fixed-technology isoquant in response to changes in relative factor prices

TABLE 9-3b. REGRESSIONS OF FERTILIZER INPUT PER HECTARE OF ARABLE LAND ON RELATIVE FACTOR PRICES: JAPAN, 1880-1980

Regression number	Coefficients of price of			Coeff. of det. ($\bar{R}^2$)	S.E.	Durbin-Watson statistics
	Fertilizer relative to land	Labor relative to land	Machinery relative to land			
(17)	−1.033 (0.347)	0.432 (0.209)	0.019 (0.487)	0.884	0.388	1.67
(18)	−1.020 (0.082)	0.427 (0.173)		0.891	0.388	1.67
(19)	−1.037 (0.093)			0.862	0.449	1.29
(20)	−1.626 (0.311)	0.496 (0.180)	0.906 (0.437)	0.909	0.345	0.63
(21)	−1.001 (0.082)	0.587 (0.190)		0.892	0.386	1.06
(22)	−1.028 (0.098)			0.844	0.477	1.07

Note: Based on quinquennial observations. Equations are linear in logarithm. Standard errors of the estimated coefficients are in parentheses.

Source: Data from Y. Hayami and V. W. Ruttan, *Agricultural Development: An International Perspective*, rev. ed. (Baltimore, Md., Johns Hopkins University Press, 1985) appendix C.

and (b) the effect of biased technical change. In addition, it is necessary to determine whether the biased technical change effect is in the same direction as the price-induced factor substitution effect.

In this section we attempt to conduct such a test using the 1880–1980 quinquennial data for U.S. and Japanese agriculture. A method of measuring biases of technical change with many factors of production was originally developed by Binswanger (1974, 1978c) using the transcendental logarithmic (translog) function. His method has found a number of applications in the analysis of agricultural production.[15] In this study, we employ the two-level constant elasticity of substitution (CES) production function, which is more robust in estimation and more clearcut in interpretation than the translog function.[16]

[15] The Binswanger method was applied to Japanese agriculture by Kako (1978) and Nghiep (1979).

[16] The two-level CES production function was originally developed by Sato (1967). This production function was first applied to the analysis of Japanese agriculture by Shintini and Hayami (1975). More recently, the two-level CES production was advocated for its relevance to the analysis of agricultural production in general by Kaneda (1982). According to Kaneda, the two-level CES function has advantages over the translog function in parsimony in parameter, ease of interpretation, computational ease, and interpolative and extrapolative robustness. For an earlier attempt to adopt the CES production function to estimate elasticities of substitution among more than two factors, see Roe and Yeung (1978).

In order to measure the effect of biased technical change, it is convenient to specify that technical change is of the factor-augmenting type. Output is assumed to be produced by n inputs $(X_1, \ldots, X_n)$ with corresponding factor-augmenting coefficients $(E_1, \ldots, E_n)$, where E_i represents the efficiency of X_i:

$$Q = f(E_1 X_1, \ldots, E_n X_n) \tag{9-1}$$

where the production function (f) is assumed to be linear homogeneous and well behaved.

When a competitive market equilibrium is assumed, the following relation can be derived (see appendix A) for accounting for changes in factor proportions:

$$\sum_{j \neq i} s_j \left(\frac{\dot{X}_i}{X_i} - \frac{\dot{X}_j}{X_j} \right) = \sum_{j \neq i} s_j \sigma_{ij} \left(\frac{\dot{P}_j}{P_j} - \frac{\dot{P}_i}{P_i} \right) + \sum_{j \neq i} s_j (1 - \sigma_{ij}) \left(\frac{\dot{E}_j}{E_j} - \frac{\dot{E}_i}{E_i} \right) \tag{9-2}$$

where the dot denotes the time derivative (hence, for example, $\dot{X}_i/X_i$ represents the growth rate of X_i); P_i is the real factor price of input i; s_i is the factor share of input i; σ_{ij} is the Allen partial elasticity of substitution representing the curvature of the fixed-technology isoquant between input i and input j, assuming optimal adjustments in other factor inputs.

The left-hand side of equation (9-2) is the weighted average of the rates of change in the proportion of factor i relative to all other factors using factor shares as weights. This term may be called a "generalized change in the factor proportion" of the ith input (GCFP$_i$). Since GCFP$_i$ can also be expressed as

$$\frac{\dot{X}_i}{X_i} - \sum_j s_j \frac{\dot{X}_j}{X_j}$$

it can be interpreted as an excess of the growth rate of X_i over the average growth rate of all factor inputs.

The first term on the right-hand side of equation (9-2) measures the contribution of changes in relative factor prices to GCFP$_i$ as it represents the effect of factor substitution along a fixed isoquant. The second term measures the contribution of biased technical change as it represents the effect of differential rates of factor augmentation among inputs. In short, the right-hand side of equation (9-2) decomposes GCFP$_i$ into the price-induced factor substitution and the biased technical change effects.

Following the Hicksian definition, factor-using bias in technical change may be defined for the many factor cases as ith factor-using,

INDUCED TECHNICAL CHANGE   269

neutral, and ith factor-saving depending on whether the second term on the right-hand side of equation (9-2) is positive, zero, or negative, respectively.

Factor-using bias can also be evaluated in terms of changes in factor shares. The rate of change in the ith factor share can be expressed as

$$\frac{\dot{s}_i}{s_i} = \sum_{j \neq i} s_j(\sigma_{ij} - 1)\left(\frac{\dot{P}_j}{P_j} - \frac{\dot{P}_i}{P_i}\right) + \sum_{j \neq i} s_j(1 - \sigma_{ij})\left(\frac{\dot{E}_j}{E_j} - \frac{\dot{E}_i}{E_i}\right) \qquad (9\text{-}3)$$

In the above equation, the rate of change in the ith factor share is decomposed into (a) the price-induced factor substitution effect (the first term on the right-hand side) and (b) the biased technical change effect (the second term). The factor-using bias can be defined in terms of the signs of the second term of equation (9-3) in exactly the same manner as for equation (9-2).

For the measurement of technical change biases we use equation (9-3). The technical change biases were measured by the changes in factor shares that would have occurred in the absence of factor price changes. First, the rate of change in the factor share for inputs that would have occurred during year t in the absence of factor price changes (b_{it}) is estimated by subtracting the first term on the right-hand side of equation (9-3) from the left-hand side. The cumulative change in a factor share due to biased technical change that would have occurred for input i from the base year (1880) up to year t (B_{it}) can be calculated as

$$B_{it} = S_{i,1880} \cdot \prod_t (1 + b_{it}) - S_{i,1880}$$

where $S_{i,1880}$ is the actual factor share in the base year. The estimates of B_{it}s are shown in table 9-4.

The first term on the right-hand side of the above equation may be called the *constant-price factor share* (S'_{it}), which is the hypothetical factor share that would have existed in year t if the factor prices had remained the same as for 1880. Note that the factor shares used for the calculation of b_{it}s were the constant-price factor shares instead of the actual shares. The index of factor-using bias for input i in year t may be calculated by

$$\frac{S'_{it}}{S_{i,1880}} \times 100$$

where $S_{i,1880}$ is the actual factor share in 1880.

For the estimation of B_{it} it is necessary to estimate the partial elasticities of substitution (σ_{ij}) and factor shares (s_i). The partial elasticities of substi-

TABLE 9-4. CUMULATIVE CHANGES IN FACTOR SHARES DUE TO BIASED TECHNICAL CHANGE IN AGRICULTURE (B_{it}) AND THE INDEXES OF FACTOR PRICES RELATIVE TO THE AGGREGATE INPUT PRICE INDEX, THE UNITED STATES AND JAPAN, 1880–1980

	United States								Japan							
	Factor-using bias (B_{it}) (percentage)				Relative factor price 1880 = 100				Factor-using bias (B_{it}) (percentage)				Relative factor price 1880 = 100			
Year	Labor (L)	Land (A)	Power (M)	Fertilizer (F)	Labor (L)	Land (A)	Power (M)	Fertilizer (F)	Labor (L)	Land (A)	Power (M)	Fertilizer (F)	Labor (L)	Land (A)	Power (M)	Fertilizer (F)
1880	0	0	0	0	100	100	100	100	0	0	0	0	100	100	100	100
1885	−0.4	−0.5	0.8	0.2	102	111	88	76	6.2	−6.8	0.2	0.4	83	134	96	95
1890	−1.0	−1.7	2.3	0.3	108	109	78	67	4.5	−7.3	0.9	1.9	81	146	86	87
1895	−3.3	−0.4	3.1	0.6	107	111	79	69	10.4	−13.9	1.8	1.6	74	177	72	84
1900	−2.9	0.9	1.2	0.8	116	107	66	69	6.4	−12.4	3.2	2.8	80	170	61	67
1905	−2.6	0.1	1.3	1.3	106	141	59	45	6.9	−14.4	3.3	4.1	77	180	64	64
1910	−2.6	−0.8	1.8	1.6	110	143	52	38	5.9	−16.8	4.6	6.1	80	191	53	48
1915	−3.6	−1.0	2.7	1.8	112	147	45	38	7.5	−21.8	5.8	8.4	77	220	48	41
1920	−6.4	−0.1	4.2	2.2	125	126	39	43	−16.4	−0.6	7.7	9.2	96	159	37	33
1925	−8.6	2.3	4.5	1.8	123	134	40	33	−12.1	−5.9	8.3	9.6	101	179	28	24
1930	−11.2	2.4	6.6	2.1	135	114	38	26	−6.0	−13.1	8.8	10.3	89	220	27	24
1935	−12.0	−0.9	10.2	2.8	118	119	51	29	−13.9	−7.5	9.2	12.0	91	203	32	22
1940	−13.0	−2.8	12.6	3.1	127	109	46	28	−28.1	4.4	10.4	13.2	110	158	27	18
1945	−21.4	1.1	17.0	3.4	159	82	31	20								
1950	−28.5	3.8	20.5	4.2	166	84	28	14								
1955	−38.4	5.4	28.0	5.2	161	91	29	11	−26.3	0.1	10.9	15.2	115	133	32	18
1960	−46.5	12.6	29.1	4.8	164	97	28	9	−18.8	−8.9	11.5	16.1	99	217	28	11
1965	−54.4	12.8	35.5	6.1	158	100	29	7	−31.1	−1.9	13.1	19.7	133	181	20	7
1970	−58.7	15.3	36.9	6.3	176	112	26	5	−40.2	5.0	14.6	20.6	169	166	14	5
1975	−61.8	16.1	37.4	8.2	181	119	24	6	−49.6	10.2	16.0	23.4	205	160	10	4
1980	−63.3	12.5	43.8	7.1	156	134	24	5	−47.3	7.9	16.0	23.4	195	183	8	4

tution were estimated with the use of the two-level CES production function (see appendixes B and C for discussion of the CES model and the factor share data).

The induced innovation hypothesis will be accepted if the factor-using bias for input i is associated with a decline in the price of input i relative to other input prices. Therefore, the indexes of input prices relative to the aggregate input price index are necessary for testing the induced innovation hypothesis. The aggregate input price index was constructed by aggregating the indexes of labor, land, power, and fertilizer prices using factor shares as weights. The relative factor price indexes were obtained by deflating the indexes of individual factor prices by the aggregate input price indexes.

The time series data for factor inputs and prices used for the analysis in this section are the same as for the analysis in the previous sections, except that the work-hour data were used exclusively for the labor variable (L). For the production function analysis, assuming market equilibrium conditions, flow measures of inputs, such as the number of work hours, are more appropriate to use than stock measures, such as the number of workers. The stock measures were still used for land (A), in terms of agricultural land area, and power (M), in terms of horsepower, because flow data were not available. The 1945–50 observations for Japan—the period of devastation due to World War II—were discarded from the analysis because government price controls and rationing precluded the possibility that the assumption of competitive market equilibrium from which our model was deduced would hold during that period.

In order to test the induced innovation hypothesis, the indexes of factor-using biases, as measured by $(S'_{it}/S_{i,1880})$, are compared with the index of relative factor prices for each individual input for the United States in figure 9-5 and for Japan in figure 9-6. In both the United States and Japan, movements in the indexes of factor-using bias are negatively associated with those of relative factor prices with only a few exceptions. The results render support to the induced innovation hypothesis. The results indicate that technical change in both countries was directed toward using (or saving) the factors which became less (or more) costly relative to other factors.

A major difference in the historical sequence of technical change biases was observed between the United States and Japan. Technical change in U.S. agriculture was biased in the labor-saving direction for the entire period in response to the continuous rise in the labor wage rate relative to other input prices. Technical change in the United States was almost neutral with respect to the use of land between 1880 and 1900, a period of relative stability in land prices. It then became land-saving until 1940, apparently induced by the rise in land prices from 1900 to 1915.

In contrast, the direction of technical change in Japan was largely labor-using and land-saving until about 1915, in response to the relative decline in the wage rate and the relative rise in the price of land. After 1915, the directions of labor and land biases were reversed in response to reversals in the movements in their relative prices. Such contrasts suggest that in the early stage of economic development, labor was relatively scarce and land relatively abundant in the United States, and that labor was relatively abundant and land relatively scarce in Japan. Thus, technology in both countries was directed to saving the relatively scarce factors—which were different in the United States and Japan.

The direction of factor-using biases in the United States and Japan tended to converge during the period after World War II. However, a major inconsistency was found in the case of land in the United States in the post-World War II period, when a land-using bias was positively associated with sharp increases in the price of land. This anomaly may have been more apparent than real because of the deficiency in land price data as a measure of the price of the factor service of land. It seems reasonable to expect that the recent inconsistency between the trends in land bias and land prices might be solved, or at least considerably reduced, when national land rent data become available.

Another possible explanation for the recent inconsistency is the difficulty of distinguishing betweeen the cause and effect of factor-using bias from *ex post* empirical data, especially for the input with inelastic supply. Given the inelastic supply of land, if technical change were very responsive to a decrease in land prices in the land-using direction, the land price would then fall less than it would in the absence of technical change. In an extreme case it might not fall at all.[17] When we consider this possibility, together with the problem of the land price data, the apparent inconsistency between the land bias and the land price trend in the United States for recent years can hardly be taken as a significant qualification of the induced innovation hypothesis.

Our estimates of factor-saving biases in U.S. agriculture are largely consistent with Binswanger's (1974b). With respect to the induced innovation hypothesis, however, he found a major inconsistency between machinery-using bias and an increase in the relative price of machinery and concluded that an innate bias in technical change existed, in a machine-using direction. His conclusion was strongly criticized by Kislev and Peterson on the grounds that the USDA machinery price index used

---

[17] This problem was pointed out by Binswanger (1978c), p. 216.

by Binswanger overestimates the price increase since it is not adjusted for quality change in machinery (Kislev and Peterson, 1981). In our case, the machinery prices were adjusted for quality change. This may explain why the inconsistency did not appear in our case.

Our results for Japan were different from Nghiep's (1979), particularly regarding his findings of labor-saving bias in technical change during the prewar period. The labor-saving bias in his case was based on labor measured in terms of the number of workers. The number of workers increased much less than the number of work hours used in our analysis. In addition, his analysis covered only the period after 1903 and did not cover the earlier period for which the labor-using bias was observed in this study.

The index of factor-using bias $(S'_{it}/S_{i,1880})$ used for the analysis in figures 9-5 and 9-6 shows the rates of change in factor shares due to biased technical change. This index is an appropriate measure of the extent to which a factor input increased relative to other inputs due to biased technical change. However, it is not an appropriate measure of the absolute effect on the agricultural production cost structure of increasing the input of that factor relative to other factors due to technical change. For example, even if the rate of increase in factor-using bias for an input as measured by $(S'_{it}/S_{i,1880})$ is high, the absolute effect of that bias on factor shares will be small if the factor share in the base period $(S_{i,1880})$ is very small. On the other hand, a modest increase in $(S'_{it}/S_{i,1880})$ may result in a large change in the corresponding constant price factor share if the factor share in the base period is large. In order to compare the absolute effects of biased technical change on the production cost structure, it is more appropriate to use the absolute change between the constant-price factor share and the actual factor share in the base period $(B_{it} = S'_{it} - S_{i,1880})$ rather than the relative change $(S'_{it}/S_{i,1880})$.

The cumulative changes in factor shares due to biased technical change are compared in figure 9-7 for the United States and Japan. The absolute effects of biased technical change on the agricultural production cost structure as measured by the cumulative changes are quite different for the two countries. In the United States, biased technical change resulted in a major increase in the share of power in total production cost and an associated decrease in the share of labor. Although the rate of increase in fertilizer-using bias was high, as measured by $(S'_{it}/S_{i,1880})$, its absolute effect on the factor-share structure was relatively small, because the initial factor share for fertilizer was very small. It seems clear that the dominant effects of biased technical change in U.S. agriculture on the production structure were labor-saving and power-using (cum machinery) throughout the whole period of analysis.

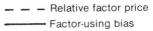
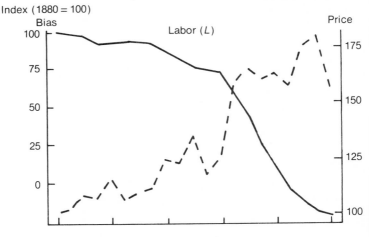
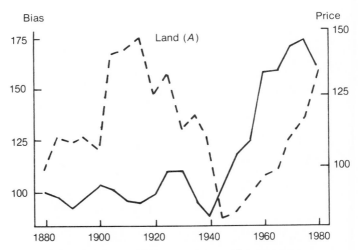

FIGURE 9-5. Individual comparisons between the indexes of factor-using biases in technical change (s'_{it}/s_i, 1880) and the indexes of factor prices relative to the aggregate input price index, the United States, 1880–1980. *Source:* data from table 9-4.

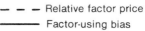
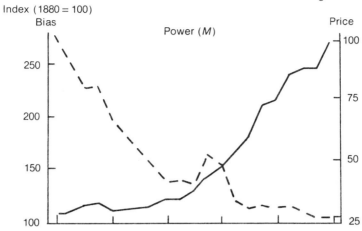
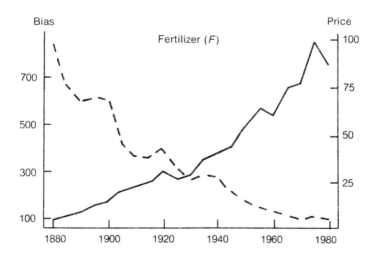

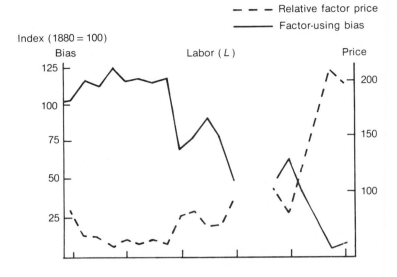
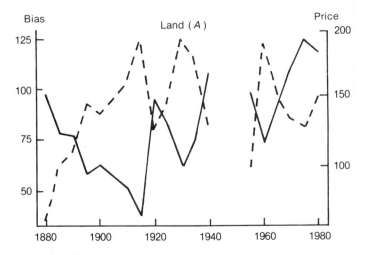

FIGURE 9-6. Individual comparisons between the indexes of factor-using biases in technical change (s'_{it}/s_i, 1880) and the indexes of factor prices relative to the aggregate input price index, Japan, 1880–1940 and 1955–1980. *Source:* data from table 9-4.

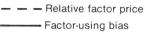
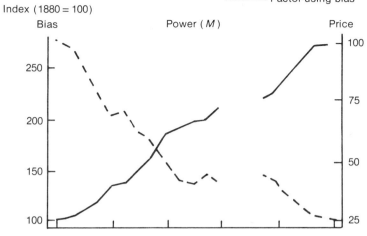
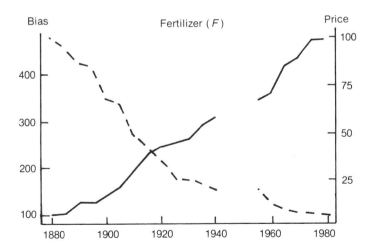

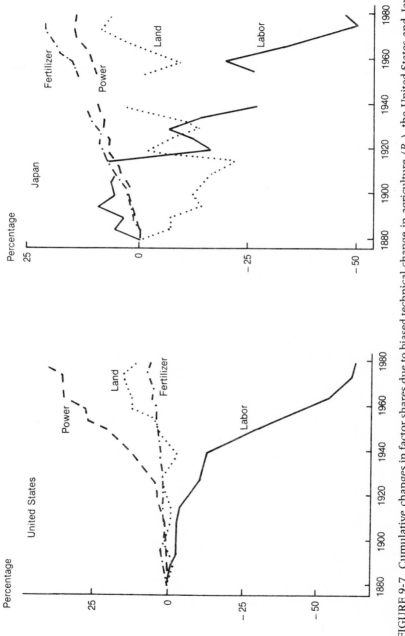

FIGURE 9-7. Cumulative changes in factor shares due to biased technical change in agriculture (B_{it}), the United States and Japan, 1880–1980. *Source*: data from table 9-4.

In contrast, in Japan the absolute effect of factor-using bias on factor shares was largest for fertilizer and second largest for power, and the factor-saving effect was the largest for labor for the entire period. However, it is noteworthy that the dominant effects of biased technical change during the early phase of economic development until about 1915 were land-saving and labor-using. It seems clear that, during the early period, in which labor was relatively abundant and land represented the major constraint on agricultural production, the major efforts of technological development in Japan were geared toward facilitating substitution of labor for land. Later, as labor wage rates rose, primarily due to increased labor demand from the nonagricultural sectors, technological development seems to have been redirected toward the labor-saving direction by facilitating substitution of power and fertilizer for labor.

GUIDING TECHNOLOGICAL CHANGE ALONG ALTERNATIVE PATHS

Despite extremely different resource endowments, both the United States and Japan were successful in sustaining growth in agricultural output and productivity over the 1880–1980 period. Common to the success of both countries was the capacity to develop agricultural technology to facilitate the substitution of relatively abundant factors for scarce factors in accordance with market price signals.

Rapid growth in agriculture in both countries could not have occurred without such dynamic factor substitution. If factor substitution had been limited to substitution along a fixed production surface, agricultural growth would have been severely limited by the inelastic supply of the more limiting factors. Development of a continuous stream of new technology, which altered the production surface to conform to long-term trends in resource endowments and factor prices, was the key to success in agricultural growth in the United States and Japan.

For both the United States and Japan, vigorous growth in the industries that supplied machinery and fertilizers at continuously declining relative prices has been an indispensable requirement for agricultural growth. Equally important were the efforts in research and extension to exploit fully the opportunities created by industrial development. Without the creation of fertilizer-responsive crop varieties, the benefits from the lower fertilizer prices would have been limited. The success in agricultural growth in both the United States and Japan seems to lie in the capacity of their farmers, research institutions, and farm supply industries to exploit new opportunities in response to the information transmitted through relative price changes.

APPENDIX 9-A.
DERIVATION OF EQUATIONS FOR MEASURING THE EFFECT OF BIASED TECHNICAL CHANGE

Assuming a homothetic production function as specified in equation (9-1) and also assuming the competitive market equilibrium, the cost function that relates production cost (C) to output and real input prices ($P_1, \ldots, P_n$) can be derived as

$$C = Q \cdot g(P_1/E_1, \ldots, P_n/E_n) \tag{9A-1}$$

where g is the unit cost function. The input prices (P_is) are divided by the factor-augmenting coefficients (E_is) because a proportional increase in E_i has the same effect as a proportional decrease in P_i. From Shephard's lemma, we obtain

$$\frac{\partial C}{\partial P_i} = X_i = \frac{Q}{E_i}\left(\frac{\partial g}{\partial P_i}\right), \quad i = 1, \ldots, n \tag{9A-2}$$

The unit cost function is related to the Allen partial elasticity of substitution (σ_{ij}) in the following way

$$\sigma_{ij} = \frac{\dfrac{\partial^2 g}{\partial P_i \partial P_j} g}{\dfrac{\partial g}{\partial P_i} \dfrac{\partial g}{\partial P_j}} \tag{9A-3}$$

The differentiation of equation (9A-2) with respect to time (t) using equation (9A-3) results in

$$\frac{\dot{X}_i}{X_i} = \frac{\dot{Q}}{Q} - \frac{\dot{E}_i}{E_i} + \sum_j s_j \sigma_{ij}\left(\frac{\dot{P}_j}{P_j} - \frac{\dot{E}_j}{E_j}\right) \tag{9A-4}$$

where the dot denotes the time derivative and s_j is the factor share of the jth input. Equation (9A-4) is the factor demand function in the growth equation form. Since $s_j \sigma_{ij}$ is the demand elasticity of input i with respect to the price of input j, and the factor demand is homogeneous of degree zero with respect to input prices, the following relation holds

$$\sum_j s_j \sigma_{ij} = 0 \tag{9A-5}$$

INDUCED TECHNICAL CHANGE   281

Equation (9-2) can be obtained by incorporating the above equation into equation (9A-4) and using the relations $\sum_j s_j = 1$ and

$$\frac{\dot{Q}}{Q} = \sum_j s_j \left(\frac{\dot{X}_j}{X_j} + \frac{\dot{E}_j}{E_j} \right)$$

Equation (9-3) can be derived by expressing the ith factor share as

$$\frac{\dot{s}_i}{s_i} = \frac{\dot{P}_i}{P_i} + \frac{\dot{X}_i}{X_i} - \frac{\dot{Q}}{Q} \tag{9A-6}$$

Because total output is fully distributed to inputs so that $Q = \sum_j P_j X_j$, the following relation holds

$$\frac{\dot{Q}}{Q} = \sum_j s_j \left(\frac{\dot{P}_j}{P_j} + \frac{\dot{X}_j}{X_j} \right) \tag{9A-7}$$

Substitution of equation (9A-7) into equation (9A-6) results in

$$\frac{\dot{s}_i}{s_i} = \sum_{j \neq i} s_j \left(\frac{\dot{P}_i}{P_i} - \frac{\dot{P}_j}{P_j} \right) + \sum_{j \neq i} s_j \left(\frac{\dot{X}_i}{X_i} - \frac{\dot{X}_j}{X_j} \right) \tag{9A-8}$$

Equation (9A-3) is obtained by inserting equation (9A-4) into equation (9A-8).

APPENDIX 9-B.
TWO-LEVEL CES PRODUCTION FUNCTION

The two-level CES production function is specified for agriculture including labor (L), land (A), machinery (M), and fertilizer (F) as four factors of production to produce output (Q) gross of intermediate products. As it is reasonable to assume that machinery is a factor substituting mainly for labor, and fertilizer is a factor substituting for land, it seems appropriate to specify the first-level equation of the two-level CES function as

$$Z_1 = [\alpha(e^{\delta_L t} L)^{-\rho_1} + (1 - \alpha)(e^{\delta_M t} M)^{-\rho_1}]^{-1/\rho_1} \tag{9B-1}$$

$$Z_2 = [\beta(e^{\delta_A t} A)^{-\rho_2} + (1 - \beta)(e^{\delta_F t} F)^{-\rho_2}]^{-1/\rho_2} \tag{9B-2}$$

where ρs are substitution parameters; α and β are distribution parameters; and δ_is ($i = L, M, A, F$) are the rates of factor augmentation for respective inputs, which are assumed to be constant for the relevant periods. It is also assumed that both functions are linear homogeneous.[18]

To use the terms of Sen (1959), Z_1 is considered an aggregate of "laboresque" inputs and Z_2 is an aggregate of "landesque" inputs. As it is reasonable to expect that the laboresque and the landesque inputs are largely separable, the relation between output (Q) and the two categories of inputs may be expressed by the second-level CES function as follows

$$Q = [\gamma(Z_1)^{-\rho} + (1-\gamma)(Z_2)^{-\rho}]^{-1/\rho} \tag{9B-3}$$

where ρ and γ are the parameters of substitution and distribution, respectively.

Whereas the direct elasticities of substitution are expressed as $\sigma_1 = 1/(\rho_1 + 1)$, $\sigma_2 = 1/(\rho_2 + 1)$ and $\sigma = 1/(\rho + 1)$, the Allen partial elasticities of substitution are expressed by

$$\sigma_{LA} = \sigma_{LF} = \sigma_{AM} = \sigma_{MF} = \sigma \tag{9B-4}$$

$$\sigma_{LM} = \sigma + \frac{1}{s_1}(\sigma_1 - \sigma) \tag{9B-5}$$

$$\sigma_{AF} = \sigma + \frac{1}{s_2}(\sigma_2 - \sigma) \tag{9B-6}$$

where s_1 and s_2 are the factor shares of the first and the second categories of inputs, respectively.

The CES production parameters are estimated using the competitive market equilibrium conditions as follows

$$\ln\left(\frac{M}{L}\right) = -\frac{1}{\rho_1 + 1}\ln\left(\frac{\alpha}{1-\alpha}\right) + \frac{1}{\rho_1 + 1}\ln\left(\frac{P_L}{P_M}\right) + \frac{\rho_1}{\rho_1 + 1}(\delta_L - \delta_M)t \tag{9B-7}$$

$$\ln\left(\frac{F}{A}\right) = -\frac{1}{\rho_2 + 1}\ln\left(\frac{\beta}{1-\beta}\right) + \frac{1}{\rho_2 + 1}\ln\left(\frac{P_A}{P_F}\right) + \frac{\rho_2}{\rho_2 + 1}(\delta_A - \delta_F)t \tag{9B-8}$$

$$\ln\left(\frac{Z_2}{Z_1}\right) = -\frac{1}{\rho + 1}\ln\left(\frac{\gamma}{1-\gamma}\right) + \frac{1}{\rho + 1}\ln\left(\frac{P_1}{P_2}\right) \tag{9B-9}$$

where P_L, P_A, P_M, and P_F are the prices of labor, land, machinery, and fertilizer, respectively, and P_1 and P_2 are the prices of Z_1 and Z_2, which are calculated by

---

[18] In equations (9-12) and (9-13), input variables are normalized to be one and time is set to be zero at the base period. In estimates of the CES function based on equations (9-18), (9-19), and (9-23), 1960 was used as the base period.

$$P_1 = (P_L L + P_M M)/Z_1 \tag{9B-10}$$

$$P_2 = (P_A A + P_F F)/Z_2 \tag{9B-11}$$

From the regression estimation of equations (9B-7) and (9B-8), we can obtain estimates of ρ_1, ρ_2, α, β, $(\delta_L - \delta_M)$, and $(\delta_A - \delta_F)$ but cannot obtain δ_L, δ_M, δ_A, and δ_F separately. Therefore, Z_1 and Z_2 cannot be measured directly and, hence, equation (9B-9) cannot be estimated. Therefore, in order to proceed to the second-level estimation, we define as

$$\hat{Z}_1 = e^{-\delta_M t} Z_1; \ \hat{Z}_2 = e^{-\delta_F t} Z_2;$$

$$\hat{P}_1 = e^{\delta_M t} P_1; \ \hat{P}_2 = e^{\delta_F t} P_2$$

which can be obtained from equations (9B-7), (9B-8), (9B-10), and (9B-11). Substitution of those relations into equation (9B-9) produces the following estimable equation

$$\ln \frac{\hat{Z}_2}{\hat{Z}_1} = -\frac{1}{\rho+1} \ln\left(\frac{\gamma}{1-\gamma}\right) + \frac{1}{\rho+1} \ln \frac{\hat{P}_1}{\hat{P}_2} + \frac{\rho}{\rho+1}(\delta_M - \delta_F)t \tag{9B-12}$$

From the regression estimates of $(\delta_L - \delta_M)$, $(\delta_A - \delta_F)$, and $(\delta_F - \delta_M)$ based on equations (9B-7), (9B-8), and (9B-12), the differential rates of factor augmentation for any pairs of inputs can be obtained.

A time dummy variable was included in the regression analysis in order to adjust for possible structural change among different phases of economic development. The dummy takes the value of zero for the 1880–1925 period and one for the 1930–1980 period in the United States, and zero for the 1880–1940 period and one for the 1955–1980 period in Japan. The time demarcation for Japan in the pre-World War II and the postwar periods should be natural considering the obvious structural change in agriculture as well as the whole economy. On the other hand, it is not clear what time demarcation may be used for the United States. Therefore, we tried several alternative demarcations and chose the one that gave the best results in terms of goodness of fit and statistical significance of estimated parameters. Through this procedure, the whole period was divided into two subperiods—before 1930 and since 1930.

For the estimation of the first-level equations (9B-7) and (9B-8), the generalized least squares (GLS) method is used. For the second-level equation, OLS is applied. Slope dummies for the relative prices were deleted from the regressions, because their coefficients were either insignificant or irrelevant in sign. The results are summarized in table 9B-1. The Allen partial elasticities of substitution derived from these results are shown in table 9B-2.

TABLE 9B-1. REGRESSIONS FOR ESTIMATING THE PARAMETERS OF THE TWO-LEVEL CES PRODUCTION FUNCTION: UNITED STATES AND JAPAN, 1880–1980

Regression number	Dependent variables	Coefficients of			Dummy variables for		Coeff. of det. ($\bar{R}^2$)	S.E. of estimates	Durbin-Watson statistics
		Intercept	Relative price	Time trend	Intercept	Time trend			
United States:									
First level:									
(R 1)	Machinery-labor ratio (M/L)	−2.121 (0.114)	0.191 (0.110)	0.037 (0.017)	2.027 (0.120)	0.334 (0.013)	—	0.083	0.98
(R 2)	Fertilizer-land ratio (F/A)	−0.174 (0.135)	0.349 (0.112)	0.183 (0.021)	0.176 (0.139)	0.073 (0.017)	—	0.102	1.39
Second level:									
(R 3)		−0.514 (0.364)	0.191 (0.276)	0.198 (0.050)	0.611 (0.367)	−0.093 (0.042)	0.997	0.099	1.18
Japan:									
First level:									
(R 4)	Machinery-labor ratio (M/L)	−1.160 (0.134)	0.111 (0.153)	−0.019 (0.023)	1.237 (0.125)	0.704 (0.052)	—	0.116	1.72
(R 5)	Fertilizer-land ratio (F/A)	0.183 (0.132)	0.182 (0.210)	0.187 (0.042)	−0.099 (0.182)	−0.110 (0.047)	—	0.157	0.71
Second level:									
(R 6)		0.055 (0.093)	0.239 (0.109)	0.219 (0.018)	−0.109 (0.116)	−0.582 (0.047)	0.996	0.080	1.20

Note: Based on quinquennial observations. For Japan, 1945–1950 observations discarded. The first-level equations are estimated by generalized least squares. The second-level equations are estimated by ordinary least squares. Standard errors of the estimated coefficients are in parentheses.

Source: Data from Y. Hayami and V. W. Ruttan, *Agricultural Development: An International Perspective*, rev. ed. (Baltimore, Md.: Johns Hopkins University Press, 1985) appendix C.

TABLE 9B-2. ESTIMATES OF THE ALLEN PARTIAL ELASTICITIES OF SUBSTITUTION

	United States		Japan	
	1880–1925	1930–1980	1880–1940	1955–1980
Allen partial elasticity of substitution:				
σ_{LM}	.191	.191	.029	.013
σ_{AF}	.777	.741	.093	.108
Other	.191	.191	.239	.239

Notes: A = land; M = power (machinery); L = labor; and F = fertilizer.

APPENDIX 9-C.
FACTOR SHARES AND AGGREGATE INPUT PRICE INDEXES IN THE UNITED STATES AND JAPAN

Factor shares for labor, land, machinery, and fertilizer in the United States were estimated by Binswanger (1978c) for 1912–1968. We extrapolated the Binswanger estimates of factor shares by the input value indexes that were the products of input quantity indexes and price indexes calculated from the data in appendix C, *Agricultural Development* (Hayami and Ruttan, 1985). The shares of labor, land, power (machinery), and fertilizer were adjusted to add up to 100. For Japan, the estimates by Yamada (1972) were used. It was assumed that the factor shares of capital and current inputs correspond to the shares of power and fertilizer, respectively. This assumption may be justified as a first approximation, considering the dominant weights of draft animals and machinery in the value of capital and of fertilizer in the value of current inputs in Japanese agriculture. It was also assumed that the factor shares for 1880 and 1980 were the same as for 1885 and 1975.

The aggregate input price index was constructed by aggregating indexes of labor, power, and fertilizer prices by using factor shares as weights. The factor share weights are changed for every five-year interval, and the indexes constructed with the different weights are chain-linked to a single aggregate input price index.

REFERENCES

Ahmad, S. 1966. "On the Theory of Induced Invention," *Economic Journal* vol. 76, no. 302, pp. 344–357.

―――. 1967a. "Reply to Professor Fellner," *Economic Journal* vol. 77, no. 307, pp. 664–665.

―――. 1967b. "A Rejoinder to Professor Kennedy," *Economic Journal* vol. 77, no. 308, pp. 960–963.

Binswanger, H. P. 1974a. "A Microeconomic Approach to Induced Innovation," *Economic Journal* vol. 84, no. 336, pp. 940–958.

―――. 1974b. "The Measurement of Technical Change Biases with Many Factors of Production," *American Economic Review* vol. 64, no. 6, pp. 964–976.

―――. 1978a. "Induced Technical Change: Evolution of Thought," in H. P. Binswanger and V. W. Ruttan, eds., *Induced Innovation: Technology, Institutions and Development* (Baltimore, Md., Johns Hopkins University Press).

―――. 1978b. "The Microeconomics of Induced Technical Change," in Hans P. Binswanger and Vernon W. Ruttan, eds., *Induced Innovation: Technology, Institutions and Development* (Baltimore, Md., Johns Hopkins University Press).

―――. 1978c. "Measured Biases of Technical Change: The United States," in Hans P. Binswanger and Vernon W. Ruttan, eds., *Induced Innovation: Technology, Institutions and Development* (Baltimore, Md., Johns Hopkins University Press).

Bronfenbrenner, M. 1944. "Production Functions: Cobb-Douglas, Interfirm, Intrafirm," *Econometrica* vol. 12, no. 1, pp. 35–44.

Brown, M. 1966. *On the Theory and Measurement of Technological Change* (Cambridge, England, Cambridge University Press).

Conlisk, J. 1969. "A Neoclassical Growth Model with Endogenously Positioned Technical Change Frontier," *Economic Journal* vol. 79, no. 314, pp. 348–362.

David, P. A. 1966. "The Mechanization of Reaping in the Ante-Bellum Midwest," in H. Rosovsky, ed., *Industrialization in Two Systems* (New York, N.Y., Wiley).

Doll, J. P., and R. Widdows. 1982. *A Comparison of Cash Rent and Land Values for Selected U.S. Farming Regions* (Washington, D.C., U.S. Department of Agriculture, Economic Research Service).

Drandakis, E. M., and E. S. Phelps. 1966. "A Model of Induced Invention, Growth and Distribution," *Economic Journal* vol. 76, no. 304, pp. 823–840.

Färe, R., and L. Jansson. 1974. "Technological Change and Disposability of Inputs," *Zeitschrift für Nationalökonomie* vol. 34, no. 3, pp. 283–290.

Fellner, W. 1961. "Two Propositions in the Theory of Induced Innovations," *Economic Journal* vol. 71, no. 282, pp. 305–308.

―――. 1962. "Does the Market Direct the Relative Factor-Saving Effects of Technological Progress?" in R. R. Nelson, ed., *The Rate and Direction of Inventive Activity* (Princeton, N.J., Princeton University Press for National Bureau of Economic Research), pp. 171–193.

―――. 1967. "Comment on the Induced Bias," *Economic Journal* vol. 77, no. 307, pp. 662–664.

Grabowski, R., and D. Sivan. 1983. "The Direction of Technological Change in Japanese Agriculture, 1874–1971," *The Developing Economies* vol. 21, no. 3, pp. 234–243.

Griliches, Z. 1957. "Hybrid Corn: An Exploration in the Economics of Technical Change," *Econometrica* vol. 25, no. 4, pp. 501–522.

———. 1958. "The Demand for Fertilizer: An Economic Interpretation of a Technical Change," *Journal of Farm Economics* vol. 40, no. 3, pp. 591–606.

Hayami, Y., and V. W. Ruttan. 1973. "Professor Rosenberg and the Direction of Technological Change: A Comment," *Economic Development and Cultural Change* vol. 21, no. 2, pp. 352–355.

——— and V. W. Ruttan. 1985. *Agricultural Development: An International Perspective*, rev. ed. (Baltimore, Md., Johns Hopkins University Press).

Hicks, J. R. 1932. *The Theory of Wages* (London, Macmillan).

Kako, T. 1978. "Decomposition Analysis of Derived Demand for Factor Inputs: The Case of Rice Production in Japan," *American Journal of Agricultural Economics* vol. 60, no. 4, pp. 628–635.

Kaneda, H. 1982. "Specification of Production Functions for Analyzing Technical Change and Factor Inputs in Agricultural Development," *Journal of Development Economics* vol. 11, no. 1, pp. 97–108.

Kennedy, C. 1964. "Induced Bias in Innovation and the Theory of Distribution," *Economic Journal* vol. 74, no. 295, pp. 541–547.

———. 1966. "Samuelson on Induced Innovation," *Review of Economics and Statistics* vol. 48, no. 4, pp. 442–444.

———. 1967. "On the Theory of Induced Invention—A Reply," *Economic Journal* vol. 77, no. 308, pp. 958–960.

Kislev, Y., and W. Peterson. 1981. "Induced Innovations and Farm Mechanization," *American Journal of Agricultural Economics* vol. 63, no. 3, pp. 562–565.

——— and W. Peterson. 1982. "Prices, Technology, and Farm Size," *Journal of Political Economy* vol. 90, no. 3, pp. 578–595.

Mowery, D., and N. Rosenberg. 1979. "The Influence of Market Demand upon Innovation: A Critical Review of Some Recent Empirical Studies," *Research Policy* vol. 8, no. 2, pp. 102–153.

Nelson, R. R. 1959. "The Economics of Invention: A Survey of the Literature," *Journal of Business* vol. 32, no. 2, pp. 101–127.

Nghiep, L. T. 1979. "The Structure and Changes of Technology in Prewar Japanese Agriculture," *American Journal of Agricultural Economics* vol. 61, no. 4, pp. 687–693.

Nordhaus, W. D. 1973. "Some Skeptical Thoughts on the Theory of Induced Innovation," *Quarterly Journal of Economics* vol. 87, no. 2, pp. 208–219.

Rosenberg, N. 1969. "The Direction of Technological Change: Inducement Mechanisms and Focusing Devices," *Economic Development and Cultural Change* vol. 18, no. 1, pp. 1–24.

Salter, W. E. G. 1960. *Productivity and Technical Change* (Cambridge, England, Cambridge University Press).
Samuelson, P. A. 1965. "A Theory of Induced Innovation Along Kennedy-Weisacker Lines," *Review of Economics and Statistics* vol. 47, no. 4, pp. 343–356.
———. 1966. "Rejoinder: Agreements, Disagreements, Doubts, and the Case of Induced Harrod-Neutral Technical Change," *Review of Economics and Statistics* vol. 48, no. 4, pp. 444–448.
Sato, K. 1967. "A Two-Level Constant-Elasticity-of-Substitution Production Function," *Review of Economic Studies* vol. 34 (April), pp. 201–218.
Scherer, F. M. 1982. "Demand Pull and Technological Invention: Schmookler Revisited," *Journal of Industrial Economics* vol. 30, no. 3, pp. 225–238.
Schmookler, J. 1962. "Changes in Industry and in the State of Knowledge as Determinants of Industrial Invention," in R. R. Nelson, ed., *The Rate and Direction of Inventive Activity* (Princeton, N.J., Princeton University Press for National Bureau of Economic Research), pp. 195–232.
———. 1966. *Invention and Economic Growth* (Cambridge, Mass., Harvard University Press).
Sen, A. K. 1959. "The Choice of Agricultural Techniques in Underdeveloped Countries," *Economic Development and Cultural Change* vol. 7, no. 3, pp. 279–285.
Shintini, M., and Y. Hayami. 1975. "Ngoyo ni okeru Yosoketsugo to Henkoteki Gitjutsushimpo" (Factor Combination and Biased Technical Change in Agriculture), in *Kindai Nihon no Keizai Hatten (Economic Development of Modern Japan)* (Tokyo, Toyokeizaishimposha).

Thirtle, C. G., 1982. "Induced Innovation in United States Agriculture" (Ph.D. dissertation, Columbia University).
———. 1985. "The Microeconomic Approach to Induced Innovation: A Reformulation of the Hayami and Ruttan Model," *The Manchester School of Economic and Social Studies* vol. 53, no. 1, pp. 263–279.
———, and V. W. Ruttan. 1987. *The Role of Demand and Supply in the Generation and Diffusion of Technical Change* (Chur, Switzerland, Harwell Academic Publishers).

van de Klundert, T., and R. J. de Groof. 1977. "Economic Growth and Induced Technical Progress," *De Economist* (The Netherlands) vol. 125, pp. 505–524.

Wan, H. Y., Jr. 1971. *Economic Growth* (New York, N.Y., Harcourt Brace Jovanovich).

Yamada, Saburo. 1982. "The Secular Trends in Input-Output Relations of Agricultural Production in Japan, 1878–1978," in Chi-ming Hou and Tzong-shian Yu, eds., *Agricultural Development in China, Japan and Korea* (Taipei, Academia Sinica).
Yeung, P., and T. Roe. 1978. "A CES Test of Induced Technical Change: Japan," in H. P. Binswanger and V. W. Ruttan, eds., *Induced Innovation: Technology, Institutions and Development* (Baltimore, Md., Johns Hopkins University Press).

10

RESEARCH, EXTENSION, AND U.S. AGRICULTURAL PRODUCTIVITY:
A STATISTICAL DECOMPOSITION ANALYSIS

ROBERT E. EVENSON

Since the pioneering studies by Schultz (1953) and Griliches (1958a,b), a large body of literature on measuring and explaining productivity change in U.S. agriculture has been produced (see chapter 3 in this volume and Norton and Davis, 1981). Several of these studies use simple benefit-cost methods. These studies usually estimate a shift in the supply curve of a commodity or commodities—often in a very simple and direct fashion—and proceed to apply discounting techniques to compute internal rates of return or benefit-cost ratios. A second class of studies in this field uses productivity decomposition methods. Some researchers use an aggregate production function framework, as outlined in chapter 2 of this volume, and include variables measuring research, schooling, and extension stocks as productivity explanation or decomposition variables. Other studies of this type utilize a two-stage procedure. In the first stage a total, or multifactor, productivity index is computed, usually for different regions and time periods. In the second stage, regression analysis is used to decompose this productivity measure by regressing it on research, extension schooling, and other variables.

In this type of study the analyst faces several definitional questions regarding the variables used. These include whether the variable should be in stock or flow form, and how it should be deflated to be comparable across observations. The existing literature uses many variable definitions but few papers provide more than a rudimentary justification for the definitions used.

This chapter addresses the definitional issues that arise in this type of analysis by relating them to the process by which productivity gains are achieved. Two studies of U.S. agricultural productivity change are discussed to illustrate applications of several of these methodological problems.

METHODOLOGY AND RESEARCH PROCESSES

Productivity decomposition can be undertaken with a production function specification or in a system of output supply and factor demand equations. A two-stage analysis is also possible. This analysis requires the calculation of productivity residuals that in turn are then regressed on decomposition variables. In all procedures the decomposition variables must be properly matched to the unit of observation. The unit in question is typically either a productivity level (relative to some base) for a farm or an aggregation of farms in a particular region in a particular time period or a change in a productivity level for the farm or aggregate in question.

The next step is to identify the location, time period, and research (or other) programs that could be expected to affect that level or change in level of productivity. Three dimensions are relevant. The first is the research specialization dimension. The second is the locational specificity on spatial dimension and the associated issue of a deflator. The third is the timing dimension.

Specialization in Agricultural Research Systems

Specialization in research systems is often characterized as falling into three classes: (a) basic or scientific research, (b) applied, technological, or mission-oriented research, and (c) development. These terms have been widely used in the general science policy literature. I find this three-level classification system inappropriate for agricultural research (and most other research as well) because it draws a sharp distinction between basic and applied research that is inappropriate. Such a distinction is inappropriate because it implies that basic research is (and should be) effectively immune from mission orientation (or demand pressures). It is also inappropriate as a characterization of actual agricultural research systems.

I find the four-level classification system depicted in figure 10-1 more appropriate. Level III, technology invention, is roughly the equivalent of mission-oriented applied research. I prefer to use the term "invention" to highlight the objectives of producing new mechanical, biological, managerial, policy, and social technology. (I do not confine the concept of invention to patentable invention.) Level IV is downstream from level III because it is derivative from level III. It includes subinvention, testing, screening, and so forth; that is, it includes what is conventionally described as "development" in "research and development."

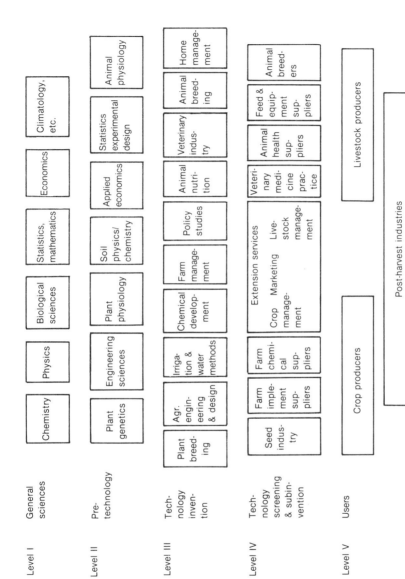

FIGURE 10-1. Classification for agricultural research systems

The important addition in my classification is level II, "pretechnology science." These activities are distinguished from level III in that they do not produce inventions of technology per se, but rather produce pretechnology scientific findings that are inputs to inventions processes. They are distinguished from the truly general sciences in that they are designed to produce pretechnology scientific knowledge rather than general scientific knowledge. In order to be effective, they must be institutionally linked to downstream inventors and they must have a capacity to respond to derived demands from downstream.

This, of course, does not mean that level I general sciences do not produce some pretechnology findings. Many of the pretechnology findings produced by level I are important, as recent experience with biotechnological advances shows. Furthermore, the general sciences are motivated to produce useful results. They are not, however, set up to respond to specific demands from below, as is the case with level II research. It should be noted further that level II researchers pay a price for their downstream demand responsiveness in that they grow more slowly, in an absolute sense, than they would if unrestricted by demand factors. A given level II research investment will accordingly produce fewer increments to knowledge, as measured by conventional standards (publications and citations of publications), than will the same investment in level III research. On the other hand, level II research products will generally have higher economic value, because they respond to economic demand (Wright and Evenson, 1980).

It is reasonable, then, to regard level II research as an input into level III research. Because it must precede level III research to be productive, it will obviously have a different time dimension. It will also have a different spatial dimension.

Specialization and Location Specificity: The Spatial Dimension

In the previous section, I noted that research conducted in one specialized unit is (or can be) of value downstream (as from level II to level III or from level III to level IV). We could say that it is transferable downstream or that it "spills" downstream (it also spills upstream). Research conducted at a particular location is similarly transferable to other locations. Each location (say, a country, or even a particular tract of land) is characterized by soil, climate, economic, and managerial environments. Linkages between these environmental characteristics of a given location and agricultural technology exist. The agronomy literature describes these links or ties as *genotype-environment interactions*. The economic meaning of these interactions is simply that the cost or profit implications of a particular technology change with the environment. That is, a given technique of production—say, a particular variety of rice—will have different cost implications in different environments. Because it is possible to tailor technology to specific environments, a given technology (or package of technologies) typically has an absolute cost or profit advantage over alternative technologies (or packages of technologies) only in a limited range of environments, as shown in figure 10-2. Panel A in figure 10-2 shows the relationship between absolute yield and an index of the water environment for three types of rice varieties. Note that all three varietal types give highest yield under good (adequate) water environments. Variety 1, a semidwarf high-yielding variety, gives the highest yields (we assume that other factors of production are at economically optimal levels). As we move to drier water-stressed environments, the upland drought-resistant variety 2 shows a comparative advantage over variety 1. As we move to locations with excess water, taller varieties such as variety 3 have the comparative advantage.

Panel B shows average variable costs associated with the three varieties in the different environments. It illustrates the point that residual land rents (prices less average variable cost) will vary by environment (peaking in the best environment), and that cost-minimizing technology will vary by location.

Panel C illustrates an improvement in the yield performance of variety 1 (from V_1 to V_1'). Note that this improvement will tend to expand the area in which V_1 type technology is superior (from E_2-E_3 to E_1-E_4). In other words, in locations between E_1 and E_2 and E_3 and E_4, the improved V_1' varietal technology will "spill in" to these locations, even though it is not tailored to these environments, as long as these regions do not improve varieties 2 and 3. If V_2 type varieties are improved, these may then spill back into the E_1-E_2 environments (and possibly beyond).

We can now see some of the factors that must be considered if we want to specify which research and inventive activity is producing technology for a specific location (this is the basic issue in the productivity-R&D specification). In locations to the left of E_1, research directed toward V_2 type varieties is relevant, but research directed toward V_1 or V_3 type technology is not—unless that research indirectly affects the productivity of V_1 type research. This is quite possible, of course. Suppose we have a single level III experiment station located in an environment between E_2 and E_3, and that it is targeting its research to this range of environments, that is, it is producing V_1 type technology. The appropriate technology generation for location E_1 is still quite complex. We can visualize it in two stages. In the short run, we can specify it for a location E_i as

$$T_i = T(\text{III}_i, P_i^*) \tag{10-1}$$

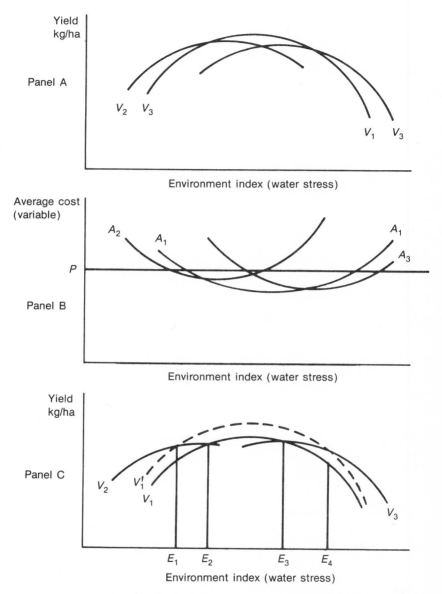

FIGURE 10-2. Relationship between an index of water environment and yields and average costs for rice

where III_i is the inventive level III research targeted to environment i and P_i^* is the stock of potential invention targeted to environment i.[1] In the long run, this stock is affected by upstream level II research and by level III research targeted to environments similar to environment i. Level II research, it may be noted, is also targeted to environments. Expression (10-2) attempts to capture this long-run process

$$P_i^* = P(c_{ij} II_j, d_{ij} III_j) \tag{10-2}$$

The c_{ij} and the d_{ij} terms are erosion coefficients, taking values between 0 and 1. (c_{ii} and d_{ii} are equal to one by definition.) They show the loss or erosion of research findings utilized in environment i when the research is targeted to environment j.

In actuality, environments are more complex than depicted in figure 10-2. The design of research system targeting may be much finer than indicated in figure 10-2, so that few locations are exclusively served by a single research station (as in the E_2-E_3 range). Most regions are subject to spill-in. The matter is further complicated by environmental concerns. Further, many research programs are designed to improve more than one commodity.

Technological research (that is, level III research) in other regions has two possible routes by which it affects technological improvements in region i. Technology produced for region j could induce invention and discovery by level III researchers in region i by disclosing information of value. Alternatively, the discovery or invention in region j might directly spill in to region i. In figure 10-2, panel C, a location in the E_1-E_2 region will be subject to spill-in from both V_1 and V_2 targeted research programs.

The specification of an appropriate R&D variable for a particular location is then twofold. First, a particular research program must be associated with the unit of observation in which productivity is measured. For example, most State Agricultural Experiment Stations (SAES), and some USDA units located in the states, may be presumed to be targeting research to the farm environments in the state. The problem in statistical work is to properly deflate these research input measures. Second, research undertaken in other research centers (other SAESs and regional or commodity-oriented USDA units) must be specified. These specifications should be sensitive to the differences in level III and II (and IV) research.

---

[1] See Kislev and Evenson (1975, chapter 8), for a specification of the search model inherent in equation (10-1).

Appropriate Deflators

Suppose we have two productivity observations (for example, states), each with a research program (SAES). Consider first only one commodity. State 1 produces twice as much of the commodity as state 2. State 1 has two distinct environmental regions of equal size. State 2 has three environmental regions, one accounting for half of its production, the other two for one quarter each. How much research in state 2 would be required to produce productivity growth equal to that in state 1 in a particular period in the future if both programs were equally productive? A proper deflator should be based on this question.

Suppose for simplicity that we have two alternative specifications of the relationship between level III research in state i, region j, III_{ij}, and productivity as production. The first, equation (10-3) is a productivity rate specification, where P_{ij} is a measure of the rate of productivity change relative to some base period

$$P_{ij} = \sum_j A_{ij} \text{III}^{\alpha_{ij}} \tag{10-3}$$

The second equation (10-3'), is a production function specification in which the dependent variable is total output or product

$$Q_{ij} = \sum_j A'_{ij} X_{ij}^{\beta_{ij}} \text{III}_{ij}'^{\alpha'_{ij}} \tag{10-3'}$$

where X_{ij} is a vector of conventional inputs.

Now if state policymakers wish to maximize the value of productivity gains subject to the costs of research stock acquisition C_{ij}, they will solve

$$\max I_i = \sum_j P_{ij} V_{ij} - \text{III}_{ij} C_{ij} \tag{10-4}$$

where V_{ij} is the value of product and

$$\max I'_i = \sum_j Q_{ij} P'_{ij} - \sum_j X_{ij} R'_{ij} - \text{III}_{ij} C'_{ij} \tag{10-4'}$$

where P'_{ij} and R'_{ij} are prices of outputs and inputs. The first-order conditions for research funding levels from equations (10-4) and (10-4') are

$$\alpha_{ij} V_{ij}/\text{III}_{ij} = C_{ij} \quad \text{or} \quad \text{III}_{ij} = \alpha_{ij} V_{ij}/C_{ij} \tag{10-5}$$

and

$$\alpha'_{ij} Q_{ij} P'_{ij} / \text{III}'_{ij} = C'_{ij} \quad \text{or} \quad \text{III}'_{ij} = \alpha'_{ij} V_{ij} / C'_{ij} \tag{10-5'}$$

The first-order conditions are the same for both specifications, except for a deflator difference (see below). They imply that if the price of research is constant across regions (or commodities) and the technology by which research activity generates productivity gains is also constant—that is, $\alpha_{ij} = \alpha_i$ for all j—then optimal research spending will be proportional to the value of the product. The price or cost of research resources is generally the same for different regions and commodities in a country and even similar in real terms across countries.[2] The technology of technology production is less likely to be the same for different commodities, however. Before exploring that issue with a more complicated model, however, it will be useful to develop the implications for a deflator from the above specifications.

Suppose that we cannot estimate equations (10-3) and (10-3') because we lack data for each region or commodity and that instead we estimate aggregate functions

$$P_i = \sum_j S_{ij} P_{ij} = \sum_j S_{ij} A_{ij} \text{III}_{ij}^{\alpha_{ij}} = \left(\sum_j A_{ij}\right)\left(\sum_j S_{ij} \text{III}_{ij}\right)^{\alpha_{ij}} \tag{10-6}$$

$$Q_i = \sum_j Q_{ij} P'_{ij} = \sum_j A_{ij} X_{ij}^{\beta_{ij}} \text{III}_{ij}^{\alpha_{ij}} = \sum_j A_{ij} (\Sigma X_i^{\beta_{ij}}) \left(\sum_j \text{III}_{ij}\right)^{\alpha_i} \tag{10-6'}$$

The term $\sum_j S_{ij} \text{III}_{ij}^{\alpha_{ij}}$ cannot be aggregated easily into a single research variable without knowledge of α_{ij}. Typically researchers simply sum research spending and ignore the fact that

$$\sum_j S_{ij} \text{III}_{ij}^{\alpha_{ij}} \neq \left(\sum_j S_{ij} \text{III}_{ij}\right)^{\alpha_{ij}} \tag{10-7}$$

Of course, alternative specifications of a linear type can avoid some of these problems. The point to be made here is that the S_{ij} term enters into equation (10-6) but not into equation (10-6'). Note that $\text{III}_{ij} = R_{ij} \sum_j \text{III}_{ij}$.

The term in equation (10-6) is $\sum_j S_{ij} R_{ij} \text{III}_i$ and can be approximated by

[2] See Judd, Boyce, and Evenson (1986) for evidence on research costs in different countries.

deflating total research spending in state i by a deflator, $\sum_j S_{ij}^2$.[3] For equation (10-6') there would be no deflator. This procedure presumes optimal research allocation.

Thus, in our example above, the deflator for equation (10-6) for state 1 would be $(\frac{1}{2})^2 + (\frac{1}{2})^2 = \frac{1}{2}$. For state 2, the deflator would be $(\frac{1}{2})^2 + (\frac{1}{4})^2 + (\frac{1}{4})^2 = \frac{3}{8}$.

The problem of aggregation can be simplified somewhat by using a simpler specification and using multiple deflators, as illustrated later in this chapter.

Many studies of R&D either do not deflate research variables (or extension variables) at all or deflate by number of farms or output. These deflation practices have little rationale behind them. One could argue that the diversity of production environments is closely related to land area. A state with twice the land area, for example, might have twice as many geoclimate regions and thus twice as many targeted programs. We know, however, from geoclimate regional data that this is not the case. Nor is the number of farms in a state a good proxy for the number of target regions. Indeed, the number of farms is itself the result of past technological change.

Research programs directed at several regions and commodities may be far from optimal, however. In one of the two studies discussed in the second section of this paper, a deflator of the following form was defined:

$$\phi = \left(1 - \tfrac{1}{2}\sum_j (C_{ij} - R_{ij})^2\right) S_{II} + (1 - S_{II}) \qquad (10\text{-}8)$$

where C_i and R_i are the shares of commodities and level III research by region and by commodity. S_{II} is the share of level II research in the budget of the state. This index equals 1 if the research budget is distributed according to the value of commodities and falls to the share of level II research if the state's research is undertaken on commodities not produced by the state.

The first term in this index can be further divided

$$\sum_j C_{ij} R_{ij} + (1 - \tfrac{1}{2}\Sigma C_{ij}^2 - \tfrac{1}{2}\Sigma R_{ij}^2) \qquad (10\text{-}9)$$

The term $\Sigma C_{ij} R_{ij}$ presumes no pervasiveness; that is, research on a commodity or region does not produce spillover to another commodity.

---
[3] This is because $R_{ij} = S_{ij}$ if research is allocated optimally.

The second term, $(1 - \frac{1}{2}\Sigma C_{ij}^2 - \Sigma R_{ij}^2)$, allows somewhat arbitrarily for spillover between research programs. This term increases as the number of commodities and regions increases and the distribution of funds and commodities becomes further apart.

The second term thus really contains two components—an opportunity for a pervasiveness component and an inefficiency component. The inefficiency effect comes from misallocation of research resources.

It may be useful to keep these two terms separate in working with a deflator. This requires a complex multiple deflator and could be pursued in the following manner.

One can use the mathematical property that

$$\frac{d_1 + d_2}{R} = \frac{d_1}{R} + \frac{d_2}{R} \qquad (10\text{-}10)$$

where d_1, d_2, and R are arbitrary variables. Within the context of this paper, this expression could be thought of as the inverse of the normal definition of a research variable. That is, instead of defining it as R/d, define it as d/R. This allows one to use a multiple deflator and estimate the relevant parameters. For example, we could define variables such as

$$\Sigma S_i^2/R \quad \text{and} \quad \Sigma(C_i - R_i)^2/R \qquad (10\text{-}11)$$

Then coefficients could be estimated for both variables.

The same idea could be employed to define environments. Suppose a rather imperfect geoclimate regional classification was possible enabling the definition of i regions in a state. One could then define two deflators:

$$d_1 = \Sigma S_{ij}^{S_{ij}} \quad \text{and} \quad d_2 = \Sigma S_{ij}^2 \qquad (10\text{-}12)$$

These deflators would allow for a more consistent estimation of the marginal product of research.

Similarly, one may have a geoclimate classification based on temperature, another based on rainfall, and so forth. Another possibility might involve the use of numbers of farms as deflators in a multiple deflator analysis.

Kislev and Evenson (1975) used a deflator defined as

$$d = \frac{n}{\sum_{i-1} |r_i - \bar{r}| + 1} \qquad (10\text{-}13)$$

where r_i is spending in the ith region, n is the number of regions, and $\bar{r}$ is average spending in the region. This deflator can be expressed as

$$d = n + V_n \tag{10-14}$$

that is, the sum of the number of regions and the variance of the size of these regions.

Borrowable Research, or Spill-in From Other Regions

The assumption of a constant α_{ij} in expression (10-3) is really an assumption about the plasticity with which nature yields her secrets. It describes the technology of producing technology. If the probability of making an additional productivity-enhancing invention, given the level of resources devoted to invention, is the same in each state and region (and we can extend the notion to commodities), the α_{ij} will be constant for all i and j.

The possibility of a constant elasticity is more likely in a somewhat richer specification. The productivity relationship could include terms for level II research in the state, "borrowable" level III research from other states, and borrowable level II research from other states. The conduct of upstream level II research in the state and in other states has a leveling effect on the productivity of level III research, because it provides findings that enhance the productivity of several downstream research programs. The conduct of and access to borrowable level III research will affect the productivity of level III research in state i in different ways. If the borrowable research program produces technology that is well matched to state i's geoclimate and economic conditions and is thus a good substitute for level III research in state i, it will lower the production elasticity of state i's research. On the other hand if it is mismatched, this research may complement the state's research by providing opportunities for adaptive inventions. Direct spill-in may occur as well.

Empirical application of specifications in which we allow for indirect productivity effects from level II research (both indigenous and borrowable) and from borrowable level III research stocks can be complex, as they require data that effectively distinguish between level II and III research. They also require a meaningful definition of borrowable research. The first issue is a matter of research definition. The CRIS data set covering public sector USDA and SAES research in the United States classifies research projects by commodity orientation, mission, and discipline and can be used to distinguish between level II and level III research. Most of the relevant level III research performed by public institutions outside the USDA-SAES system (universities) can almost certainly be identified as well. More difficult is the task of obtaining data on level II research activities by private firms.

The issue of borrowability or "spatial pervasiveness" has not been considered extensively in the literature. Traditionally, borrowability has been said to exist when research is performed in neighboring states (or countries) or when research is performed in states or countries doing more than an average level of research on the commodity.

Both of these definitions are inconsistent with the logic of the literature on technology transfer inhibitions. The genotype environment interactions depicted in figure 10-2 are the relevant determinants of technology flow or spill-in from one region to another. These soil and climate factors do not coincide well with neighboring states, although consistency in terrain and so forth do provide some coincidence. Genotype environment interactions do not coincide with the decisions of states regarding high levels of research spending or of research intensities. The argument that states spending relatively high amounts are targeting their research to states spending low amounts is also not tenable.

Clearly we require a means for comparing regions on the basis of factors that inhibit or interact with technology. Soil and climate factors clearly do and we have geographic classifications of these factors. One usable classification for U.S. agriculture is based on the sixteen regions reported in the 1957 *Yearbook of Agriculture*. These regions are further delineated into forty-one subregions that are defined as relatively homogenous soil and climate regions. Borrowable research programs can be defined as programs in similar regions or subregions (see below).

The Time Dimension

Research processes at all levels have two characteristics of relevance to the time-shape dimension. First, research is a time-consuming process. Second, it is a naturally stochastic process, in that research experiments and research initiatives have uncertain outcomes. The expected time shape is essentially the distributed lag specification showing which research investments in the past are relevant to the production of the new technology being implemented in the current period. This lag relationship can be divided into several components:

a. the lag between investment in level II research and the discovery of level II findings

b. the lag between investment in level III research and the implementation (that is, adaption) of level II findings
c. the lag between investment in level III research and the production of inventions
d. the lag between inventions and market development
e. the lag between technology availability and its implementation by farmers.

All productivity-increasing technology cannot be traced backward to identify each of these lags. Some inventions cannot be easily traced to level II findings and for these the lag components a and b are not relevant. Similarly, for some subinvention lags, components a, b, c, and d may not exist.

We have little in the way of theory to guide us in specifying the shape of these time weights. In general, we expect the length of the lag to depend on the type of research, with longer lags as we move upstream from level IV to III to II. The actual shape of the lag depends on the form of the productivity variable. In some formulations the observed productivity may be in the form of a level of productivity relative to some historical period or relative to other cross-section observations. In other formulations the observation may be a rate of change (say, over the past year). Some formulations are mixtures.

Figure 10-3 illustrates two simple time shapes. Panel A shows the expected effect of an investment in time t on the level or productivity (as measured in the intercept terms of a Cobb-Douglas production function, for example) in the future. Shape 1 shows the case where technology is invented, developed, and adapted beginning in year $t + i$. By year $t + i$ the technology is fully adapted and productivity levels have risen by M units and this level is permanently maintained. Shape 2 shows a case where the technology deteriorates after adaption. This often happens with biological technology, as, for example, when newly released crop varieties induce changes in pest populations and disease pressures.

Panel B shows the same time shapes when the dependent variable is the rate of change in productivity. As may be readily noted the shapes and the implied definitions of the research variable are very different.

The literature is often quite confused on this matter, particularly when the dependent variable is an index number. In time series data an index number is a percentage change from the base period level. If the time series is short and the base period is in the middle of the series, it is not really a level series, even though it may be treated as such. If it is a cross-section with all states having a common base the problem is ex-

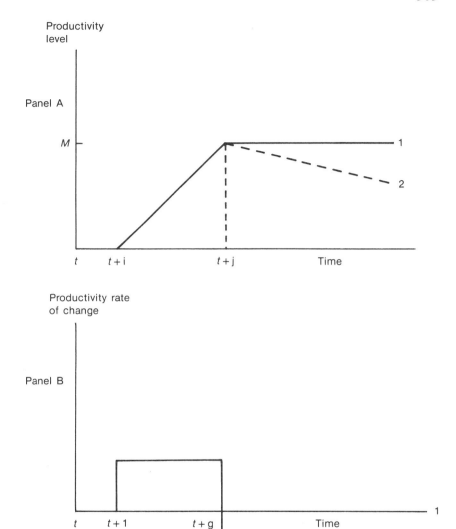

FIGURE 10-3. Relationship between productivity and alternative time dimensions for invention and adoption of technology

acerbated. Care should be taken in studies of level effects to have a reasonably long time series.

Short time series are probably best suited to rate of change specifications. Note that in case 1 this required lagged research data only for the $t-i$ to $t-j$ periods, whereas the levels regression required long time series (even where technology deteriorates). (Note that in a productivity specification, we reverse the time dimension in figure 10-3 and ask what sum of past research investments are effecting the current productivity level or rate of change.)

THE EVENSON-WELCH STUDY (1974)[4]

The Evenson-Welch study defined the research variable as follows. First state research expenditures were allocated by commodity to subregions (and regions) within the state on the basis of the share of the subregion in the state's production of each commodity. Data were available on research expenditures by commodity for five livestock commodities (beef, dairy, swine, poultry, and sheep) plus a general livestock category. For crops, sixteen crop categories plus a general crops research category were used.[5] County data from the 1964 *Census* were utilized to allocate research data to subregions.

For each state and for each commodity, then, a research "stock" variable was defined by cumulating expenditures in years prior to 1964 utilizing the time-shape inverted V weights estimated by Evenson (1968). For each commodity i, in state j, subregion k, let E_{ijk} be the research stock in 1964. A similar subregion stock, $E_{ijk}{}^{ss}$, can then be defined as

$$E_{ijk}{}^{ss} = \sum_j E_{ijk} - E_{ijk}$$

The similar subregion stock is the sum of the research stocks for this commodity in similar subregions outside the state. A comparable similar region stock, $E_{ijk}{}^{sr}$, can also be defined for each commodity.

[4] Both the Evenson-Welch and the Evenson studies use the geoclimate classification as delineated in the 1957 *Yearbook of Agriculture* as a basis for defining spatial pervasiveness. One of the earliest studies of this genre, by Latimer and Paarlberg (1965) concluded that the degree of pervasiveness was sufficiently great to defeat their efforts to identify the effect of research and production.

[5] They include barley, corn and sorghum, cotton, flax, forestry and forest products, fruits, hay, oats, peanuts, potatoes, rice, soybeans, sugarbeets, sugarcane, tobacco, vegetables, and wheat.

We now aggregate across commodities to obtain a state-specific crop research variable as follows

$$CR_j = \sum_i \phi_{ij}\left(\sum_k S_{ijk}(E_{ijk} + \theta_c)E_{ijk}^{ss}\right)$$

where

$$\phi_{ij} = 1 - \tfrac{1}{2}(C_{ij} - r_{ij}^2)$$

A similar specification for a livestock research variable is defined.

In the above specification, θ_c is a spatial pervasiveness parameter to be estimated by iterative methods. The term S_{ijk}, the share of total crop sales of commodity i in subregion k, is a deflator that implicitly deflates the research stocks by the number of subregions in the state. The ϕ_{ij} term is a commodity pervasiveness term, equal to one if the share of sales of the commodity, C_{ij}, in the state is equal to the share of the commodity research expenditures, r_{ij}.[6] This implies that general research is completely pervasive over all crop or livestock categories, but that applied research is only partially pervasive.[7]

The production function specification used in the study estimated separate production functions for crop and livestock production. A procedure for optimally allocating labor between crop and livestock production was employed (see Evenson and Welch, 1974).

The results, summarized in table 10-1, support hypotheses that agricultural output is affected both by research within the state in which output is observed and by research in similar subregions (or regions) of other states. The estimates of pervasiveness in crop research are not affected by the specification pervasiveness in livestock research and estimates for livestock pervasiveness are independent of the crop specification.

For crop research, when the borrowable stock is defined for regions, no evidence of pervasiveness is found. Statistical results for state-specific

---

[6] This index ranged between .99 in Illinois and .25 in Washington for crop research, and for livestock reached a low of .77 in New York and a high of .99 in New Jersey. The unweighted mean for the forty-eight coterminous U.S. states is .57 for crop research and .93 for livestock.

[7] If commodity-oriented research were not pervasive across commodities, the appropriate index of ϕ_{ij} would be $C_i r_i$. Any increase in research expenditure would be confined to the commodity. Our index ϕ can be decomposed into $C_i r_i$ plus a pervasiveness term because

$$1 - \tfrac{1}{2}(c_i r_i)^2 = c_i r_i + (1 - \tfrac{1}{2}c_i^2 - \tfrac{1}{2}r_i^2)$$

TABLE 10-1. ESTIMATES OF EFFECTS OF CROP AND LIVESTOCK RESEARCH ON FARM PRODUCTION WITH ALTERNATIVE ASSUMPTIONS OF SPATIAL PERVASIVENESS

Research	Estimated production elasticity for research index	Coefficient estimated divided by standard error[a]	Variance of the estimated equation[a]
A. Crop research			
State expenditures only	0.0138	8.93	0.2189
State + 0.25 similar subregions	0.0211	14.18	0.2182
State + 0.50 similar subregions	0.0244	16.60	0.2179
State + 0.75 similar subregions	0.0280	19.21	0.2174
State + 0.90 similar subregions	0.0290	19.96	0.2172
State + 1.00 similar subregions	0.0299	20.66	0.2171
B. Livestock research			
State expenditures only	−0.0367	−1.77	0.2219
State + 0.25 similar subregions	−0.0269	−1.09	0.2257
State + 0.50 similar subregions	−0.0115	−0.45	0.2277
State + 1.00 similar subregions	0.0002	0.08	0.2281
State + 0.50 similar regions	0.0445	1.23	0.2248
State + 0.75 similar regions	0.0675	1.81	0.2207
State + 0.90 similar regions	0.0756	2.03	0.2206
State + 1.00 similar regions	0.0821	2.28	0.2171

Note: These estimates utilize the education adjusted standard quantity labor variable definition.
[a] These are asymptotic normal statistics.

Source: Based on results from R. E. Evenson and F. Welch, "Research, Farm Scale, and Agricultural Production," Economic Growth Center, Yale University, 1974.

research are consistently superior to specifications in which some fraction of similar region research is added. On the other hand, when pervasiveness is restricted to subregions, evidence of pervasiveness emerges. Using the similar subregion specification for crop research, the production elasticity estimate, its statistical significance, and the likelihood of the sample all rise as the pervasiveness index is increased. We consider this strong evidence that (a) crop production is increased by increased research activity and (b) that there is a significant spillover between similar geoclimate subregions of the United States.

The evidence for livestock research is less convincing. Notice first that when the research index is restricted to within state expenditures there is no evidence of a significant positive effect. As horizons broaden to similar subregions, the point estimate of the production elasticity changes from negative to positive, but the variance of the estimated equation increases. When pervasiveness is expanded to the regional level, the estimates improve. In particular, when 90 to 100 percent of the expenditures in similar regions is added to expenditures within the state, the effect of livestock appears positive and significant. It appears that livestock research products are pervasive over a broad geographical base.

We estimate that both crop and livestock research programs significantly affect technical efficiency. The pervasiveness estimate is that crops research is highly pervasive over geoclimate subregions as defined in this study. The nature of our estimation procedure is sufficiently crude that we have captured only the major features of pervasiveness. It is plausible that some research results are more broadly pervasive, some less so. Our estimates for livestock research are somewhat less consistent from a statistical perspective. We estimate that pervasiveness is greater for livestock, extending over geoclimate regions.

These results imply, of course, that producers in a given state appropriate only part of the research products produced by the state experiment station and that they benefit from research done in other states. For crop research we estimate that one third of the production effect of research will be realized by producers in the state producing the research. For livestock research we estimate that only 15 percent of the effect will be realized by producers in the state of origin of the research.

We can compute an estimate of the marginal contribution to crop and livestock production of an added increment to the research stocks. A $1.00 increase in the crop research stock in a typical state produced by a sequence of prior annual expenditures (appropriately weighted by time-shape weights) in 1964 would have produced an increment to farm production of $8.21, one third of which would have been realized in the state. The comparable estimate for livestock research is $11.85, 15 percent of which would have been realized in the state.[8] By assuming a pattern of annual investments sufficient to produce a $1.00 increment to research stocks given the implicit time logic, an internal rate of return to a marginal investment in crop research of 55 percent is computed. For livestock research the figure is slightly higher, although it may be plausible to presume a longer lag for livestock research.

The Evenson Study (1984)

The Evenson productivity decomposition study entailed two stages. First, an analysis of the combined time shape and contiguity pattern of applied agricultural research was undertaken. Secondly, a more complete decomposition analysis was undertaken.

---

[8] A $1 increase in the stock in a given state actually increases the total national research stock by $3 for crops and $6.96 for livestock.

The procedure adopted for the time shape-contiguity analysis is a partial correlation scanning procedure for a general research variable of the form[9]

$$A(a,b,c) + \alpha SA(a,b,c) + \beta RA(a,b,c)$$

where A is the within-state applied research stock; SA the stock in similar subregions outside the state; and RA the stock in similar regions (which include the subregions) outside the state. The parameters a, b, and c refer to alternative time shapes: a is the time period of rising linear weights; b, the time period of constant weights; and c, the period of declining linear weights. The parameters α and β represent pervasiveness and measure the extent to which research in contiguous or similar regions contributes to state productivity growth.

Table 10-2 reveals the results of a partial correlation scanning analysis across varying time-shape and contiguity parameters. The analysis is undertaken for Northern states (Northeast, Corn Belt, and Lake States regions), Southern states (Appalachian, Southeast, and Delta regions), and Western states (Northern Plains, Southern Plains, Mountain, and Pacific regions). The highest partial correlation for the Northern states is for the variable constructed with a seven-year lag from investment to peak effect, a further eight-year constant lag, and a fifteen-year period of declining weights. The contiguity weight is half of the similar subregions outside the state, and the research variable is deflated by the number of commodities and subregions in the state.

The estimated time-shape weights for the Southern states were 5, 6, and 11, and the contiguity weight was .25 of the similar subregions. Note that very little difference exists between the Northern and Southern regions. The Western region shows the same pattern in the subregion weight as the other regions. However, the contiguity weight is .25 of similar regions (which include the subregion) indicating a somewhat broader range of technology transferability.[10]

---

[9] The partial correlation scanning proceedure is based on a proposition by Theil (1964), who showed that a structured search for the highest partial correlation or lowest residual sum of squared errors by iterating over a, b, and c is in fact a nonlinear least squares procedure for estimating a, b, and c. The use of partial correlation coefficients instead of residual sums of squared errors as the criterion is simply a matter of convenience.

[10] An appropriate standard error was computed for these weights (see Evenson, 1980). It showed that the shortest lags $R(3, 4, 7)$ were significantly different from the longest lags $R(15, 20, 25)$ for all weights α and β.

TABLE 10-2. TIME-SHAPE AND CONTIGUITY ESTIMATES:
U.S. AGRICULTURE, 1948–1971

	Partial correlation coefficients, controlling for scaling parameter, business cycles, and education								
	$\alpha, \beta = 0$	$\alpha = .25$	$\alpha = .5$	$\alpha = .75$	$\alpha = 1$	$\beta = .25$	$\beta = .5$	$\beta = .75$	$\beta = 1$
Northern States									
R (3, 4, 7)	.135	.324	.304	.284	.273	.224	.224	.219	.218
R (3, 4, 11)	.145	.321	.323	.303	.289	.225	.224	.222	.220
R (5, 6, 11)	.165	.339	.338	.314	.297	.234	.230	.226	.223
R (5, 6, 15)	.161	.323	.343	.325	.308	.229	.228	.227	.224
R (7, 8, 15)	.167	.326	.346	.327	.309	.234	.234	.231	.228
R (7, 8, 19)	.158	.304	.342	.331	.305	.227	.231	.239	.227
R (7, 8, 25)	.145	.277	.286	.266	.249	.278	.219	.218	.216
R (11, 12, 25)	.140	.274	.282	.267	.246	.273	.218	.217	.215
R (15, 20, 25)	.122	.221	.222	.202	.187	.221	.206	.206	.205
Southern States									
R (3, 4, 7)	.456	.487	.481	.474	.468	.266	.184	.107	.078
R (3, 4, 11)	.451	.484	.483	.478	.473	.395	.203	.143	.107
R (5, 6, 11)	.460	.490	.488	.482	.476	.310	.207	.146	.109
R (5, 6, 15)	.451	.483	.485	.482	.478	.328	.232	.171	.131
R (7, 8, 15)	.451	.483	.485	.482	.478	.329	.233	.172	.133
R (7, 8, 19)	.442	.475	.481	.480	.477	.337	.250	.190	.149
R (7, 8, 25)	.429	.465	.470	.469	.466	.464	.216	.157	.118
R (11, 12, 25)	.436	.471	.475	.471	.469	.471	.215	.155	.116
R (15, 20, 25)	.418	.452	.459	.458	.456	.452	.210	.151	.112
Western States									
R (3, 4, 7)	.224	.234	.201	.171	.150	.268	.240	.203	.101
R (3, 4, 11)	.237	.252	.230	.203	.181	.293	.253	.225	.208
R (5, 6, 11)	.248	.261	.238	.203	.186	.302	.258	.230	.212
R (5, 6, 15)	.253	.268	.254	.230	.207	.318	.278	.248	.226
R (7, 8, 15)	.257	.273	.257	.232	.208	.328	.280	.260	.228
R (7, 8, 19)	.258	.275	.266	.244	.222	.323	.292	.240	.238
R (7, 8, 25)	.295	.272	.254	.225	.199	.271	.286	.254	.233
R (11, 12, 25)	.259	.272	.251	.221	.193	.272	.283	.250	.229
R (15, 20, 25)	.257	.267	.245	.213	.184	.267	.295	.261	.240

Table 10-3 reveals the results of productivity decomposition analysis for U.S. agriculture for the 1948–71 period. The general specification is

$$\text{TFP} = a_0(\text{ED})^{a_1}(\text{EXTECON})a_2 + a_3\,\text{ED}_{(AR)}\,a_4 + a_5\,\text{BR} + a_6\,\text{EXTECON}_{(e)}\,a_7\,\text{PL} + a_8\,\text{BC} + a_{9i}\,Z_i$$

where TFP is the total factor productivity index; ED is an index of years of school completed by farm operators. It is constructed from Census data and utilizes weights developed in a study by Welch (1970); EXTECON is a composite variable based on extension expenditure plus

TABLE 10-3. PRODUCTIVITY DECOMPOSITION:
U.S. AGRICULTURE, 1948–1971

Independent variables	Dependent variable: ln (TFP) specifications[a]				
	(1)	(2)	(3)	(4)	(5)
Constant	4.69	4.25	4.73	4.77	4.86
ln (AR)		0.04237	0.0174		
		(0.00997)	(0.0085)		
ln (AR) south				0.03309	0.03407
				(0.00856)	(0.00086)
ln (AR) north				0.01187	0.00991
				(0.00848)	(0.00861)
ln (AR) west				0.01874	0.01882
				(0.00887)	(0.00903)
ln (ED)		0.3143	0.1770	0.3540	0.3731
		(0.0404)	(0.0362)	(0.0426)	(0.0419)
ln (EXTECON)		−0.000276	−0.0388	−0.0394	−0.0514
		(0.01176)	(0.0099)	(0.0097)	(0.0104)
ln (EXTECON)[b] (ED)		−0.01223	−0.00659	−0.0116	−0.0120
		(0.00242)	(0.00206)	(0.0021)	(0.0021)
ln (AR)[b] (EXTECON)		0.1314 D-5	0.1730 D-5	0.1821 D-5	0.1962 D-5
		(0.0260 D-5)	(0.0230 D-5)	(0.0230 D-5)	(0.0227 D-5)
ln (AR)[b] (BR)		0.2054 D-7	0.0171 D-6	0.2061 D-6	0.2166 D-6
		(0.8300 D-7)	(0.0737 D-6)	(0.0710 D-6)	(0.0705 D-6)
ln (AR[b] GRAD)					0.000247
					(0.000071)
ln (AR[b] SCALE)					−0.543 D-7
					(0.600 D-7)
Productivity scaling factor (PL)		−0.000136	−0.00014	−0.00016	−0.00016
		(0.000030)	(0.000034)	(0.00003)	(0.00003)
Business cycle index (BC)		0.34509	0.2486	0.2297	0.2237
		(0.0200)	(0.0180)	(0.0176)	(0.0176)
1957–63 south dummy	0.165		0.158	0.076	0.075
1957–63 north dummy	0.118		0.074	0.102	0.102
1957–63 west dummy	0.156		0.136	0.113	0.112
1964–71 south dummy	0.308		0.246	0.136	0.132
1964–71 north dummy	0.246		0.115	0.128	0.124
1964–71 west dummy	0.286		0.192	0.152	0.149
R^2	0.484	0.413	0.618	0.573	0.651
R^2 (ADJ)	0.481	0.409	0.613	0.569	0.646

[a] Specifications are described in text.
[b] Standard errors are in parentheses.

expenditures on production-oriented economic (farm management) and applied engineering research;[11] AR is the applied (level III) research stock variable constructed with table 10-2 weights; BR is an index of level II research constructed with time shape (a, b, c) weights of $(11, 12, 25, \alpha = .25)$ for Southern states, $(15, 20, 25, \alpha = .5)$ for Northern states, and $(15, 20, 25, \beta = .5)$ for Western states. These weights were estimated in a partial correlation scanning analysis similar to that reported in table 10-2. BR is undeflated; PL is a productivity scaling factor for states; BC is a business cycle index designed to capture the productivity effectiveness of the business cycle. It is constructed as the ratio of two moving averages of real farm income. Productivity gains are expected to be higher in the trough phases of the business cycle than in the peak phases because of adjustment pressures; and Z_i are region and time dummies.

The specifications reported in table 10-3 demonstrate the effect of adding the region-time dummy variables and of estimating separate coefficients for the three major regions of the country for the research variable. Specification 1 is included to show how much of the change in total factor productivity is associated with the region and time dummy variables. It also allows a relatively simple comparison of the proportion of the growth in total factor productivity change explained by the research and related variables.

The second specification is included to show the effects of the decomposition variables and to enable the reader to assess the effect of adding the region-time dummy variables in specification 3. An experiment in which a simple time trend variable replaced two region-time dummies was conducted. The results were essentially the same as those obtained for specification 3.

Specification 3 provides the basic decomposition results. The negative coefficients for the extension variable and the extension-education interaction variable do not mean that the marginal product of extension on education is negative. The negative extension-education effect is to be expected. It shows that extension or adult education is a substitute for formal schooling in terms of its effect on farmer efficiency. In states with high levels of farmer schooling, extension activities have a smaller impact. The positive (and highly significant) research-extension interaction term shows these activities to be complements. We would expect extension to be more productive the higher the level of research activity in a

---

[11] The EXTECON variable has geometrically declining time shape weights. The weights are .5 for the current period, .25 for $t - 1$, .125 for $t - 2$, and so forth.

given state. The positive applied (level II) research term also indicates that higher levels of level II research increase the productivity of applied level III research. Thus, level II research in the agricultural experiment stations is productive through its effect on the productivity of applied research.

The fourth specification estimates separate coefficients for the applied research (AR) variable for the three major regions of the study: South, North, and West. This extension shows that regional differences have existed. In particular, the southern states have realized faster rates of productivity growth and it appears that at least part of this is due to the research system. Note that in specification 3, which imposed a single AR coefficient, the time variable in the South accounts for almost 80 percent of the change in total factor productivity from the beginning of the period until the ending period. In specification 4 this proportion falls to less than 50 percent. In all three regions the variables account for 50 percent or less of the explanation of productivity growth in specification 4.

The fifth specification extends the analysis farther in an attempt to explore whether experiment station characteristics have an effect on the productivity of agricultural research. Two variables, a measure of the scale of the main experiment station (measured as number of scientists) and a measure of the size of graduate programs associated with the main experiment station (number of Ph.D.s granted annually in the 1950s, denoted as GRAD in table 10-3) were interacted with the applied research variable. The results suggest that the size of the associated graduate program positively affects research productivity, but that scale *per se* does not.

The productivity scaling variable has the expected sign and can be interpreted as an indicator of economic slack, in that states with relatively low scaling parameters have more potential for productivity growth: they have more catching up to do and catching up requires fewer resources than leading requires. The business cycle variable also indicates that as farm income falls in a cycle total factor productivity rises. As the farm income cycle reaches a boom phase, total factor productivity slows down.

Table 10-4 summarizes the marginal contribution of schooling, research, and extension to productivity. The total gains are divided into those occurring in a state making the investment and those spilling into other states. These estimates show high marginal rates of return to investment in schooling, extension and economic research, applied level III research, and pretechnology science (level II) research.

TABLE 10-4. COMPUTED MARGINAL CONTRIBUTION OF CHANGES IN RESEARCH, EXTENSION, AND EDUCATION STOCKS, 1948–1971 (dollars)

Change in farm production due to	Appropriated by state	Transferred to other states	Total
One year of primary schooling			
Specification (3)	120		120
Specification (4)	260		260
$1,000 added to extension applied economics stock			
Specification (3)	2,947		2,947
Specification (4)	2,173		2,173
$1,000 added to scientific research stock			
Specification (3)	755	1,585	2,330
Specification (4)	1,450	3,050	4,500
$1,000 added to applied research stock			
Specification (3)	6,820	5,180	12,000
Specification (4)			
South	14,100	7,100	21,000
North	5,070	6,530	11,600
West	8,270	3,930	12,200

CONCLUDING REMARKS

In this paper I have attempted to discuss empirical specification problems in the context of a model of the process of technology generation. Many early studies failed to deal with these problems in a consistent manner. Many current studies ignore them.

As noted earlier in this book and emphasized throughout the remaining chapters, we are now entering a new phase of productivity analysis that relies on duality-based systems of output supply and factor demand equations. This work is promising in many respects because it allows us to estimate biases of research and related inputs on both the output and input sides. These models are just as demanding of proper specification as the primal models and we are unlikely to achieve proper interpretation of results without more careful development of variables, including the R&D variables.

Several important issues have been left out of this discussion. Perhaps the most serious omission is that of simultaneity problems in productivity decomposition. To date, we have no strong evidence of simultaneity problems in these studies, but the issue has not been fully explored. Several studies have shown that research and extension investment responds to levels of production and to some extent to productivity changes. It is shown in chapter 12, however, that current investment responds to past productivity change and current productivity change responds to past research investment due to the dynamics of the system. These dynamic relations imply that the simultaneity bias problem is not serious.

Finally, a note on total factor productivity indexes. There is much confusion in the literature over what they measure. They are affected by the treatment of human capital, scale economy adjustments, and capital stock construction. They are also affected by incorrect weights, as when the real value of a marginal product is not equal to a factor price. These problems plague efforts to compare total factor productivity series from different studies, as noted by Capalbo in chapter 5, but they are not fundamentally a serious problem for decomposition studies, because decomposition studies can incorporate variables that can pick up errors due to incorrect weights and other problems, and because a flawed productivity index will not necessarily cause a bias in the estimate of the impact of a particular decomposition variable. (The flaw may be uncorrelated with the decomposition variable.)

REFERENCES

Binswanger, H. P., and R. E. Evenson. 1978. "Technology Transfer and Research Resource Allocation," in H. P. Binswanger and V. W. Ruttan, eds., *Induced Innovation* (Baltimore, Md., Johns Hopkins Press).

Boyne, D. H. 1962. "Changes in the Real Wealth Position of Owners of Agricultural Assets" (Ph.D. dissertation, University of Chicago).

Bredahl, M., and W. Peterson. 1976. "The Productivity and Allocation of Research: U.S. Agricultural Experiment Stations," *American Journal of Agricultural Economics* vol. 58, no. 4, pp. 684–692.

Evenson, R. E. 1968. "The Contribution of Agricultural Research and Extension to Agricultural Production" (Ph.D. dissertation, University of Chicago).

———. 1971. "Economic Aspects of the Organization of Agricultural Research," in W. Fishel, ed., *Resource Allocation in Agricultural Research* (Minneapolis, Minn., University of Minnesota Press).

———. 1984. "Technical Change in U.S. Agriculture," in R. Nelson, ed., *Government and Technical Change: A Cross Industry Analysis* (New York, N.Y., Pergamon Press).

———, and F. Welch. 1974. "Research, Farm Scale, and Agricultural Production" (New Haven, Conn., Economic Growth Center, Yale University).

———, R. W. Herdt, J. C. O'Toole, W. R. Coffman, and H. E. Kauffman. 1978. "Risk and Uncertainty as Factors in Crop Improvement Research," Research Paper No. 15 (Los Banos, Philippines, International Rice Research Institute).

Griliches, Z. 1958a. "The Demand for Fertilizer: An Economic Interpretation of Technical Change," *Journal of Farm Economics* vol. 40, no. 3, pp. 591–606.
———. 1958b. "Research Costs and Social Returns: Hybrid Corn and Related Innovations," *Journal of Political Economy* vol. 66, no. 5, pp. 419–431.

Judd, M. A., J. K. Boyce, and R. E. Evenson. 1986. "Investing in Agricultural Supply," *Economic Development and Cultural Change* vol. 35, no.1, pp. 77–113.

Kislev, Y., and R. E. Evenson. 1975. *Agricultural Research and Productivity* (New Haven, Conn., Yale University Press).

Landau, D., and R. E. Evenson. 1974. "Productivity Change in U.S. Agricultural States," Economic Center Discussion Paper No. 198 (New Haven, Conn., Yale University).

Latimer, R., and D. Paarlberg. 1965. "Geographic Distribution of Research Costs and Benefits," *Journal of Farm Economics* vol. 57, no. 2, pp. 234–241.

Norton, G. R., and J. S. Davis. 1981. "Evaluating Returns to Research, A Review," *American Journal of Agricultural Economics* vol. 63, no. 4, pp. 685–699.

Schultz, T. 1953. *Economic Organization of Agriculture* (New York, N.Y., McGraw-Hill).

Theil, H. 1964. *Economic Forecasting Policy*, chapter VI (Amsterdam, North-Holland).

Welch, F. 1970. "Education in Production," *Journal of Political Economy* vol. 78, no. 1, pp. 35–59.

Wright, B., and R. E. Evenson. 1980. "An Evaluation of Methods for Examining the Quality of Agricultural Research," in *An Assessment of the United States Food and Agricultural Research System* vol. II (Washington, D.C., Office of Technology Assessment).

11

ENDOGENOUS TECHNOLOGY AND THE MEASUREMENT OF PRODUCTIVITY*

YAIR MUNDLAK

In macroeconomic analysis, the technology of the economy is summarized by a production function. Yet the production function, strictly speaking, is a microeconomic concept. It has a relatively clear meaning when it specifies a well-defined process, such as the production of a crop under well-defined conditions. But there are many crops and environmental conditions. Consequently, we observe many production functions in agriculture. In order to explicate the meaning of the agricultural production function, we must first consider the issue of aggregation, a subject that has been discussed at length in the literature. The problem of aggregation in agriculture differs from the common aggregation problem, however, in that the set of aggregated functions is endogenous to the economic system.

The question raised here is how to represent and measure technology of a sector (or any other level of economic activity) when output is produced by using more than one technique. The chapter is divided into six parts. We begin by presenting the conceptual framework. In the sections that follow, we develop the aggregate production function relevant for empirical analysis; examine the issue of estimation; study the endogeneity of technology; and examine the state variables relevant for empirical analysis. Finally, some implications for future research are identified.

THE CHOICE OF A TECHNIQUE

Each technique can be described by a production function, which is associated with an input requirement set. Technology (T) is defined as the collection of all possible techniques. In symbols,

$$T = \{F_j(x)\} \tag{11-1}$$

---

*This paper draws on Mundlak (1983).

where $F_j(x)$ is the production function associated with the jth technique. The technology defines an input requirement set obtained by convexification of the input requirement sets of the individual techniques.

These concepts are illustrated in figure 11-1, where the technology consists of two production functions, represented by their unit isoquants. The input requirement set of each technique is bounded from below by its isoquant. The input requirement set associated with T contains all the convex combinations of the individual input requirement set. To obtain it, we note that there exists a cost line with a slope $\tilde{\omega}$ that is tangent to both isoquants. Let the inputs be capital (K) and labor (L), then corresponding to $\tilde{\omega}$ we have threshold capital-labor ratios $\tilde{k}_j = k_j(\tilde{\omega})$, $j = 1, 2$ determined by the tangency of the cost line and the two isoquants. Let k be the overall capital-labor ratio. Then for $k \geq \tilde{k}_2$ the isoquant associated with T is identical with $Y_2 = 1$. Similarly, for $k \leq \tilde{k}_1$, it is identical with $Y_1 = 1$. For $\tilde{k}_1 \leq k \leq \tilde{k}_2$, it is given by the segment MN along the tangent line.[1]

A technological change is defined within this framework as a change in the technology T. The main objective of empirical analysis is to infer something about the technology from the data. The data can reveal information only about techniques that were actually implemented. For instance, if the two techniques described in figure 11-1 represent two varieties of wheat—Y_1 representing the traditional technique and Y_2 representing the modern technique—it is clear that when the capital-labor ratio in the economy is below the threshold level ($\tilde{k}_1$), then only the traditional variety will be employed, even though the modern variety is available. The data in this case do not reveal any information about the modern variety. We thus make a distinction between technology (T) and implemented technology (IT), which consists only of techniques actually implemented.

The choice of technique is made at the firm level. To simplify the analysis, we deal with a single-period optimization and single-output production functions. We distinguish between fixed (b) and variable (v)

---

[1] The concept of a technique is very general. Techniques can be associated with products. The assumption made at some point in the foregoing analysis that the various techniques produce the same product can cover the multiproduct case by defining the output by its value. Thus, the isoquants of figure 11-1 will represent one dollar's worth of output. Moving from firms to the industry, firms themselves can be represented by techniques. This will require an extension of the optimization framework by including in the constraints a variable specific to the firm and introducing alternative costs for the fixed resources. Then, if the optimization is solved by the market, the exit and entry of firms will be one aspect of the choice of techniques. The consequences of such a choice on the aggregate production function follows the pattern developed above.

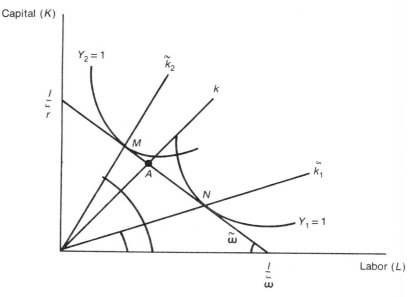

FIGURE 11-1. Choice of technique

inputs and assume for simplicity that the fixed inputs have no alternative cost. The optimization problem can then be described as maximizing

$$L = \sum_j p_j F_j(v_j, b_j) - \sum_j w v_j + \lambda \left(b - \sum_j b_j \right) \quad (11\text{-}2)$$

such that $F_j(\cdot) \epsilon T$, where p_j is the price of the product of technique j; w is the vector of factor prices; and b is the constraint on Σb_j.[2] The Kuhn-Tucker necessary conditions for a solution are

$$L_{vj} = p_j F_{vj} - w_j \leq 0 \quad (11\text{-}3)$$

$$L_{bj} = p_j F_{bj} - \lambda \leq 0 \quad (11\text{-}4)$$

$$\sum_j (L_{vj} v_j + L_{bj} b_j) = 0 \quad (11\text{-}5)$$

$$v_j \geq 0 \qquad b_j \geq 0 \quad (11\text{-}6)$$

$$L_\lambda = \Sigma b_j - b \leq 0 \quad (11\text{-}7)$$

$$\lambda L_\lambda = 0 \quad (11\text{-}8)$$

[2] A similar formulation is used by Glenn Johnson (1972). His formulation also includes salvage values for the constraints. This addition is not essential for the present discussion.

where L_{vj}, F_{vj}, L_{bj}, F_{bj}, and L_λ are vectors of the first partial derivatives. The solution gives

$$v_j^*(s), b_j^*(s), \lambda^*(s)$$

where s represents the exogenous variables of this problem, to be referred to as the *state variables*

$$s = (b, p, w, T)$$

The solution thus depends on the available technology T, on the constraint b, and on the products and variable input prices. The solution determines both the techniques used and the level of their use, as determined by the optimal allocation of fixed inputs b_j^* and variable inputs v_j^*.[3] This can be seen by rearranging equations (11-3) through (11-5)

$$0 = \sum_j (p_j F_{vj} - w_j) v_j + \sum_j (p_j F_{bj} - \lambda) b_j$$

Due to equation (11-6), when either equation (11-3) or equation (11-4) is negative, then $v_j^* = 0$ and $b_j^* = 0$. The implemented technology is the collection of all implemented techniques and it can be described formally by

$$IT(b, p, w, T) = \{F_j(v_j, b_j) | F_j(v_j^*, b_j^*) \neq 0, F_j \epsilon T\} \qquad (11\text{-}9)$$

The optimal output of technique j is $y_j^* = F_j(v_j^*, b_j^*)$. The implemented technology, IT, is a subset of T. As such, the envelope of IT is in general not the same as the envelope of T. The difference, of course, is due to the constraints encountered by the firm. Put differently, when the constraints are binding a constrained optimum is inferior to an unconstrained one.

For any set of state variables, equation (11-9) describes a well-behaved technology. Consequently, a profit function can be derived:

$$\pi(s) = \sum_j p_j F_j(v_j^*(s), b_j^*(s)) - \Sigma w_j v_j^*(s)$$

The various theorems dealing with the duality between the profit function and the production function hold true conditional on s. Specifically,

---

[3] The number of implemented techniques is related to the number of constraints, or the dimensionality of b. This is a familiar property in linear models. However, in this formulation no limit is set on the number of state variables except that it is finite.

$IT(s)$ is dual to $\pi(s)$ and vice versa.[4] Using Hotelling's lemma, it is possible to derive factor demand at the technique level, $v_j^*(s)$, by

$$-\frac{\partial \pi(s)}{\partial w} = v_j^*(s)$$

The aggregate input demands are $v^*(s) = \Sigma v_j^*(s)$ and $b^*(s) = \Sigma b_j^*(s)$. Similarly, the supply of output of technique j is given by

$$y_j^*(s) = \frac{\partial \pi(s)}{\partial p_j}$$

and the aggregate value supply is given by

$$y^*(s) = \sum_j p_j y_j^*(s)$$

AGGREGATION OVER TECHNIQUES

From the foregoing discussion, it is clear that for any given value of the state variables, the techniques to be implemented and the intensity of their implementation—as determined by the endogenous quantities, that is, inputs and outputs, associated with those techniques—are determined simultaneously. The crux of the analysis stems from the fact that at any point more than one technique is used.

The data are generally aggregate in the sense that there is no differentiation of inputs and outputs by techniques. It is therefore important to examine the relationships between aggregate output and aggregate input. To simplify the discussion, it can be assumed that all the techniques produce the same product. Let x represent the vector of inputs, and $x^* = x(s)$ its optimal level. Then total optimal output is given by

$$F(x^*, s) = \sum_j y_j^*(s) \qquad (11\text{-}10)$$

It should be noted that the production function (11-10) is defined conditional on s. Variations in s cause a joint change in x^* as well as in

---

[4] It is important to note that the exploitation of this property in empirical analysis is restricted by the fact that s varies over the sample. Thus, strictly speaking, each point in the sample comes from a different profit function, which in turn describes a different technology.

$F(x^*, s)$. This is the main difference between the present approach and conventional analysis. In the latter, changes in prices generate a spread of points on a given production function and as such are important for identifying the function. Under our approach, changes in prices generate not only changes in inputs and outputs but a different set of implemented functions as well.

The aggregate production function can then be thought of as an approximation to equation (11-10) in a specific way. For equation (11-10) to be a production function, x^* should be disconnected from s. This can be done by allowing for a discrepancy between observed (x) and optimal (x^*) inputs; we can express the observed output as

$$\Sigma y_j \cong F(x, s) \tag{11-11}$$

Strictly speaking, $F(x, s)$ need not be a function, as x can be allocated to the various techniques in an arbitrary way. Only when we have an allocation rule leading to x^* can uniqueness be achieved. However, holding s constant, the implemented technology is determined. Consequently, the difference between x and x^* produces information on that technology. This provides a key to the identification and estimation of the aggregate production function.

Following Fuss, McFadden, and Mundlak (1978), $F(x, s)$ can be approximated, using a weak assumption, by a set of functions:

$$F(x, s) \cong \sum_{i=1}^{m} a_i h_i \equiv g(x, s) \tag{11-12}$$

where $g(x, s)$ is the approximating function; a_i are parameters; and h_i are known functions. Expanding $F(x, s)$ about x^* and omitting the argument (x^*, s) wherever ambiguity does not result, we set

$a_0 = F(x^*, s)$, $h_0 = 1$;
$a_1 = \nabla F(x^*, s)$ is the gradient of $F(x^*, s)$;
$h_1 = (x - x^*)$ is the discrepancy between the optimal and actual vector of inputs; and
$2a_2 = \nabla^2 F(x^*, s)$ is the Hessian matrix of F evaluated at x^*.

It then follows that

$$g(x, s) = F(x^*, s) + (x - x^*)' \nabla F(x^*, s)$$
$$+ (x - x^*)' \nabla^2 \frac{F}{2} (x^*, s)(x - x^*) \tag{11-13}$$

Rearranging terms,

$$g(x,s) = \Gamma(x^*,s) + x'B(x^*,x,s) \tag{11-14}$$

where

$$\Gamma(x^*,s) = F(x^*,s) - x^{*\prime}\nabla F(x^*,s) + x^{*\prime}\frac{\nabla^2 F(x^*)x^*}{2} \tag{11-15}$$

$$B(x^*,x,s) = \nabla F(x^*,s) - \nabla^2 F(x^*,s)x^* + \frac{\nabla^2 F(x^*s)}{2}x$$

$$= \frac{\partial g(x,s)}{\partial x} - \frac{\nabla^2 F(x^*,s)}{2}x \tag{11-16}$$

If the variables are originally in logs, then equation (11-14) has the form of a Cobb-Douglas function, with one major difference: the coefficients are functions and not constants. This, of course, is the main feature of the present approach. As x^* changes, so do Γ and B. Variations in the state variables affect Γ and B directly as well as through their effect on x^*. In turn variations in x^* affect the input-output combinations. If this model is an accurate description, a constant coefficient production function would fail to explain all sources of variation in productivity. Such variation in productivity would be incorrectly interpreted as random by the researcher who fails to take account of the state variables that determine the implemented technology.

Failure to account for endogenous technology can lead to difficulties in the estimation of production functions. Consider the efficiency frontier approach, which involves the estimation of the production function under the assumption that the function is indeed an envelope.[5] The results derived above show that this objective cannot be achieved. At best, it is possible to estimate the envelope of the implemented technology and not of the technology. But that envelope varies with the state variables, and the question then is, what is the meaning of estimating an envelope ignoring such variations.

Equation (11-14) indicates that the coefficients can vary either with variations in the state variables—as reflected by x^*—or with x. The literature on production functions deals with the latter. For instance, the translog production function (Christensen, Jorgenson, and Lau, 1973) can be derived from equation (11-14) by setting x^* to be identically zero. Thus, although $B(0,x)$ is not a constant, it is invariant to variations in the state variables that affect the implemented technology in that it is implicitly assumed that the observations are generated by a

---
[5] For discussions of this approach, see for instance the *Journal of Econometrics*, May 1980.

well-defined production function, which is not the case. This may explain the frequent failure of empirical analysis to obtain the concavity consistent with the second-order condition for profit maximization. Such a failure is serious when the first-order conditions are used in the estimation procedure, as is actually the case in such studies. This reflects the fact that the process of formulation and estimation of production functions has not yet reached a satisfactory stage. One direction of research aimed at correcting the situation has been to introduce higher degree polynomials to approximate the production function.[6] It is true that sufficiently high degree polynomials will approximate any function to a desired degree of accuracy. The fact that aggregate technology is a function of the state variables, however, raises questions about the use of higher polynomial functions to represent production technologies. The function that this method intends to estimate does not exist. In principle, there may be as many functions as there are sample points. The key to estimation and interpretation is to take the endogeneity of the technology into account, as is discussed in the next section.

IDENTIFICATION AND ESTIMATION

The key to the estimation of $g(x,s)$ as defined by equation (11-12) is the discrepancy between observed and optimal quantities. Such a discrepancy occurs at two levels. First is the error made by firms in correctly determining $x^*(s)$. Second is the error of specification arising from the fact that the model simplifies reality; the optimal value consistent with the model is not necessarily the same as that viewed by the firm. It should be noted that it is the existence of a discrepancy that is utilized here and as such it is independent of its actual distribution. Thus $x - x^*$ can be white noise and still help in the identification.

Thus, the estimation of $g(x,s)$ requires the estimation of $\Gamma(x^*,s)$ and $B(x^*,x,s)$, which are unknown functions in s, and the unobserved variable x^*. There is no point in trying to determine x^* separately from Γ and B. The procedure is to consider Γ and B as composite functions in s and to expand functions in terms of s. Denoting the vector of state variables by s, we can then write

$$\Gamma(x^*,s) \cong \pi_{00} + s'\pi_{10} + s'\pi_{20}s \qquad (11\text{-}17)$$

$$B(x^*,x,s) \cong \pi_{01} + \pi_{11}\begin{pmatrix} s \\ x \end{pmatrix} + Q(s,,x) \qquad (11\text{-}18)$$

[6] See Galant (1982).

The πs represent coefficients: π_{00} is a scalar; π_{10}, π_{01} are vectors; and π_{20}, π_{11} are matrices whose orders are obvious from the equations. $Q(s,x)$ is a quadratic term, not spelled out as it is likely to be omitted as explained below.

Combining equations (11-17), (11-18), and (11-14) we obtain

$$g(x,s) \cong \pi_{00} + s'\pi_{10} + s'\pi_{20}s + x'\pi_{01} + x'\pi_{s1}s + x'\pi_{x1}x \tag{11-19}$$

where π_{11} is now decomposed in an obvious way to $\pi_{11} = (\pi_{s1}, \pi_{x1})$. $Q(s,x)$ is omitted from equation (11-19), as its multiplication by x' gives third-degree terms that are unlikely to be empirically relevant. Such an omission simplifies the discussion but does not change it in a substantive way.

Even in its present form, equation (11-19) contains potentially too many terms. Its direct estimation is thus likely to yield imprecise results. The number of variables can be reduced by using principal component techniques. This procedure was followed by Mundlak and Hellinghausen (1982) in applying some of these concepts to cross-country analysis of agricultural productivity.

Additional information on the first derivatives can be derived from the factor shares. Let g and x represent logs of output and inputs, respectively, and let θ be the vector of factor shares. Then under equilibrium

$$\frac{\partial g(x,s)}{\partial x} = \theta \tag{11-20}$$

Differentiating equation (11-19) and using equation (11-20), we can write

$$\theta = \pi_{01} + \pi_{s1}x \tag{11-21}$$

Using equation (11-18) without $Q(s,x)$,

$$B = \theta - \pi_{x1}x \tag{11-22}$$

We can then write the production function as

$$y = \Gamma + x'(\theta - \pi_{x1}x) \tag{11-23}$$

The system to be estimated consists of equations (11-23), (11-21), and (11-17). Adding error terms to these equations, the system can be estimated by any system method, such as FIML or one of its approximations.

The dependence of θ on x indicates that the production technology exhibits a varying elasticity of substitution. The variation of θ with s indicates that different techniques may have different factor shares, that is, the techniques may vary in their factor intensity. Empirically, the variations of the state variables may explain to a large degree the variations in factor shares.

Equation (11-21) was obtained subject to the equilibrium condition in equation (11-20). This condition can be modified to allow for distortion in the factor market. This can be done by adding a term $s'\pi_{30}x$ to Γ

$$\Gamma(x^*s) \cong \pi_{00} + s'\pi_{10} + s'\pi_{20}s + s'\pi_{30}x \qquad (11\text{-}24)$$

in which case $\partial g(\ )/\partial x$ will differ from θ by $s'\pi_{30}$. This allows for systematic deviations from equilibrium in the factor market that are related to the state variable. Such a distortion could be made to depend on other variables as well, or even on x itself, following a similar approach. This approach makes it possible to use the information conveyed in the factor shares without imposing the equilibrium conditions. The system to be estimated will now consist of equations (11-24), (11-21), and (11-17).

ENDOGENEITY OF TECHNOLOGY

In what sense is the technology endogenous? The foregoing discussion has shown explicitly that the implemented technology is endogenous in that it depends on the state variables. This is the only aspect of technology that is actually observed. Like any other observed economic variable, it is determined by supply and demand. The supply of new techniques is represented by T. Together, T and the remaining state variables determine the implemented technology. The question is, to what extent is T endogenous? The process of generating new techniques is a subject that has been studied broadly and is not dealt with here in any comprehensive way. One important aspect of it should be brought up explicitly, however, because of its important repercussions for our analysis. The present framework facilitates an insight into Hicks' view of induced innovations being labor saving.

Following Danin and Mundlak (1979), it can be shown that capital accumulation results in the employment of capital-intensive techniques and, conversely, that the introduction of capital-intensive techniques requires capital accumulation. Turning to figure 11-1, even if it is available, technique 2 will not be used as long as $k < \tilde{k}_1$. As the capital-labor ratio goes up and passes the threshold level $\tilde{k}_1$, technique 2 will be introduced, and its utilization will increase with k. For $k > \tilde{k}_2$, only technique 2 will be employed. This analysis can be viewed as dealing with a given industry, say agriculture, in isolation. It assumes that agriculture has two techniques, traditional and modern, and as the capital-labor ratio in agriculture increases, the traditional variety will lose ground to the modern variety. It can be shown that the same result is obtained within an equilibrium analysis for the economy as a whole.[7] The implication is that the introduction of capital-intensive techniques is subject to capital constraints and consequently the rate of adoption of the technique depends on capital accumulation.

New techniques are generated by firms, private or public, that allocate resources to research and development. For a given state of science, a choice can generally be made in determining the research strategy. For the purpose of our discussion, the key variable is the capital intensity of the new techniques. As we have shown, capital accumulation generates demand for capital-intensive techniques. Thus, the producers of techniques should aim at the development of capital-intensive, rather than labor-intensive, techniques. However, overshooting is counterproductive. Because the rate of implementation depends on the rate of capital accumulation, the threshold level of the new techniques should not be too high or the market for such techniques will be limited.

In the absence of a new capital-intensive technique, capital accumulation increases the capital-labor ratios, thus increasing real wages and decreasing the real rental rate on capital. Thus, the owners of capital will be interested in investing their capital in techniques that prevent the rate of return from falling. This generates the demand for the capital-intensive techniques.

For the purpose of simplification, we have dealt with two techniques: traditional and modern. The appearance of additional techniques can be handled in a similar fashion. One particular case—that of neutral technical change (NTC) in the modern technique—is worthy of examination here. As it has been argued that the process of capital accumulation generates a demand for capital-intensive techniques, then—other things being equal—this demand will be realized through the development of the NTC to be implemented on the modern techniques. In a more detailed framework, the cost of producing and changing techniques, as well as the required research time, should be introduced. If the required time is significant, by the time the research is completed the traditional technique may no longer be of any importance. Therefore, effort will be directed at increasing the productivity of the modern techniques. This consideration has a dynamic aspect. With time, the modern techniques

---

[7] See Danin and Mundlak (1979).

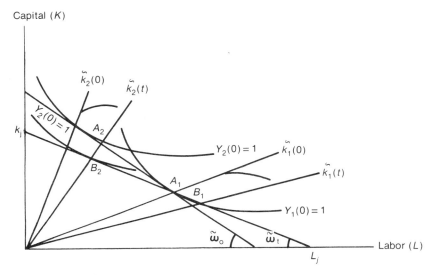

FIGURE 11-2. Neutral technical change and choice of technique

become traditional and, therefore, have already been worked on so that the easy gains may already have been made and additional gains may be subject to increasing cost. Thus, both from the demand side and the supply side, it is likely that the effort of improving an existing technique will be aimed at the modern techniques.

An improvement in the productivity of a technique should increase the degree of its utilization. In part, this can be illustrated graphically in figure 11-2. The initial techniques are represented by $Y_1(0)$ and $Y_2(0)$, with threshold values $\tilde{k}_1$ and $\tilde{k}_2$. Neutral technical change in the modern technique shifts its unit isoquant to $Y_2(t) = 1$. The threshold values decline accordingly to $\tilde{k}_2(t)$ and $\tilde{k}_1(t)$. For any value of k, the relative importance of the traditional variety declines.[8] The net effect of this change is labor saving and thus can be expressed as a decline in the labor share at any level of k. This is believed to be the situation in agriculture where, on a net basis, we observe labor-saving technical change. From this discussion, it emerges that the technology set will be expanding with capital accumulation. To be sure, capital is viewed here in a comprehensive way, in that it includes capital accumulated in research, education, and other forms of human capital. As such, it represents both the supply and the demand of new techniques.

[8] This can be shown analytically, but we omit here the technical details.

It is interesting to compare this approach to the introduction of new techniques with that of Solow's embodied technical change. If the new techniques are embodied in new investment, then the rate of their introduction depends on gross rather than on net investment, as is the case in the present framework. However, the empirical application of this distinction is not immediate. This is discussed in more detail in Mundlak (1984). The foregoing discussion suggests that net investment and depreciation are included in the empirical analysis as state variables with expected positive effects on productivity. This does not tell the whole story, however. A broader framework should also allow for a cost of adjustment that increases with investment. This partial effect of investment on output due to the cost of adjustment is negative. Thus, the expected net effect of investment is ambiguous.

EMPIRICAL IMPLEMENTATION

The primary goal of empirical analysis is the identification of the state variables, which can be classified as those representing technology, constraints, and prices.

The technology variables should represent the movement of the frontier, or what is commonly understood as technical change. Sometimes there are observable indicators of such progress, such as the proportion of the area sown by high-yielding varieties. Such variables themselves are endogenous within the economic system, as discussed below. In general, the set of techniques T is a function of the overall stock of capital in the economy, including the various facets of human capital. The rate of implementation of the new techniques may also be determined by gross investment, although the sign is ambiguous.[9] Per capita GNP can be used as a measure of comprehensive capital (Mundlak and Hellinghausen, 1982). Given that this measure is subject to cyclical fluctuation, its historical peak values are better indicators of T.[10]

The technology constraints are easier to identify. In the cross-country analysis of agricultural productivity, the constraints included the basic resource endowment, such as labor and land, as well as measures of the physical environment. In the study of Argentina's sectoral growth (Cavallo and Mundlak, 1982), the constraints also included the share of agriculture in total credit, as this share reflected a supervised program.

---

[9] This approach is taken by Coeymans and Mundlak (1983) in the study of sectoral growth in Chile.
[10] See Mundlak (1984).

It is possible to endogenize some of the constraints. Doing so will allow the state variables to be determined by the economy. The state variables will then determine productivity, which affects the state variables. This is basically the approach that has been followed by Cavallo and Mundlak (1982) in the study of sectoral growth in Argentina and by Coeymans and Mundlak (1983) in the study of sectoral growth in Chile.

Prices are generally observable. However, what matters for production decisions are not actual prices but expected prices, which are not observed. Thus, the analysis should be extended to explain expectation formation. When dealing with an industry, prices are likely to be endogenous. Therefore, to obtain a complete model, the production sector should be analyzed together with product demand and factor supply.[11] Thus, when demand is expected to be weak due to cyclical variations, one would expect a lower output. If resources are not adjustable instantaneously, such a downward cycle will cause a decline in measured productivity in conventional studies; in the presented framework, the state variables will explain such deviations from the frontier.

AGENDA FOR FUTURE RESEARCH

This analysis has sought to endogenize productivity and formulate it in a way that has empirical relevance. The previous section reviewed some directions taken in recent research. This research, however, constitutes only a first step. It is, therefore, useful to summarize the main directions that should be followed in future research.

a. Integration of the analysis of product demand and factor supply into the analysis of productivity. As indicated above, this approach integrates cyclical variations in productivity analysis. It also generates the necessary link with which to analyze sectoral growth within the framework of economic growth. If the sector is important, then the study of its factor or product market cannot be isolated. Thus, when agriculture constitutes an important sector of the economy, the off-farm migration of labor or the sectoral allocation of investment is interdependent with developments in the rest of the economy.

b. Integration of technological uncertainty and price uncertainty into the analysis of productivity. The main thrust of our analysis has been on the growth aspects of changes in technology; we have ignored the question of uncertainty associated with the supply and demand of new techniques. Thus, even when a technique is available, its implementa-

---

[11] See Johnson (1950).

tion is not immediate because of uncertainty considerations (Griliches, 1957). Any empirical analysis should be concerned with this aspect. A related issue is that of the greater variability associated with new techniques relative to existing techniques (Barker, Gabler, and Winkelman, 1981; Hazell, 1984; and Mehra, 1981).

c. Adoption of a multiperiod framework. The investments associated with the generation of new techniques as well as with their implementations should be evaluated within a multiperiod model. The rate of interest will play an important role, suggesting the possibility of extending the analysis by endogenizing the rate of interest.

REFERENCES

Barker, R., E. C. Gabler, and D. Winkelman. 1981. "Long-Term Consequences of Technological Change on Crop Yield Stability," in A. Valdes, ed., *Food Security for Developing Countries* (Boulder, Colo., Westview Press).

Cavallo, D., and Y. Mundlak. 1982. *Agriculture and Economic Growth in an Open Economy: The Case of Argentina*. Research Report No. 36 (Washington, D.C., International Food Policy Research Institute).

Christensen, L. R., D. W. Jorgensen, and L. J. Lau. 1973. "Transcendental Logarithmic Production Frontiers," *Review of Economics and Statistics* vol. 55, no. 1, pp. 284–285.

Coeymans, J., and Y. Mundlak. 1983. "Productividad Endogena y la Evolucion de la Produccion y Empleo Sectoral en Chile." Paper presented at the Fourth Latin American Congress of the Econometric Society, Santiago, Chile, July.

Danin, Y., and Y. Mundlak. 1979. "The Introduction of a New Technique and Capital Accumulation," Working Paper 7909 (Rehovat, Israel, The Center for Agricultural Economic Research).

Fuss, M., D. McFadden, and Y. Mundlak. 1978. "Survey of Functional Forms in Economic Analysis of Production," in M. Fuss and D. McFadden, eds., *Production Economics: A Dual Approach to Theory and Applications* (New York, N.Y., North-Holland).

Gallant, A. R. 1982. "Unbiased Determination of Production Technologies," *Journal of Econometrics* vol. 20, no. 2, pp. 285–323.

Griliches, Z. 1957. "Hybrid Corn: An Exploration in the Economics of Technological Change," *Econometrica* vol. 25, no. 4, pp. 501–522.

Hazell, P. 1984. "Sources of Increased Instability in Indian and U.S. Cereal Production," *American Journal of Agricultural Economics* vol. 66, no. 3, pp. 302–311.

Johnson, D. Gale. 1950. "The Nature of the Supply Function for Agricultural Products," *The American Economic Review* vol. 40, no. 3, pp. 539–564.

Johnson, Glenn. 1972. "Mathematical Notes on the Neoclassical and Modified Neoclassical Theory of the Firm," in G. L. Johnson and C. L. Quance, eds., *The Overproduction Trap in U.S. Agriculture* (Baltimore, Md., Johns Hopkins University Press).

Mehra, S. 1981. *Instability in Indian Agriculture in the Context of the New Technology*. Research Report No. 25 (Washington, D.C., International Food Policy Research Institute).

Mundlak, Y. 1983. "Lectures on Agriculture and Economic Growth, Theory and Measurement (Part I)." Mimeograph.

———. 1984. "Capital Accumulation, the Choice of Techniques and Agricultural Output" (Washington, D.C., International Food Policy Research Institute). Mimeograph.

———, and R. Hellinghausen. 1982. "Intercountry Comparison of Agricultural Productivity—Another View," *American Journal of Agricultural Economics* vol. 64, no. 4, pp. 664–672.

Solow, R. M. *Capital Theory and the Rate of Return* (Amsterdam, North-Holland).

12

DYNAMICS, CAUSALITY, AND AGRICULTURAL PRODUCTIVITY

JOHN M. ANTLE

This chapter explores the dynamic structure of agricultural production processes and the implications of this structure for the measurement and explanation of agricultural productivity. Underlying this approach is the view that changes in agricultural productivity occur over time as producers respond to the evolution of complex and interrelated natural, economic, technological, and institutional phenomena. Modern theories of economic development, such as Hayami and Ruttan's version of induced innovation, are consistent with this dynamic view of agricultural development.

Empirical agricultural productivity research, in contrast, is usually based on static models, as evidenced by the preceding chapters in this volume and the studies reviewed in chapter 3. Static production, cost, and profit function models are used to measure technological change and to draw inferences about the explanation of technological change. However, given that theory portrays technological change as a dynamic process, it can be questioned whether contemporaneous correlations between variables such as factor cost shares and factor prices can be used to explain the processes generating technological change in agriculture. It can be argued more generally that most of the existing empirical research has provided evidence that time trends and other measurable variables, such as agricultural research, human capital, relative prices, and agricultural policies are correlated to some degree with agricultural productivity. However, it is unclear what the observed correlations between measured output and time and other variables imply about the causal mechanisms governing agricultural productivity.

As Zellner (1984, p. 72) has emphasized, analysis of causal relations involves "the delicate and difficult work of integrating statistical techniques and subject matter considerations"; that is, it requires both theory and measurement. Measurement alone, involving either contemporaneous or temporal correlations, cannot be used to infer causation according to the accepted philosophical and econometric definitions. This chapter attempts to integrate a theory of dynamic production relations with measurement methods in order to draw inferences about the causal relations among factors affecting agricultural productivity.

The literature on the temporal dimension of production can be classed into three generations of dynamic models (Berndt, Morrison, and Watkins, 1981), beginning with the Nerlovian partial adjustment models. The more recent models, based on firms' dynamic optimizing behavior, can be traced to the investment literature of the 1960s. Lucas (1967) introduced the time dimension into production using the cost-of-adjustment hypothesis that investment disrupts production and hence imposes costs on the firm. More recently, Kydland and Prescott (1982) introduced the time dimension in production through the assumption that investment occurs over time. Long and Plosser (1983) introduced a dynamic Cobb-Douglas model that introduced lagged outputs to explain real business cycles. Using a similar kind of model, Antle (1983a,b) suggested that agricultural production is fundamentally dynamic because it depends on biological processes and because farmers make production decisions sequentially over time.

The theoretical explanations in this literature all concern the microdynamics of production at the individual firm or farm level. Aggregate agricultural productivity at a point in time is a function of the individual producer's behavior within the existing natural, economic, technological, and institutional environment. This environment itself is not static. When data are aggregated over production units and time, the macrodynamics of agricultural production emerge, involving both the microdynamics of the individual producer and the evolution of the environment within which the individual producer operates.

These observations suggest that it is necessary to begin with the microdynamics of production to understand the dynamic processes governing agricultural productivity. The first section of this chapter discusses the dynamic structure of farm-level production and the effects of aggregation. The second section addresses the causal relations in agricultural production implied by its dynamic structure. The third section then considers macrodynamics within a model of technological change; implications for econometric measurement of aggregate technology; and the explanation of technological change. The fourth section analyzes the implications of the dynamic structure of production for analysis of agricultural policies, including research investment policy and government intervention in agricultural markets. The final section discusses the implications of the analysis for future research on agricultural productivity.

DYNAMICS, STOCHASTICS, AND AGGREGATION

Virtually all intraseasonal agricultural production processes are multistage, that is, they involve a sequence of inputs applied over time to a sequence of intermediate production stages. These intermediate stages lead to a final salable output. Thus, intraseasonal agricultural production is characterized by output dynamics that are represented by a sequence of production functions in the autoregressive form

$$q_t = f_t[x_t, q_{t-1}, q_{t-2}, \ldots, q_{t-j_t}, \epsilon_t], \qquad t = 1, \ldots, T \tag{12-1}$$

where q_t is output, x_t is an input vector, ϵ_t is a random error, and T is the number of stages. Interseasonal production may also exhibit output dynamics. This is exemplified by crop rotation, in which productivity in one season depends on the acreages and types of crops grown in preceding seasons. The fundamental property of processes with output dynamics is that they are time recursive. Thus, production in period t depends on current and past inputs and production disturbances: for example

$$\begin{aligned} q_t &= f_t[x_t, q_{t-1}, \epsilon_t] \\ &= f_t[x_t, f_{t-1}[x_{t-1}, q_{t-2}, \epsilon_{t-1}], \epsilon_t] \\ &= h_t[x_t, x_{t-1}, \ldots, \epsilon_t, \epsilon_{t-1}, \ldots] \end{aligned}$$

An example of intraseasonal output dynamics is irrigation water that is applied over time at each stage of plant growth (see Antle and Hatchett, 1986).

Production may also be dynamic because a sequence of inputs affects output. In this case the process is characterized by input dynamics in the moving average form

$$q_t = f_t[x_t, x_{t-1}, \ldots, x_{t-j_t}, \epsilon_t] \tag{12-2}$$

An example of interseasonal input dynamics is fertilizer carryover (see Ackello-Ogutu, Paris, and Williams, 1985). More generally, all types of investment can be represented as input dynamics.

Within a given season with $t = 1, \ldots, T$ stages, a risk-neutral farmer's objective function can be written as

$$\max_{\{x_t\}} E\left[p_T q_T - \sum_{t=1}^{T} r_t x_t \right] \tag{12-3}$$

where $E[\cdot]$ denotes the mathematical expectation operator; p_T is the price of final output q_T; and r_t is the input price. However, many of the farmer's decisions affect production both within and across seasons. In the case in which the farmer's decisions span S seasons, this multiperiod decision problem is

$$\max_{\{x_t\}} E\left[\sum_{s=1}^{S}\left(p_{T_s}q_{T_s} - \sum_{t=1}^{T_s} r_t x_t\right)\right] \quad (12\text{-}4)$$

This nesting of decisions is exemplified by the row-crop farmer in the Sacramento Valley who rotates summer crops of processing tomatoes and corn. Intraseasonal output dynamics are associated with land preparation, planting, cultivating, irrigating, and harvesting operations. Interseasonal output dynamics are associated with soil fertility, diseases, and the crop rotation. In the longer run, all processes exhibit input dynamics in the form of physical capital investment. Both output and input dynamics are characteristic of many other types of agricultural production, including perennial crops, livestock, and poultry. Thus, at the farm level, most production processes are likely to exhibit both input and output dynamics.

The farmer's decision making is intimately related to the structure of the production process. The general solution to a dynamic production problem such as equation (12-3) or (12-4) is given by the dynamic programming algorithm. Let $x^t = (x_1, \ldots, x_t)$ and $q^t = (q_1, \ldots, q_t)$, ignoring the distinction between stages and periods for notational simplicity. Define ψ_t as the parameter vector of the decision maker's subjective joint probability distribution function of current and future outputs, output prices, and input prices. The general form of the structural input demand functions implied by the solution to the multistage or multiperiod decision problem, when they exist in closed form, is

$$x_t = x_t[x^{t-1}, q^{t-1}, \psi_t, r_t], \quad t = 1, \ldots, T \quad (12\text{-}5)$$

Equation (12-5) is the form of the solution for a technology with both input and output dynamics. If the technology had only input dynamics, q^{t-1} would not appear in equation (12-5); if the technology had only output dynamics, x^{t-1} would not appear in equation (12-5). Note that equation (12-5) can be interpreted as a reduced-form equation, since x^{t-1} and q^{t-1} are predetermined and ψ_t and r_t are exogenous.

A complete dynamic production model is a system of production functions such as equation (12-1) or (12-2) and input demand functions (12-5). One remarkable property of dynamic production models of this type is their recursive, or triangular, structure.[1] Processes characterized by output dynamics are recursive systems of both production functions and input demand functions, whereas processes with input dynamics are recursive only in the input demands (lagged outputs do not appear in equation (12-2)). This recursive structure is central to the problem of econometric estimation of dynamic production models, and proves to be critical to the analysis of causality in section 2.

It must be emphasized that the dynamics of agricultural production are intimately related to the time-dependent nature of the production process itself. This contrasts with the cost-of-adjustment theory underlying many dynamic production models. The cost-of-adjustment model is based on the idea that investment imposes costs on the firm in the form of reduced output (see Berndt, Morrison, and Watkins, 1981, for a survey of the literature). Since net investment can be expressed as the change in the capital stock, the cost-of-adjustment model is derived from a production function that is a special case of input dynamics given in equation (12-2) (as shown in chapter 2). Thus, the cost-of-adjustment model does not exhibit output dynamics, and therefore cannot represent many of the dynamic relations in agricultural production at the farm level.

Aggregation at the Firm Level

Productivity analysis—even at the firm level—must utilize data that have been aggregated to some degree. This raises the question of the dynamic structure of production after data aggregation. Consider first farm level data aggregated over a growing season into total inputs and outputs for the season. For simplicity, assume there is only one (scalar) input x_t applied in each stage. Using equation (12-5), we calculate the total input use in the season to be

$$X = \sum_{t=1}^{T} x_t[x^{t-1}, q^{t-1}, \psi_t, r_t] \qquad (12\text{-}6)$$

Substituting equation (12-1) for q^t and equation (12-5) for x^{t-1} in equation (12-6) gives the final-form equation

$$X = X[\psi^T, r^T, \epsilon^{T-1}]$$

---

[1] These systems are referred to as recursive without specifying the structure of the error covariance matrix. If the covariance matrix is not diagonal, then the system is either block recursive or triangular.

showing that inputs aggregated across stages depend on the history of price expectations, input prices, and intermediate stage production shocks over the season. A production function defined in terms of seasonally aggregated inputs and production disturbances is

$$Q = f[X, \epsilon^T] \tag{12-7}$$

Therefore, the production model consisting of the production function (12-7) and input demand functions (12-6) is also recursive, because inputs explain final output and not vice versa. The final-form equation below equation (12-6) shows that inputs are correlated with final output, but are not functions of final output.

Aggregating across seasons to obtain annual data in the presence of interseasonal output dynamics gives aggregate inputs that depend on previous seasons' final outputs. For example, let annual farm output be measured as an output index $Y[Q_{T_1}, \ldots, Q_{T_s}]$, where Q_{T_s} is final output in season s. With interseasonal output dynamics

$$Q_{T_s} = f_{T_s}[X_{T_s}, Q_{T_s-1}, Q_{T_s-2}, \ldots, \epsilon_{T_s}] \tag{12-8}$$

so annual output aggregated over seasons can be expressed as

$$Y = Y[X_{T_1}, \ldots, X_{T_s}, \epsilon_{T_1}, \ldots, \epsilon_{T_s}] \tag{12-9}$$

where X_{T_s} is the input vector for season s aggregated across operations in season s. Solution of maximization problem (12-4) shows that X_{T_s} depends on prices, price expectations, and previous seasons' inputs and production disturbances

$$X_{T_s} = X_{T_s}[X^{T_s-1}, \epsilon^{T_s-1}, \psi_t, r_t] \tag{12-10}$$

Observe that the production model composed of equations (12-9) and (12-10) is recursive in inputs. Therefore, as in the case of input aggregation within a season, output aggregation across seasons preserves the recursive structure of production models.

The behavioral interpretation of these findings should be emphasized. Production at the farm level is recursive because of the logical time ordering of events in the production process: inputs must be chosen before output can be realized. Aggregation over time at the farm level causes inputs and outputs to be correlated because some inputs and outputs are functions of the same production disturbances. However, it is not possible for inputs employed during a given time interval to be functions of final output realized at the end of that period. Thus, the recursive structure of production is preserved in aggregation across production stages and seasons.

Production as a Stochastic Process

In modern production theory, technology is represented in terms of the production possibilities set, which defines the set of technologically feasible production plans. Each production plan defines an output (or vector of outputs) that is obtainable from a given input vector. This formulation of the production problem does not account for the presence of random events in the production process. As in the previous section, output uncertainty can be modeled by introducing random terms into the production function. More generally, uncertainty can be introduced into the production problem by defining the firm's outputs and prices as stochastic processes satisfying certain conditions (analogous to the convexity properties of neoclassical technologies) that define the technological and economic conditions facing the firm.

Outputs and prices may be described in terms of a joint probability distribution. To simplify notation and the discussion to follow, we assume that firms are price takers and prices and outputs are assumed to be statistically independent. These assumptions allow the firm's stochastic technology to be defined separately from price distributions, in a manner analogous to the neoclassical theory of the firm, in which the production function is distinct from prices. In the more general case, in which output and prices are correlated, one must distinguish the output distribution conditional on prices from the marginal output distribution. In the discussion of aggregation that follows, it will become apparent that this assumption of price and output independence does not limit the generality of the results for the purposes of this study.

The firm's output q_t in period t; its vector of fixed and variable inputs x_t; and its output price p_t are defined on nonnegative, bounded subsets S_q, S_{x_t}, S_p of R, R^{n_t}, and R. The firm formulates expectations based on an information set ω_t that is defined on a subset S_{ω_t} of the appropriately dimensioned Euclidean space. In the case of static production, the production process is defined in terms of the conditional probability distribution function $F[q_t|x_t] : S_q X S_{x_t} \to [0, 1]$. The distribution of output prices is defined as $G[p_t|\omega_t] : S_p X S_{\omega_t} \to [0, 1]$, where X denotes the Cartesian product. In place of the neoclassical regularity conditions, we assume that the probability distribution function is continuous and twice differentiable for all $x_t \in S_{x_t}$ such that the choice problems (12-3) or (12-4) have unique interior solutions.

The dynamic production functions described above (equations (12-1) and (12-2)) can be related to the properties of the output distribution function. In the case of input dynamics, the distribution of output is $F[q_t|x^t]$, where the convention $x^t = (x_{t-j_t}, \ldots, x_{t-1}, x_t)$ is employed (note that j_t, the number of lagged input vectors conditioning the distribution of output, may vary with t). For output dynamics, the stochastic process generating output is $F[q_t|x_t, q^{t-1}]$, where $q^{t-1} = (q_{t-k_t}, \ldots, q_{t-2}, q_{t-1})$. In general, the production process may exhibit both input and output dynamics, in which case $\{q_t\}$ is generated by

$$F[q_t|x^t, q^{t-1}] \qquad (12\text{-}11)$$

Stochastic Production and Technological Change[2]

In the neoclassical model, technological change is interpreted as a shift in the production function that enables the firm to obtain more output per unit of input or per unit of cost. When production is described as a stochastic process, technological change causes the output distribution function (12-11) to change over time. Time series theorists define a stochastic process as *stationary* if its distribution function does not depend on time, and as *nonstationary* if its distribution function varies with time (Doob, 1953; Granger and Newbold, 1977). Thus, we can define technological change in terms of the stationarity properties of the stochastic process generating output. Let the time dependence of the conditional distribution function be denoted by a time subscript

$$F_t[q_t|x^t, q^{t-1}] \qquad (12\text{-}12)$$

F_t represents a distribution that can change over time either in terms of form or in terms of the parameters of the distribution.

For the purpose of defining the stationarity properties of the distribution of output, the conventional concept of stationarity, based on unconditional distributions, requires a modification. The stochastic process q_t is said to be *conditionally stationary* if and only if $F_{t+j}[q_{t+j}|x^{t+j}, q^{t-1+j}] = F[q_t|x^t, q^{t-1}]$, $x^{t+j} = x^t$, and $q^{t-1+j} = q^{t-1}$ for all t and for all $j > 1$. The stochastic process q_t is said to be *conditionally nonstationary* if and only if $F_{t+j}[q_{t+j}|x^{t+j}, q^{t-1+j}] \neq F[q_t|x^t, q^{t-1}]$, $x^{t+j} = x^t$, and $q^{t-1+j} = q^{t-1}$ for at least one t and one $j > 1$. In words, conditional stationarity (nonstationarity) means that, for given values of the conditioning variables, the distribution of q_t is (is not) time invariant. Note that the distinction

---

[2] This section and the following one are based on Antle (1986).

between stationarity and conditional stationarity is needed because q_t may not be stationary in the conventional sense (because the moments of its conditional distribution vary with x^t and q^{t-1}) even if it is conditionally stationary. The distinction between nonstationarity and conditional nonstationarity is needed because conditional nonstationarity requires that the form or the parameters of the distribution depend on time.

These definitions allow the neoclassical concept of technological change, in terms of a dated production function, to be generalized to the dynamic, stochastic case in terms of the dated conditional output distribution function. Intuitively, technological change in the stochastic case must mean that the process generating output changes over time, and technological progress must mean that over time more output can be obtained from a given resource bundle than was possible in the past. It is reasonable to assume that if rational firms adopt innovations, observed technological change is, in fact, technological progress. For example, in the static risk-neutral case in which the firm's objective function involves only the mean of the output distribution, the implemented production processes would be expected to satisfy non-retrogression in the mean, that is, at constant factor prices cost per unit of expected output is nonincreasing

$$r_t x_t / E_t[q|x_t] \leq r_t x_{t-1} / E_{t-1}[q|x_{t-1}] \qquad (12\text{-}13)$$

However, in the dynamic, stochastic case represented by equation (12-4), consideration of the mean alone may not be sufficient to infer technological progress, for two reasons. First, the firm's objective function is nonlinear in future periods' outputs and therefore depends on higher moments of future periods' outputs even if the firm is risk neutral (Antle, 1983a). If the firm is risk averse, the objective function is nonlinear in both current and future periods' outputs. Second, the general representation of production as a stochastic process suggests that technological progress may involve beneficial changes in higher moments of the output distribution. The general representation (equation (12-12)) of the output distribution makes clear that the analysis of technological change may be a multidimensional problem, involving more than the mean of the output distribution.

It is not possible to estimate statistically the parameters of a non-stationary process (strictly speaking, of a nonergodic process; see, for example, Parzen, 1962, chapter 3) with single realizations of a time series, although estimation may be possible with pooled cross-section and time series data. The time series literature typically discusses conditions for stationarity of linear autoregressive moving average (ARMA) processes (for example, Granger and Newbold, 1977) and considers transformations of the data (by taking differences, for example) that eliminate nonstationarity under certain assumptions. For some purposes, the modeling of output as an ARMA process may be adequate.

However, in contrast to conventional time series analysis, the aim here is to measure and explain the conditional nonstationarity of the output series—that is, technological change—using theory and observable variables. Our approach is to hypothesize that the time dependence of the distribution F_t can be explained in terms of a vector of explanatory variables z_t suggested by economic theory.

Generally, the z_t are economic variables that the firm knows in period t but that it does not know in future periods and thus must forecast along with prices and output. For example, human capital theory suggests that a firm's adoption of technology may depend on its manager's human capital; the induced innovation theory suggests that relative prices influence the type and rate of technological change; thus z_t could contain human capital variables and prices. Alternatively, z_t could be a time trend, under the assumption that technological change is an exogenous process correlated with time. Thus, if theory implies that the time dependence of the output distribution is explained by z_t, there exists a conditionally stationary process F satisfying

$$F_t[q_t | x^t, q^{t-1}] = F[q_t | x^t, q^{t-1}, z_t] \text{ for all } t \qquad (12\text{-}14)$$

This condition is important in attempting to model technological change because it allows the definition of technological change in terms of the conditional nonstationarity of the output series to be transformed into a theory that explains technological change in terms of a conditionally stationary series.

Aggregation Across Firms

Productivity analysis typically involves the use of data that have been aggregated across firms as well as across stages of production and across seasons at the firm level. It is thus necessary to relate the stochastic processes defining firm-level production to their aggregate counterparts for the industry. The aggregation problem is analyzed here by adapting Stoker's (1982) formulation of the general aggregation problem to the present context. To illustrate the approach, we consider aggregation of the stochastic process (equation (12-12)). The principal issue is whether the structural and stationarity properties of the microeconomic process are preserved in aggregation. For simplicity, assume all firms produce with the same technology F_t in each period; this assumption can be relaxed without loss, as noted below.

Aggregation is interpreted here as the adding up of attributes of individuals in the population to obtain summary statistics for the population that cannot be differentiated by the individual outcomes that are aggregated. Therefore, aggregate analysis must be conducted in terms of the parameters that define the population and its behavior. These parameters are the distribution of prices firms face when decisions are made and the distribution of firm attributes. The ith firm's attributes are defined in terms of the information set ω_t^i it uses to make production decisions. For example, in equation (12-5) we see that the ith firm's attributes are its input and output histories x^{it} and q^{it-1} and its expectations parameters ψ_t^i, so $\omega_t^i = (x^{it}, q^{it-1}, \psi_t^i)$. From the econometrician's point of view, ω_t^i is unobservable and can be interpreted as a draw from the distribution $A(\omega|\theta_t)$, where $\theta_t \in \Theta$ (Θ is an appropriately dimensioned subspace of Euclidean space) is the parameter vector characterizing the population in period t. Define

$$\xi_t \equiv E(\omega|\theta_t) = \int \omega dA(\omega|\theta_t) = b_1(\theta_t)$$

From equation (12-5) it follows that x_t^i, the ith firm's input vector, is a random variable and the expected input vector for the population is

$$\mu_t \equiv E[x|r_t, \theta_t] = \int \delta_t[r_t, \omega]dA(\omega|\theta_t) = b_2(r_t, \theta_t)$$

The aggregate analog of ω_t^i is defined as the aggregate information set Ω_t that contains the aggregate counterparts of the firm's information, for example, a subset of the histories of aggregate quantity and price data.

Two assumptions are required for the derivation of the aggregation analysis

A.1. The functions $\xi_t = b_1(\theta_t)$ and $\mu_t = b_2(r_t, \theta_t)$ are invertible such that the functions $\theta_t = h_1(\xi_t)$ and $r_t = h_2(\mu_t, \theta_t)$ exist.

A.2. The aggregate information set Ω_t; the aggregate input vector X_t; and aggregate output Q_t are averages of ω_t^i, x_t^i, and q_t^i in the sense that they satisfy

$$\plim_{n_t \to \infty} \Omega_t = \theta_t, \quad \plim_{n_t \to \infty} X_t = \mu_t, \quad \text{and} \quad \plim_{n_t \to \infty} Q_t = E_t[q|r_t, \theta_t]$$

where n_t is the number of firms in period t. Define the ith firm's input and output histories as x^{it} and q^{it-1}. The conditional expectation of aggregate output Q_t can be written in terms of all firms' current and past inputs and past outputs

$$E_t[Q|x^{1t},\ldots,x^{n_t t}, q^{1t-1},\ldots,q^{n_t t-1}]$$
$$= \int\ldots\int Q_t\, dF_t[q|x^{1t}, q^{1t-1}]\ldots dF_t[q|x^{n_t t}, q^{n_t t-1}]$$

This conditional expectation cannot be inferred from aggregate data, because it requires knowledge of individual firms' inputs and outputs. Expected aggregate output can be expressed independently of individual firm decisions and output realizations by averaging the above expression conditional on factor prices r_t and the population parameters θ_t

$$E_t[Q|r_t, \theta_t] = \int E_t[Q|x^{1t},\ldots,x^{n_t t}, q^{1t-1},\ldots,q^{n_t t-1}] dA(\omega|\theta_t) \qquad (12\text{-}15)$$

Using assumption A.1 it follows that aggregate output can be expressed in terms of μ_t and θ_t as

$$E_t[Q|r_t, \theta_t] = E_t[Q|h_2(\mu_t, h_1(\xi_t)), h_1(\xi_t)]$$
$$= M_t[\mu_t, \theta_t] \qquad (12\text{-}16)$$

By assumption A.2 and equation (12-10) it follows that for large n_t,

$$E_t[Q|r_t, \theta_t] = M_t[\mu_t, \theta_t] \simeq M_t[X_t, \Omega_t] \qquad (12\text{-}17)$$

Similar arguments show that other moments of Q_t, such as its variance and autocovariances, can also be expressed as approximate functions of X_t and Ω_t for large n_t. By the same line of reasoning, if a closed-form solution (12-5) exists, the aggregate factor demand functions can be expressed as

$$X_t = \Delta_t[r_t, \Omega_t] \qquad (12\text{-}18)$$

It follows that, under condition (12-14), mean aggregate output and the factor demand functions can be written in terms of undated functions of X_t, Ω_t, r_t, and z_t:

$$M_t[X_t, \Omega_t] = M[X_t, \Omega_t, z_t] \qquad (12\text{-}19)$$
$$\Delta_t[r_t, \Omega_t] = \Delta[r_t, \Omega_t, z_t] \qquad (12\text{-}20)$$

Comparing the microeconomic and aggregate models, we reach two important conclusions. First, both the firm-level and aggregate models have a recursive or triangular structure. Thus, the recursive structure of the microeconomic model is preserved in aggregation. Second, although the structure of the microeconomic and aggregate models is similar, there is a fundamental difference between the two models. Equation (12-17) shows that the stochastic process generating aggregate output is a function of current inputs and the information with which the firms' expectations are formed (histories of inputs, outputs, and prices). Mean aggregate output depends on these variables because, as equation (12-15) shows, the distribution of aggregate output depends on the distribution of firms' inputs and outputs, which in turn depend on prices and the population parameters. In contrast, the stochastic process generating an individual firm's output depends only on current inputs, and the histories of past input and output quantities; other elements of ω_t^i, such as prices, do not enter into equation (12-11) or (12-12). In the microeconomic model, prices and other elements of ω_t^i can condition the firm's output distribution only if they are contained in z_t and equation (12-14) is assumed. These distinctions between the microeconomic and aggregate output processes play a central role in the identification of the aggregate model.

The assumption that all firms use the same technology F_t can be relaxed without altering the results of the aggregation analysis. For example, let technological change be modeled in terms of the parameters of the function (12-12). Then the expectation (as in equation (12-15)) can be taken over individual firms' inputs, outputs, and parameters representing technology. Again, the aggregate expectation of output is defined in terms of prices and population parameters, and the remaining aggregation results follow. Thus the properties of the aggregate production function described by Mundlak in chapter 11 can be obtained from this formal aggregation analysis. Mundlak argues that the aggregate production function depends on current inputs and a set of state variables measuring prices, fixed capital, and technology. In Mundlak's model this result flows from the assumption that the technology implemented by each firm depends on its state variables. Letting the firm's state variables be Ω_t shows that either expectations formation as discussed here or different rates of technology adoption as discussed in chapter 11 imply that the aggregate production process depends on the variables contained in Ω_t.

CAUSALITY IN PRODUCTION

Zellner (1984) discusses the various definitions of causation in philosophy and economics. It is beyond the scope of this chapter to consider the various definitions in detail. Instead, two recent approaches to causality will be used to explore causality in production. These two approaches are referred to as *structural* and *nonstructural* by Cooley and LeRoy (1985).

The structural approach is based on the traditional econometric concepts of endogeneity, identification, and structural form. What will be referred to here as structural causality is defined by Jacobs, Leamer, and Ward (1979, p. 403) as meaning that in an equation system, "y does not cause x" when "the disturbance in the y equation is never transmitted to x."[3] More specifically, consider the following example

$$q_t = \theta_1 x_t + \theta_2 z_t + \beta_{11} q_{t-1} + \beta_{12} x_{t-1} + \beta_{13} z_{t-1} + e_t \qquad (12\text{-}21)$$

$$x_t = \gamma_1 q_t + \gamma_2 z_t + \beta_{21} q_{t-1} + \beta_{22} x_{t-1} + \beta_{23} z_{t-1} + \delta r_{1t} + u_{1t} \qquad (12\text{-}22)$$

$$z_t = \tau_1 q_t + \tau_2 x_t + \beta_{31} q_{t-1} + \beta_{32} x_{t-1} + \beta_{33} z_{t-1} + \delta r_{2t} + u_{2t} \qquad (12\text{-}23)$$

where Greek symbols denote parameters, and e_t, u_{1t}, and u_{2t} are independent errors. Thus, q does not structurally cause x, for example, if $\gamma_1 = \gamma_2 = \beta_{21} = \beta_{23} = 0$.

Equations (12-21), (12-22), and (12-23) can be interpreted as a production model in which q_t is output in period t; x_t and z_t are inputs chosen at the beginning of period t; and r_{1t} and r_{2t} are the input prices normalized by output price, so that equation (12-21) is a production function and equations (12-22) and (12-23) are input demand functions. By interpreting the variables as logarithms, a system of this form can be derived from a dynamic Cobb-Douglas model, as in Antle (1983b). Viewed as a production model, it is known *a priori* that inputs x_t and z_t are chosen before q_t is realized, so $\gamma_1 = \tau_1 = 0$ with probability one. However, it is not necessarily true *a priori* that $\beta_{21} = \beta_{31} = 0$, that is, that output in the previous period does not affect current input decisions. Indeed, it was argued in the previous section that output dynamics are present in most agricultural production, implying $\beta_{21} \neq 0$ and $\beta_{31} \neq 0$, so that causality does run from output to inputs.

In terms of causality from inputs to output, x does not structurally cause q when the marginal productivity of both current and past inputs is zero. If the production process does not exhibit input dynamics, then $\beta_{12} = \beta_{13} = 0$, but positive marginal productivity of x_t and z_t implies

---

[3] There is a remarkable similarity in the models and terminology used by Jacobs, Leamer, and Ward to define causality and the model and terminology used by Mundlak and Hoch (1965) in their classic article on estimation of the Cobb-Douglas production function. Mundlak and Hoch considered the case of production disturbances being transmitted from output to inputs, thus clearly recognizing the importance of structural causality in production function estimation.

$\theta_1 \neq 0$ and $\theta_2 \neq 0$. Thus, either nonzero marginal productivity or input dynamics imply inputs structurally cause output.

The nonstructural approach to causality is due primarily to Granger (1969, 1980), who emphasized the existence of feedback and the direction of the flow of time in economic relationships. Granger's concept of causal ordering is based on "the notion that absence of correlation between *past* values of one variable X and that part of another variable Y which cannot be predicted from Y's own past implies absence of causal influence from X to Y" (Sims, 1972, p. 544). Whereas structural causality is based on identification of structural parameters, Granger causality is based on correlations between variables observed at different points in time.[4] Jacobs, Leamer, and Ward (1979) note that Granger's definition of causality implies that q does not cause x in a system such as equations (12-21), (12-22), and (12-23) if the coefficient of q_{t-1} in the reduced form equation of x_t is zero. It is easily shown that this reduced form coefficient is $(\gamma_1 \beta_{11} + \gamma_2 \beta_{31} + \beta_{21})$. Note that to infer Granger causality, therefore, it is not necessary to identify the structural parameters $\gamma_1, \gamma_2, \beta_{11}, \beta_{21}$, and β_{31}.

Using system (12-21)–(12-23), Granger causality can be related to the dynamic properties of production systems. First, observe that output dynamics occur when β_{11}, β_{21}, and β_{31} are nonzero. Thus, if the process does (not) exhibit output dynamics, then q does (not) Granger cause x or z. Similarly, if the process does not exhibit input dynamics, it can be shown that neither x nor z Granger cause q. Second, observe that the condition $(\gamma_1 \beta_{11} + \gamma_2 \beta_{31} + \beta_{21}) = 0$ does not imply that any of the individual parameters is necessarily zero. It follows that if q does not Granger cause x, it cannot be inferred that the production process does not exhibit output dynamics. But if q does Granger cause x, it can be inferred that output dynamics exist. Similar results hold for input dynamics.

The discussion above shows that output (input) dynamics exist if and only if outputs (inputs) Granger cause inputs (outputs). Moreover, the absence of output (input) dynamics implies outputs (inputs) do not Granger cause inputs (outputs), but the converse is not generally true. The econometric implication of these results is that the estimates of reduced form dynamic factor demand models cannot be used to draw conclusive inferences about the dynamic structure of production processes. For example, Nerlovian partial adjustment models have been

---

[4] As Zellner (1984) has emphasized, Granger's concept of causality, based on temporal correlations without reference to theory, conflicts with definitions used by philosophers and other economists. The temporal correlations of the type discussed by Granger can be used in conjunction with theory to draw inferences about causation, and are used here in that manner.

used to explain output and inputs as functions of lagged outputs, inputs, and prices. These models can be interpreted as reduced-form or final-form equations and therefore the lack of correlation between reduced-form variables cannot be used to reject the null hypothesis of production dynamics. Only estimates of the structural production function parameters can be used to reject the hypothesis that a production process does not exhibit either input or output dynamics.

The causal relations between inputs and outputs can be translated into causal relations between quantities (either input or output) and prices. For example, recursive substitution of the demand functions in equation (12-5) to obtain the final-form equation shows that x_t generally depends on the parameters of expected future price distributions and on past prices. When the individual producer is a price taker, this means that causality runs from prices to quantities in microeconomic data. However, as data are aggregated, price exogeneity may be questionable. If market prices are endogenous in aggregated data, they may depend on past market quantities if the market takes more than one period to adjust to exogenous shocks. Thus, aggregate data may exhibit feedback between quantities and prices.

THE DYNAMICS OF INNOVATION

The macrodynamics of agricultural productivity involve the farm-level dynamics discussed earlier as well as the dynamics of the natural, economic, technological, and institutional environment within which the individual producer operates. Some factors that affect productivity, notably the natural environment, are clearly exogenous to the economic system, but influence it by determining the relative scarcity of resources and the productivity of a given technology. Although the state of technology may be exogenous to the economic system, relative prices may influence the direction of technological change, as argued by Hicks (1932). More recently, Hayami and Ruttan (1971) hypothesized that relative prices could influence public sector institutions, such as agricultural research organizations, and thus affect the state of technology.

Hayami and Ruttan emphasize that induced innovation is a dynamic process. They describe technical change in agriculture as "a dynamic response to the resource endowments and economic environment in which a country finds itself at the beginning of the modernization process" (Hayami and Ruttan, 1971, p. 26). In their analysis of changes in factor proportions during the twentieth century in Japan and the United States, Hayami and Ruttan conclude "that such changes in input mixes represent a process of dynamic factor substitution accompanying changes in the production surface induced primarily by changes in relative factor prices" (1971, p. 133). Following the logic of the Hayami-Ruttan theory, this section outlines some elements of a stylized dynamic model of technological change, relates it to the dynamic structure of agricultural production, and discusses the implications for the empirical explanation of technological change.

The innovation process can be thought of as a sequence of interrelated investments. It begins with additions to the stock of basic scientific knowledge, and is proceeded by the transformation of that knowledge into technological (applied) knowledge, the embodiment of technological knowledge in physical inputs, and finally the diffusion of the inputs and related knowledge to producers. Thus, the innovation process involves a sequence of individuals and institutions. The induced innovation hypothesis states that this process is influenced by relative factor prices such that the resulting technology saves relatively scarce resources and uses relatively abundant resources.

A model of the innovation process begins with the stock of basic scientific knowledge, K_t.[5] Assume K_t evolves over time as

$$K_{t+1} = \delta_t K_t + k_t + \kappa_t \qquad (12\text{-}24)$$

where δ_t is a depreciation rate (reflecting knowledge obsolescence); k_t is systematic investment in basic research; and κ_t is a random term representing scientific discoveries that are not explained by purposeful investment. The induced innovation hypothesis implies that prices may influence basic research, but research has not addressed this question. Noneconomic forces (for example, World War II) clearly affect both the amount and type of investment in basic research.

The Hayami-Ruttan theory is more closely related to the subsequent stages of the innovation process, in which basic knowledge is transformed into on-farm technology. The next step in the innovation process is the transformation of basic knowledge into applied or technological knowledge. The equation of motion for technological knowledge, T_t, is

$$T_{t+1} = \rho_t T_t + I_t[R_t, K_t, r_t, \psi_t] + \tau_t \qquad (12\text{-}25)$$

where ρ_t is a depreciation rate; I_t is gross investment as a function of research funding R_t; K_t is basic knowledge; r_t is factor prices; ψ_t is price

[5] The stock of knowledge is difficult to define and more difficult to measure. For some creative work in this direction, see Evenson and Kislev (1975) and Evenson's chapter in this book.

expectations; and τ_t is a random term. In this formulation, the induced innovation hypothesis enters through the presence of r_t and ψ_t in the gross investment term I_t. If T_t and I_t are thought of as vectors whose elements measure stocks of knowledge in various technological fields, then the induced innovation hypothesis can be interpreted as stating that ψ_t determines how applied research funding R_t is utilized to transform basic knowledge K_t into specific types of technological knowledge T_{t+1}.

The stock of technological knowledge is next embodied in physical inputs by private industry and public institutions, and diffused to producers. This process is hypothesized to depend on T_t and price expectations ψ_t. Technology diffusion also may be constrained by such factors as farmers' human capital, government spending on technology diffusion (extension), and transportation and communication infrastructure. We denote these factors by D_t. Let the on-farm technology be expressed as a function $A_t[T_t, \psi_t, D_t]$. Now a general model of on-farm productin can be written as the production function

$$q_t = f[x^t, q^t, A_t, \epsilon_t] \tag{12-26}$$

and the input demand functions

$$x_t = x[x^{t-1}, q^{t-1}, \psi_t, r_t, A_t] \tag{12-27}$$

where A_t represents the state of on-farm technology. Note that the existence of either input or output dynamics makes the input demand functions recursive, so that a final-form demand equation system can be written as

$$x_t = x[r^t, \epsilon^t, \psi^t, A^t] \tag{12-28}$$

to show that the input decisions in period t are functions of the histories of prices, price expectations, technologies, and production shocks.

The structure of the model represented by equations (12-26) and (12-27) has important implications for testing the induced innovation theory and for modeling agricultural production. First, observe that the final-form factor demand equations (12-28) depend on the histories of prices, expectations, and technologies if the production process is dynamic, whether or not the induced innovation hypothesis is true. That is, the final form has the same general structure whether or not w_t and ψ_t influence I_t and A_t, because lagged prices also enter the final form through input and output dynamics. Thus, the final form generated by an induced innovation model is observationally equivalent to the final form generated by a model with exogenous technological change; the two models therefore cannot be distinguished in the final form. For example, consider the alternative hypothesis that the evolution of T_t and A_t is exogenous, and that the technology is nonhomothetic, such that factor proportions respond to relative factor prices in the same qualitative manner as they would if T_t and A_t were functions of prices. Both models could generate time series that would be consistent with the reduced form equation (12-28). Therefore, if production dynamics cannot be ruled out *a priori*, final-form factor demand equations cannot be used to test the induced innovation hypothesis. However, note that if production is static, the induced innovation model generates a final form involving lagged prices, due to equation (12-25), whereas the model of exogenous technological change implies a final form without lagged prices. The final form, therefore, could be used to test the induced innovation hypothesis under the assumption of static production.

The above results have strong implications for econometric measurement of agricultural productivity and testing of hypotheses about the structure of technology and technological change with farm-level data. Consider the use of dual cost or profit functions for these purposes. Dual functions and their derivatives have been advocated for econometric studies because they directly yield estimation equations with only exogenous variables on the right-hand side (Lau, 1978), that is, they directly yield factor demand and product supply equations in final form. Given the above results on final form equations, it is clear that the dual approach is of limited usefulness for identifying parameters needed to differentiate between theories of technological change and dynamic production, unless sufficiently strong *a priori* structure is imposed on the technology. To illustrate, we consider one element x_{it} of the final-form system (12-28). Multiplying x_{it} by its price and dividing by total variable profit gives a system of profit share equations

$$\pi_{it} = \pi_i[r^t, \epsilon^t, \psi^t, A^t], \qquad i = 1, \ldots, n \tag{12-29}$$

Alternatively, consider the system of equations

$$\pi_{it} = \pi_i[r_t, t] \tag{12-30}$$

System (12-30) (or its cost share analog) has been used in the studies cited by Capalbo and Vo to measure biased technological change and technology structure. System (12-30) can be derived from equation (12-29) by assuming static nonstochastic production, static expectations, and $A_t = t$. If the true process generating the data were given by equation (12-29), the interpretation of the time trend in system (12-30) as representing technological change could lead to spurious inferences about the structure of the technology and the nature of technological change. Moreover, if the induced innovation hypothesis is entertained so that A_t is a function of past prices and price expectations, the profit shares can be expressed as a system depending on r^t, ϵ^t, and ψ^t. Estimation of this final-form system clearly could not be used to differentiate the effects on profit shares of production dynamics from those of induced innovation.

In contrast, a structural production model can be used to distinguish the hypotheses of induced versus exogenous technological change in the presence of production dynamics. The production function (12-25) shifts over time as a function of the on-farm technology, which, according to the induced innovation theory, is a function of the history of prices and other variables influencing the innovation process. The induced innovation theory also implies that factor proportions should be functions of the history of prices, given current prices, price expectations, and the histories of input use and output. Therefore, the structural form of a dynamic production model can be used to test the induced innovation hypothesis by inferring causality from past prices to present productivity.

These conclusions regarding the identification of technological change with a structural model are valid only for a model of the individual firm, however. To see this, observe that the aggregate versions of equations (12-26) and (12-27) are given by equations (12-19) and (12-20). Let $z_t = A_t$ and define Ω_t as the vector containing the aggregates of x^t, q^t, and ψ_t. Then if the state of technology A_t is determined by variables distinct from Ω_t, the aggregate structural model can be identified if conventional identification criteria are satisfied. However, suppose that the induced innovation theory holds in a pure form such that $A_t = (r_t, \psi_t)$; that is, the state of technology is hypothesized to be only a function of prices and price expectations. Then equations (12-19) and (12-20) become

$$M_t[X_t, \Omega_t] = M[X_t, \Omega_t, z_t] = M[X_t, \Omega_t]$$

$$\Delta_t[r_t, \Omega_t] = \Delta[r_t, \Omega_t, z_t] = \Delta[r_t, \Omega_t]$$

thus demonstrating that the aggregate structural model would be underidentified, because the dated model is observationally equivalent to the undated model. It can be concluded that technological change can be identified in an aggregate model only if sufficient identifying restrictions are imposed on firms' expectations and on the process generating technological change. For example, in the description of the innovation process above, it was suggested that A_t may depend not only on prices and price expectations, but also on exogenous factors that determine investment in basic research and institutions that transmit technological information to firms. In addition, it could be hypothesized that firms and researchers form expectations based on different subsets of the existing price information. For example, firms could take a shorter-run perspective on expectations formation, and researchers could take a very long-run view. Under these conditions, A_t could be distinct from Ω_t and the aggregate structural model could be identified.

In conclusion, the analysis of a dynamic production model with technological change shows that both the dynamic structure of production and the dynamics of innovation may lead to observationally equivalent final forms. Therefore, unless production dynamics can be ruled out *a priori*, which seems unreasonable, final-form parameter estimates cannot be used to differentiate between production dynamics and alternative induced models of technological change. If production dynamics cannot be ruled out *a priori*, it is necessary to identify the structural parameters in the model to test the induced innovation theory.[6] Aggregation analysis shows that the aggregate structural model can be identified and used to test the induced innovation theory only if sufficient identifying restrictions are imposed on firms' expectations and the process generating technological change.

POLICY AND PRODUCTIVITY

Two sets of issues are considered in this section: the measurement of the productivity of investments in agricultural research, human capital, and physical infrastructure; and the relation of agricultural price and production policies to productivity.

Measuring Productivity of Public Sector Investments in Agriculture

The innovation model outlined in the previous section shows that the evolution of on-farm technology is a function of the histories of agricultural investments, both private and public. If the demand for and supply

---

[6] Similar results on reduced-form equations of dynamic models have been noted in the rational expectations literature. For example, Eckstein (1984) finds that a dynamic rational expectations model and a Nerlovian partial adjustment model give observationally equivalent reduced form acreage response equations. However, the rational expectations model's structural form implies overidentifying restrictions that can be used to differentiate the two models.

of public sector investments is a function of agricultural productivity, then they are not exogenous to production and causality between them and productivity is bidirectional. Under such a scenario, would measured static productivity of the investments be biased?

To explore this issue, suppose agricultural research spending in year t in region i, R_{it}, is a function of the region's history of average land productivity, $P_i^{t-1} = (P_{it-1}, P_{it-2}, \ldots)$. Consider the estimation of a regional production function using pooled regional cross-section and time series data. Equations (12-24) and (12-25) suggest that a production function for region i could be specified as

$$Q_{it} = f_i[x_i^t, Q_i^{t-1}, R_i^{t-1}, D_{it}, \epsilon_{it}] \qquad (12\text{-}31)$$

where the technology index A_{it} has been substituted out of the model, and the unobservable histories of the variables K_t, ψ_t, and τ_t are subsumed in the error term ϵ_{it}. The measured productivity of R_i^{t-1} would be biased if it is correlated with the unmeasured effects represented by the error term ϵ_{it} in the production function. Such correlation could be due to autocorrelated outputs, but this seems unlikely because annual agricultural output autocorrelations typically are small and of low order. Correlation between R_i^{t-1} and ϵ_{it} is much more likely to be due to omitted variables or measurement error. For example, the stock of basic knowledge in region i, K_{it}, may also be a function of past productivity. Since the $K_{it-1}, K_{it-2}, \ldots$, are subsumed into the error term of equation (12-31), ϵ_{it} and R_i^{t-1} would be correlated and the measured productivity of R_i^{t-1} would be likely to be biased upward. Another likely possibility is that the regions in the sample differ in land quality, and hence in their past productivity. If the data are not accurately adjusted for quality differences, then ϵ_{it} and R_i^{t-1} would be correlated. The marginal productivity of agricultural research would be biased upward, because high levels of research would be associated with high levels of productivity, regardless of true research productivity.

The preceding analysis shows that bias in the measured research productivity is not due to the dynamic structure of production or to the causal relations between research and productivity *per se*. In a correctly specified model with accurately measured inputs, ϵ_{it} and R_i^{t-1} are not correlated if output is not autocorrelated, whether causality between productivity and research is unidirectional or bidirectional. Therefore, the biases that exist are most likely the result of common measurement and specification problems. The same analysis could be conducted for other public sector investments, including human capital and physical infrastructure. The recursive structure of the dynamic production model ensures that the productivity of these investments is unbiased as long as output is not autocorrelated, the model is not misspecified, and input measurement error is not severe.[7]

Causality is important to the measurement of the productivity of public sector investments because these investments themselves are often interrelated. These investments have a logical time ordering in the process of agricultural growth: development of the physical infrastructure is a necessary precondition for development of certain other forms of physical and institutional infrastructure, such as education, extension, and marketing systems. Diffusion of agricultural technology also depends on related physical and institutional investments. Thus, agricultural investments have a logical causal ordering in their relation to each other and to productivity.

The pervasiveness of these interrelations between public sector investments means that it is difficult to measure the distinct productivity effects of agricultural research, human capital, and related physical and institutional infrastructure. For example, in attempting to estimate a production function similar to equation (12-31) with aggregate U.S. time series, we found research, human capital, and physical infrastructure data to be highly correlated. It has been argued elsewhere (Antle, 1984) that the interdependence of agricultural investments is so pervasive that the meaningfulness of discussing the productivity of individual investments is brought into question, suggesting that only the productivity of sets of interrelated investments can be measured. This conclusion, if true, has important implications for research that has used production models to measure the productivity of agricultural research, human capital, and other public sector investments. Because many such studies include only one or two of the many variables that would be required to represent the full set of interrelated investments, it seems probable that the measured productivity of these investments is biased, most likely upwards.

Production and Price Policy

The model of innovation outlined in the previous section shows that innovation can come from both the demand and supply sides of the technology market. Induced innovation can be initiated through the

---

[7] In this connection Chaudhri's (1979) study of education and agricultural productivity is worth mentioning. Chaudhri argued that education is an endogenous variable and estimated a static aggregate production model in which education was endogenous. However, it is clear that education (a measure of the stock of farmers' human capital) in year t cannot be a function of agricultural output in t. Thus, Chaudhri's model contradicts the logic underlying the recursive structure of production, and must be misspecified.

effects relative prices have on the demand for technology that saves scarce resources, and subsequent supply response by public institutions and private firms. Technological change can also come about through exogenous shifts in the supply of a particular type of technology, without being induced by prices. One of the important policy questions is, therefore, to what degree do agricultural price and production policies influence the innovation process? The answer to this question clearly depends on the validity of the induced innovation theory. Price policies have the potential to influence productivity through the demand side only if induced innovation is a significant factor in the innovation process. Public subsidization of research and diffusion can stimulate the supply side of the technology market whether or not induced innovation is operative.

The model of induced innovation shows that the transformation of basic knowledge into on-farm technology is influenced by price expectations. Therefore, a necessary condition for agricultural policies to affect productivity is that price policies must alter the long-run price expectations of farmers on the demand side and of institutions on the supply side of the technology market. The implication—very much in the spirit of the rational expectations theory—is that erratic or unanticipated government policies do not modify expectations and hence do not influence technological change, whereas established long-term policies or credible announced changes in policy may influence price expectations and thus alter the direction and rate of technological change.

Specific examples of both unanticipated and long-term policies are the Payment-in-Kind (PIK) program and the milk price support program. PIK was announced as a temporary program and thus would not be expected to alter long-term expectations. In contrast, the long-term, stable character of milk price supports appears to have significantly influenced producers' demand for capital-intensive milking technology and related biological research, as well as the allocation of both public and private research resources. For example, the California Milk Advisory Board provides several hundred thousand dollars of private research funding for dairy research. The University of California recently established a multimillion-dollar teaching and research center in Tulare County, California. An important policy question is the degree to which this kind of commitment of research resources has been influenced by the milk price supports.

The milk price support example is also of interest because it suggests that relative output prices may influence the direction of innovation just as much as input prices. Induced innovation theorists have emphasized the role of input prices in influencing scarce-factor-saving technology.

But there seems to be nothing in the logic of the induced innovation theory contradicting the possibility that relative product prices influence the overall relative rates of technical change in different industries or in specific inputs in certain industries. In the case of dairy production, it seems likely that increasing relative wage rates encouraged mechanized milking. But it seems equally plausible that dairy mechanization continued beyond what would have been profitable without government intervention because of output price and tax policies.

Production and related nonprice policies may also influence the direction of innovation by altering the perceived opportunity costs of resources and technologies. Production policies such as PIK clearly affect product and input prices just as direct price support programs do. But other policies only indirectly related to production may also have significant effects on opportunity costs. One example is pesticide regulation and the commitment of research resources to integrated pest management (IPM). Modern chemical pesticides were a spinoff of scientific developments during World War II (see Flint and van den Bosch, 1981). Advances in related biological technology, mechanical technology, and increases in wage rates helped stimulate the development and diffusion of the new chemical technology. But as pollution and health externalities became a serious concern, government regulations and restrictions began to increase the on-farm opportunity costs of pesticides, and public institutions began to respond to society's demand for less chemical intensive agricultural technologies. IPM, because it emphasizes the use of all available methods of pest control, biological and chemical, was seen as a means to reduce agricultural pesticide pollution. It is not yet clear that government policy has raised the opportunity cost of chemical control of pests to the point that major technological change is forthcoming.

EMPIRICAL EVIDENCE ON THE DYNAMIC STRUCTURE OF U.S. AGRICULTURAL PRODUCTION

This section draws upon the econometric methods and analysis developed by Antle (1986) to investigate the dynamic structure of U.S. agricultural production in the twentieth century. With adequate data, a complete structural production model could be specified and estimated for this purpose. However, using the aggregate time series data that are available, a degrees-of-freedom problem would arise if the specification of the output function were flexible (that is, translog) and included current inputs, and lagged inputs, outputs, and prices. Alternatively, a subset of the equations in a complete structural model can be investigated. This latter approach is pursued here.

Consider the first-order conditions to a dynamic optimization problem such as equation (12-4), expressed as $g_{it}[x_t, \omega_t] - r_{it} = 0$ for the choice of input x_{it} (recall ω_t denotes the decision maker's information set, upon which expectations at time t are based). Thus

$$r_{it}/r_{jt} = g_{it}[x_t, \omega_t]/g_{jt}[x_t, \omega_t] \equiv \phi_{ijt}[x_t, \omega_t] \qquad (12\text{-}32)$$

where $\phi_{ijt}[x_t, \omega_t]$ can be interpreted as a dynamic generalization of the marginal rate of technical substitution between x_{it} and x_{jt} because $g_{it}[x_t, \omega_t]$ takes into account the effects of x_{it} on current as well as future productivity. Hicks' two-factor measure of biased technical change can be generalized to the dynamic case using equation (12-32). Consider the hypothesis that a set of variables A_t causes the bias in technological change. Thus, define $\phi_{ij}[x_t, \omega_t, A_t] \equiv \phi_{ijt}[x_t, \omega_t]$. The bias in technological change due to $a_{kt} \in A_t$ is measured as

$$B_{ijk} \equiv \partial \ln \phi_{ij}/\partial \ln a_{kt} \qquad (12\text{-}33)$$

As in the static case (see chapter 2), these pairwise bias measures can be aggregated using cost shares c_{jt} into a cost share measure:

$$B_{ik} \equiv \partial \ln c_{it}/\partial \ln a_{kt} = \sum_{j \neq i} c_{jt} B_{ijk} \qquad (12\text{-}34)$$

The B_{ik} can be used to measure and test alternative explanations of biased technological change. More generally, the dynamic properties of the technology, measured as the effect on c_{it} of any exogenous or predetermined variable in the model, can be measured in this manner.

It was argued earlier in this chapter that the process of aggregation must be accounted for in productivity analysis. Following results presented above, the aggregate analog of equation (12-32) is

$$r_{it}/r_{jt} = \Phi_{ij}[X_t, \Omega_t, A_t] \qquad (12\text{-}35)$$

where X_t is the vector of aggregate inputs; Ω_t is the aggregate analog of the firm's information set ω_t; and A_t is a vector of variables hypothesized to explain technological change.

The identification issues raised earlier can be interpreted in terms of equation (12-35). If equation (12-35) were solved for the reduced form, variables contained in Ω_t and A_t would be introduced into the equation through X_t. Thus, it would be impossible using the reduced form to differentiate the indirect effects of A_t on Φ_{ij} through X_t from the effects of A_t on the bias in technological change. Therefore, to identify the effects of A_t on the bias in technological change, it is necessary to estimate the structural form.

Earlier it was shown that the aggregate structural model may be underidentified. If the variables contained in A_t are also contained in the information sets firms use to formulate expectations, then it will not be possible to differentiate the effects of A_t from those of Ω_t. Put differently, if A_t were a subset of Ω_t, then the model

$$r_{it}/r_{jt} = \Phi_{ij}^* [X_t, \Omega_t]$$

would be observationally equivalent to equation (12-35) and (12-35) would be underidentified. Therefore, to identify the model, it is necessary to impose the restriction that A_t and Ω_t are distinct sets.

For estimation of equation (12-35), Ω_t was specified as unweighted moving averages of lags 1 through 5 of inputs, outputs, and input prices normalized by the output price. Following the induced innovation hypothesis, A_t was specified as unweighted moving averages of normalized input prices for years 6 through 10 and 11 through 15. This structure for A_t and Ω_t was used to obtain an identified model. This specification is justified by evidence that suggests that agricultural research takes more than five years to have an impact on productivity (Evenson, 1968); it also is consistent with Binswanger's (1978) evidence on the lags between relative price trends and biases. The five-year moving averages can be interpreted as measuring the long-run trend in relative prices that should be relevant to expectations formation and technological innovation.

Equation (12-35) was specified in log-linear form. In this form, the lagged price variables' coefficients provide estimates of the B_{ijk}, and the B_{ik} can then be calculated using equation (12-31). To account for the endogeneity of contemporaneous input quantities on the right-hand side of equation (12-35), instrumental variables were created by regressing input quantities on Ω_t, A_t, measures of agricultural research and transport and communications infrastructure, and time dummy variables for 1930–39, 1940–49, 1950–59, 1960–72, 1973–74, and 1975–78. Under the assumption that the factor demand equations are identified by the shift variables in the factor supply equations, the instrumental variables estimator produces consistent parameter estimates.

The 1910–78 data used in the analysis are described in detail in Antle (1984). Four aggregate inputs were used: machinery, chemicals, labor, and land. Since $\Phi_{ij} = 1/\Phi_{ji}$, three regressions are required to estimate the parameters of the biases in equation (12-34). Parameter estimates for the regressions using labor as numéraire are presented in table 12-1. Table 12-2 summarizes the bias measures.

TABLE 12-1. PARAMETER ESTIMATES OF DYNAMIC MARGINAL RATE OF TECHNICAL SUBSTITUTION, U.S. AGRICULTURE, 1910–78

		Equation		
Independent variable		(1) r_A/r_L	(2) r_C/r_L	(3) r_M/r_L
Labor	(t)	−0.653	2.643	0.141
		(−1.40)	(2.90)	(0.37)
	$(t-1,\ldots,t-5)$	−0.302	−0.873	−0.541
		(−1.72)	(12.54)	(−3.78)
Land	(t)	−3.718	1.039	−0.513
		(−1.47)	(0.21)	(−0.25)
	$(t-1,\ldots,t-5)$	1.612	3.823	2.499
		(4.97)	(6.04)	(9.48)
Machinery	(t)	−0.488	−1.729	−0.548
		(−0.77)	(−1.40)	(−1.07)
	$(t-1,\ldots,t-5)$	0.072	−0.890	−0.356
		(0.49)	(−3.09)	(−2.96)
Chemicals	(t)	−0.162	−1.856	0.810
		(−0.76)	(−4.48)	(4.69)
	$(t-1,\ldots,t-5)$	−0.294	−0.615	−0.347
		(−2.61)	(−2.80)	(−3.79)
Labor Price	$(t-1,\ldots,t-5)$	0.144	0.044	0.268
		(1.35)	(.21)	(3.09)
	$(t-6,\ldots,t-10)$	0.341	0.146	0.166
		(4.21)	(0.92)	(2.52)
	$(t-11,\ldots,t-15)$	0.055	0.180	0.103
		(0.32)	(0.54)	(0.74)
Land Price	$(t-1,\ldots,t-5)$	0.022	0.171	−0.014
		(0.20)	(0.79)	(−0.15)
	$(t-6,\ldots,t-10)$	0.064	0.215	1.74
		(0.56)	(0.97)	(1.89)
	$(t-11,\ldots,t-15)$	0.213	−0.365	0.101
		(1.92)	(−1.69)	(1.12)
Machinery Price	$(t-1,\ldots,t-5)$	−0.531	−0.411	−0.651
		(−3.58)	(−1.42)	(−5.40)
	$(t-6,\ldots,t-10)$	−0.590	−0.140	−1.037
		(−2.11)	(−0.26)	(−4.57)
	$(t-11,\ldots,t-15)$	−0.385	−0.371	−0.481
		(−1.97)	(−0.97)	(−3.02)
Chemicals Price	$(t-1,\ldots,t-5)$	0.307	−0.245	0.490
		(1.84)	(−0.75)	(3.61)
	$(t-6,\ldots,t-10)$	0.711	−0.202	1.044
		(2.14)	(−0.31)	(3.86)
	$(t-11,\ldots,t-15)$	0.181	0.596	0.331
		(1.09)	(1.83)	(2.44)
Output	$(t-1,\ldots,t-5)$	0.158	1.460	0.41
		(0.77)	(3.65)	(2.46)
Time		0.058	0.284	0.201
		(1.87)	(4.69)	(7.99)
R^2 adjusted		0.985	0.991	0.996
Durbin-Watson statistic		2.24	2.33	2.48

Notes: Based on log-linear specification of equation (12-22). r_i is the price of factor i. A = land, L = labor, C = chemicals, M = machinery, as defined in Antle (1986). t-statistics are in parentheses.

Source: J. M. Antle, "Aggregation, Expectations, and the Explanation of Technological Change," *Journal of Econometrics* vol. 33, no. 1, pp. 213–236, 1986.

TABLE 12-2. AVERAGE BIASES IN TECHNOLOGICAL CHANGE
AND OUTPUT AND TIME TREND EFFECTS, U.S. AGRICULTURE, 1910–1978

Price biases	Lags		
	1–5 years	6–10 years	11–15 years
Own-price biases			
Machinery	−0.323	−0.607	−0.237
Chemicals	−0.440	−0.648	0.433
Labor	−0.109	−0.139	−0.050
Land	0.011	−0.009	0.150
Cross-price biases			
Machinery/labor	0.159	0.126	0.053
Chemicals/labor	−0.066	0.007	0.130
Land/labor	0.035	0.202	0.005
Cost share	Output effect	Time trend effect	
Machinery	0.189	0.119	
Chemicals	1.239	0.202	
Labor	−0.221	−0.081	
Land	−0.063	−0.023	

Note: Based on equation (12-24) and table 12-1, calculated at the sample means of the data.
Source: J. M. Antle, "Aggregation, Expectations, and the Explanation of Technological Change," *Journal of Econometrics* vol. 33, no. 1, pp. 213–236, 1986.

Several findings are apparent from tables 12-1 and 12-2. First, the lack of significant positive or negative first-degree autocorrelation in the residuals indicates that the variables in the model adequately represent the systematic trend in the aggregate marginal rate of technical substitution. For example, if technological change were biased towards machinery and against labor, the marginal rate of technical substitution of machinery for labor would be increasing over time at given factor proportions; failure to account for this trend over time would introduce positive autocorrelation into equation 3 in table 12-1. Second, there is evidence of input dynamics, output dynamics, and exogenous technological change, as lagged output is significant at the 5 percent level in two equations, lagged inputs are significant as a group in all three equations, and the time trend is significant in all equations. Third, there is evidence that lagged prices play a role, as the price variables are significant as a group in all three lag groups in all equations. Thus, there is evidence that firms use lagged prices to form expectations, or that technological change is endogenous, or both. Under the assumption that the longer lags represent the effects of induced innovation, these results can be interpreted as providing evidence of induced innovation.

A more rigorous test of the induced innovation explanation of technological change involves the signs of the lagged price biases. Note that if a_{kt}

is a lagged price, the pairwise bias effects B_{ijk} defined in equation (12-33) measure the effect of the lagged price on the marginal rate of technical substitution between inputs i and j. These pairwise effects can be averaged into the overall measure B_{ik} that can be interpreted as the effect of the lagged price on the cost share c_{it}. If the induced innovation theory is true, an increase in the relative price of factor i, *ceteris paribus*, should encourage adoption of innovations that are factor-i saving, and should therefore lead to a decrease in the B_{iji} and thus to a decrease in B_{ii}. By the same reasoning, if innovation in U.S. agriculture has been factor-j saving, the effect of lagged price of factor j on other inputs' cost shares should be positive.

Table 12-2 shows that there is one positive own-price bias for years 1 through 5, none for years 6 through 10, and two for years 11 through 15. The parameter estimates in table 12-1 show that the latter two positive own-price biases are due to statistically significant parameter estimates on land and chemicals price terms that violate the implications of the induced innovation theory. Table 12-2 also shows that only one of the cross-price biases—for years 1 through 5—violates the implications of the induced innovation hypothesis. Thus, under the assumption that the moving averages for lags 6 through 10 and 11 through 15 represent the effects of induced innovation, the data contradict the implications of the induced innovation hypothesis in terms of the 11 through 15 year own-price biases for chemicals and land, but the remaining own-price and cross-price biases are consistent with induced innovation.

These results suggest that the evidence from the U.S. aggregate time series data is not entirely supportive of the induced innovation hypothesis. One implication of this finding is that care should be taken in interpreting residual bias measures, such as those reported in Binswanger (1978) or Antle (1984), as providing evidence in support of induced innovation. However, the proper interpretation of these results is that the joint hypothesis of five-year price expectations formation and the induced innovation hypothesis were not supported. One possible explanation for the positive own-price biases for land and chemicals in table 12-2 is that the expectations assumption used to identify the model is invalid.

Table 12-2 also contains estimates of the effects on cost shares attributable to the time trend and the moving average of lagged output. These numbers can be interpreted as measures of the effects of exogenous factors (represented by the time trend) and scale changes (represented by output) on cost shares. These results show that both effects led to larger shares for chemical and machinery inputs and smaller shares for labor and land. Considered with the other evidence in table 12-2, these findings also raise questions about the interpretation of the residual biases in technological change. If, as these findings suggest, lagged relative factor prices do not solely explain the residual bias in technological change, then there must be more to technological change than is revealed by the induced innovation theory. Thus these findings present a challenge to researchers to develop alternative explanations of technological change that are consistent with the data.

IMPLICATIONS FOR RESEARCH

The preceding sections suggest that productivity change in agriculture is a complex dynamic process. Measurement and explanation of this process requires an understanding of how individual producers, private industry, and public institutions respond to incentives to innovate. Econometric productivity measurement can contribute to our understanding of this complex process, but it seems that adequate data will rarely be available for estimation of many of the behavioral relations involved. Therefore, careful description and analysis of the unquantifiable institutional and behavioral components of the system will always be needed to understand productivity change and evaluate policy options. Given this caveat, the analyses in this chapter suggest several possible directions for fruitful research.

1. The measurement and explanation of farm-level technological change is especially important for analysis of farm-level decision making, and can provide information on the behavioral lags attributable to microdynamics. The results regarding the recursive structure of farm-level production greatly simplify estimation. Considering the results of the aggregation analysis, we know that the availability of farm-level time series data would allow researchers to study the dynamics of the innovation process without confronting the identification problems that arise due to aggregation. Unfortunately, farm-specific time series data are rarely available.

2. The relation between microdynamics and aggregate dynamics has received little attention (one exception is Day, 1984). It was shown above that the recursive structure of farm-level production is preserved in aggregation. This fact can be used to study the causal relationships between aggregate quantities and prices, and to study aggregate production dynamics.

3. It was shown that bidirectional causality between agricultural research, human capital, and infrastructure investments does not bias the measured productivity of these investments. However, the causal inter-

relations between these investments themselves is not well understood and is important in identifying the productivity of individual investments.

4. A host of questions related to the dynamics of innovation need to be answered before the relation of agricultural price and production policies to agricultural productivity can be fully understood. First, the dynamic model developed above suggests that an understanding of farmers' expectations may be important to the evaluation of agricultural policy. A second related need is for a better understanding of the dynamics of innovation. For policy analysis, evidence is needed on the period of time required for relative price changes to be translated into productivity changes through the induced innovation process. There is a need for estimates of what might be called "the price elasticity of the supply of innovations." Casual empiricism suggests, for example, that producers adjust their factor proportions rapidly to price changes, to the degree possible with the existing technology; but that public institutions adjust more slowly and less predictably to relative price changes. Research into the relative speed of adjustment of the various organizations involved in the innovation process could provide much insight into the dynamics of technological change.

5. The welfare implications of the links between policy and productivity remain largely unexplored. For example, in the case of the milk price support policy, given the technology that was in place when the policy was instituted, income was apparently transferred from consumers and taxpayers to certain asset owners and producers. However, it is less clear what the welfare implications are as the milk price support began to influence the allocation of research resources and dairy productivity. This effect can be interpreted as an additional distortion caused by the policy, but it is different from a conventional market distortion because it also generates scientific research that presumably increases social welfare in the long run.

REFERENCES

Ackello-Ogutu, C., Q. Paris, and W. A. Williams. 1985. "Testing a von Liebig Crop Response Function Against Polynomial Specifications," *American Journal of Agricultural Economics* vol. 67, no. 4, pp. 873–880.

Antle, J. M. 1983a. "Incorporating Risk in Production Analysis," *American Journal of Agricultural Economics* vol. 65, no. 5, pp. 1099–1106.

———. 1983b. "Sequential Decision Making in Production Models," *American Journal of Agricultural Economics* vol. 65, no. 2, pp. 282–290.

———. 1984. "Measuring Returns to Marketing Systems Investments for Agricultural Development," in *Agricultural Markets in the Semi-Arid Tropics* (Patancheru, India, International Crops Research Institute for the Semi-Arid Tropics).

———. 1986. "Aggregation, Expectations, and the Explanation of Technological Change," *Journal of Econometrics* vol. 33, no. 1, pp. 213–236.

———, and S. A. Hatchett. 1986. "Dynamic Input Decisions in Econometric Production Models," *American Journal of Agricultural Economics* vol. 68, no. 4, pp. 939–949.

Berndt, E. R., C. J. Morrison, and G. C. Watkins. 1981. "Dynamic Models of Energy Demand: An Assessment and Comparison," in E. R. Berndt and B. C. Field, eds., *Modeling and Measuring Natural Resource Substitution* (Cambridge, Mass., MIT Press).

Binswanger, H. P. 1978. "Measured Biases of Technical Change: The United States," in H. P. Binswanger and coeditors, *Induced Innovation: Technology, Institutions, and Development* (Baltimore, Md., Johns Hopkins University Press).

Chaudhri, D. P. 1979. *Education, Innovation, and Agricultural Development* (London, Croom Helm).

Cooley, T. F., and S. F. LeRoy. 1985. "Atheoretical Macroeconometrics: A Critique," *Journal of Monetary Economics* vol. 16, no. 3, pp. 283–308.

Day, R. H. 1984. "Micro-Macro Dynamics and Complicated Economic Behavior." Paper presented at Southern Regional Research Project S-180 Meeting, New Orleans, March.

Doob, J. L. 1953. *Stochastic Processes* (New York, Wiley).

Eckstein, Z. 1984. "A Rational Expectations Model of Agricultural Supply," *Journal of Political Economy* vol. 92, no. 1, pp. 1–19.

Evenson, R. E. 1968. "The Contribution of Agricultural Research and Extension to Agricultural Production" (Ph.D. dissertation, University of Chicago).

———, and Y. Kislev. 1975. *Agricultural Research and Productivity* (New Haven, Conn., Yale University Press).

Flint, M. L., and R. van den Bosch. 1981. *Introduction to Integrated Pest Management* (New York, N.Y., Plenum Press).

Granger, C. W. J. 1969. "Investigating Causal Relations by Econometric Models and Cross-Spectral Methods,"*Econometrica* vol. 37, no. 3, pp. 424–438.

———. 1980. "Testing for Causality: A Personal Viewpoint," *Journal of Economic Dynamics and Control* vol. 2, no. 4, pp. 329–352.

———, and P. Newbold. 1977. *Forecasting Economic Time Series* (New York, N.Y., Academic Press).

Hayami, Y., and V. W. Ruttan. 1971. *Agricultural Development: An International Perspective* (Baltimore, Md., Johns Hopkins University Press).

Hicks, J. R. 1932. *The Theory of Wages* (London, Macmillan).

Jacobs, R. L., E. E. Leamer, and M. P. Ward. 1979. "Difficulties with Testing for Causation," *Economic Inquiry* vol. 17, no. 3, pp. 401–413.

Kydland, F. E., and E. C. Prescott. 1982. "Time to Build and Aggregate Fluctuations," *Econometrica* vol. 50, no. 1, pp. 1345–1370.

Lau, L. J. 1978. "Applications of Profit Functions," in M. Fuss and D. McFadden, eds., *Production Economics: A Dual Approach to Theory and Applications* vol. 1 (Amsterdam, North-Holland).

Long, J. B., and C. I. Plosser. 1983. "Real Business Cycles," *Journal of Political Economy* vol. 91, no. 1, pp. 39–69.

Lucas, R. E., Jr. 1967. "Adjustment Costs and the Theory of Supply," *Journal of Political Economy* vol. 75, no. 4, pp. 321–334.

Mundlak, Y., and I. Hoch. 1965. "Consequences of Alternative Specifications in Estimation of Cobb–Douglas Production Functions," *Econometrica* vol. 33, no. 4, pp. 814–828.

Parzen, E. 1962. *Stochastic Processes* (San Francisco, Holden-Day).

Sims, C. A. 1972. "Money, Income, and Causality," *American Economic Review* vol. 62, no. 4, pp. 540–552.

Stoker, T. M. 1982. "The Use of Cross-Section Data to Characterize Macro Functions," *Journal of the American Statistical Association* vol. 77, no. 378, pp. 369–380.

Zellner, A. 1984. *Basic Issues in Econometrics* (Chicago, University of Chicago Press).

13
INCORPORATING EXTERNALITIES INTO AGRICULTURAL PRODUCTIVITY ANALYSIS*

SANDRA O. ARCHIBALD

Economic growth in the U.S. agricultural sector has resulted largely from productivity increases driven primarily by technological change.[1] Gains in productivity due to technology have long been recognized; only recently have the negative effects of technology begun to attract attention. The buildup of salinity in soils from continuous irrigation and inadequate drainage; soil erosion; the development of resistance in pests to chemical pesticides; and the depletion and contamination of aquifers from irrigation and fertilizer application are some of the obvious examples.

As noted in earlier chapters of this volume, productivity analysis has not explicitly incorporated production externalities—the unpriced, unintended "products" from widespread adoption and cumulative use of a particular production technology—into the measurement of growth. Environmental policy has been used to reduce potentially harmful effects of agricultural technology on the environment and natural resource base, but its consequences for productivity and technical change have not been fully identified.

The intertemporal and external effects of technology, along with the effects of mitigating environmental and resource policy, should be included in analyses of gains from technology adoption for several reasons. First, excluding production externalities can overstate (understate) productivity gains from technology, as some resource costs (benefits) are not counted. Secondly, as public policy moves in the direction of requiring producers to bear more of the total costs of production and to internalize externalities, the total, or social, costs and benefits from technology must be determined. Thirdly, as interest focuses on the long-run profitability of technology, the biological and physical sustainability of

---

*The author gratefully acknowledges the constructive comments and suggestions of Gordon A. King, C. O. McCorkle, Jr., Anne E. Peck, and anonymous referees.

[1] See chapters 3, 5, 9, and 10 in this volume, and Farrell (1981).

technology becomes critical. Fourthly, examination of the effects of production externalities on productivity growth may reveal inconsistencies between agricultural commodity programs and environmental policies. Finally, evidence that existing regulatory policy has adversely affected productivity growth rates underscores the need for further analysis.[2]

Models that explicitly incorporate externalities must be developed in order better to measure and explain agricultural productivity. Pittman (1983) provides a revision to the multilateral productivity indexes that enables the researcher to include undesirable outputs as well as desirable outputs in the productivity measures. In this chapter a dynamic model of agricultural production that incorporates intertemporal and externality effects is developed and linked to the analysis of long-run productivity. This model provides a means of analyzing the effects of environmental policy on future productivity levels. In the second section environmental externalities in agriculture and the nature of existing regulatory policy to mitigate them are discussed. The appropriateness of conventional production models and productivity measures to incorporate these externalities is treated in the third section, along with suggestions for an alternative joint production model. In the fourth section, a dynamic production model is developed for a sector that jointly produces intended agricultural output, cotton, and an unintended intertemporal externality, pesticide resistance. The production relations for cotton and pesticide resistance provide the dynamic constraints for an intertemporal economic optimization problem. Optimal control theory is then used to derive dynamic measures of productivity and social cost under alternative policy regulations. Although this chapter concentrates on the pesticide resistance problem for a cotton-producing region in California, parallels to other regions and commodities, as well as to other environmental and resource externality problems, exist.

EXTERNALITIES AND REGULATION IN AGRICULTURE: IMPLICATIONS FOR PRODUCTIVITY

National environmental regulation affecting agriculture has focused primarily on pesticide hazards and water pollution. Regulatory objectives have included protection of consumers, farm workers, and wildlife from hazardous chemicals, and minimization of sediment and nutrient pollution in waterways.

---

[2] There is evidence in the nonagricultural sector that environmental regulations are contributing factors to observed declines in productivity growth rates (see Kendrick, 1980; Portney, 1981; and Kopp and Smith, 1981).

Standards on technology and its use have been the principal regulatory mechanism employed. In cases such as livestock and feedlot operations, where the source of the externality can be identified, standards have been imposed directly on the externality output, for example, by limiting runoff. In cases of potentially hazardous technologies, product standards and use restrictions are imposed directly on the technology, as is the case with feed additives and growth stimulants. Resource regulations, such as those directed at soil conservation, regulate land use with conservation set aside requirements and tillage practices, and are often preconditions to participation in commodity programs. The 1985 farm bill requires compliance with soil conservation provisions in order to participate in farm programs, for example.

Pesticide regulations restrict or prohibit the use of particular pesticides, set specific quality standards for chemicals, and establish tolerance levels for chemical residues on output. These regulations alter the production possibilities set and the supply curve through their effect on production costs. It was expected that these regulatory constraints would, in the long run, result in adoption of an environmentally less damaging technology that would substitute information (on pest levels and their potential economic damage), cultural practices, and biological control for pesticides. Regulation would thus reduce pesticide use over time, and decrease the potential for development of pesticide resistance.[3] However, evidence is accumulating that technology has not substituted for chemical control to the degree expected (Dover, 1985) and that pesticide resistance is increasing (Brattsten and coauthors, 1986). If new chemicals are not forthcoming, the productivity of current pesticides can be expected to decrease as higher doses are required to achieve previous control levels. Significant changes in output mix and location of production or both also may result.

Several other resource issues related to agricultural production are also of concern, foremost among them the potential for declining productivity of physical and biological resources in agricultural uses. For example, cultivation practices have been linked to excessive erosion rates, potentially leading to declining soil productivity over time. This issue is particularly complex in that decline in the productivity of soil can be compensated for with purchased inputs or new tillage practices that actually augment soil productivity (at some cost) and mask any real declines in the productivity of the resource. Crosson (1983) studied the

---

[3] Regulation may not only have failed to reduce pesticide use but may have actually spurred resistance development by restricting the range of materials available and raising the costs of new product development (Council for Agricultural Science and Technology, 1983).

effects of soil erosion on productivity, finding that the effect on major crops was small but negative and significant. Estimated reductions were, however, more than compensated for by technical change and input substitution. Another example is groundwater supplies, which are gradually declining over time. Competition for surface water supplies in many areas in the United States is likely to reduce quantities of surface water available to agriculture in the future (National Research Council, 1982). Although the productivity effects and adjustments associated with changes in resource quality and quantity could be substantial, they are not included in any current measures of agricultural productivity.

Productivity Effects of Alternative Regulatory Policies

When externalities exist a competitive economy will, in general, no longer ensure a Pareto-optimal allocation of resources, because the externality-producing firm imposes damages or costs on others that it does not consider in its profit-maximizing decisions.[4] For production decisions to be Pareto optimal in the presence of externalities, marginal social benefits must be equated to marginal social costs. The latter are equal to the sum of the firm's marginal private costs and the marginal externality (user) costs. That is, social optimality will result only if the external costs of technology are considered in production decisions (internalized). For agricultural production, most of the externalities occur over long time horizons and thus the above optimality conditions must hold intertemporally as well.

Regulation of agricultural technology to internalize externalities (for example, pesticide use) can be expected to increase costs as higher cost inputs are substituted, or lead to reduced output when such substitutes are not available. The extent to which productivity is affected depends upon opportunities for input substitution, possibilities for "abatement" technology, and the specific regulatory mechanism employed. When no abatement possibilities exist and the externality being generated is proportional to the agricultural output produced, production will fall if producers are made to internalize the externalities. In figure 13-1, (p_0, q_0) is the price-output combination that results from a competitive equilibrium when marginal external costs (MEC) are excluded. If MEC are added to marginal private costs (MPC) so that the producer faces marginal social costs (MSC), the socially efficient output will be q_2 and output price will be p_2. Although overall welfare improves (by the area shaded above MSB and below MSC) as output falls initially to q_1 but

---

[4] See Just, Hueth, and Schmitz (1982) for a full discussion of externalities.

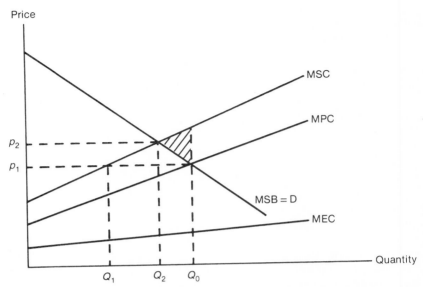

FIGURE 13-1. Price-output combinations under competitive and socially efficient scenarios

adjusts to q_2 as output price rises, both direct and indirect productivity effects can be expected.

The productivity effects depend upon the structure of technology. If production is homogeneous of degree one (constant returns to scale), a reduction in agricultural output would result in a proportionate decline in inputs and total factor productivity (TFP) would remain unchanged. On the other hand, if technology is not scale-neutral, the change in input use may not be proportionate and TFP could change. In the longer run, the effects on productivity may be traced through the changes in profitability. If profit declines as a result of the regulations, adjustments in cropping patterns and factor use can be expected. Assuming producers were allocating production optimally before regulation, changes in response to regulations should result in a loss in sectoral productivity.

By contrast, if abatement technology is feasible and producers are forced to adopt it, output may increase relative to q_2. With abatement technology marginal private costs increase but marginal external costs will also be less than without it. At the social optimum, marginal social costs with abatement technology will be below the marginal social cost when the abatement technology is not feasible; the optimum quantity produced will be higher. The productivity effects, however, depend upon the cost of the abatement technology or the productivity of new inputs. Assuming that farmers did not voluntarily adopt the abatement technology, productivity is presumably less than at q_0 but may be greater than or less than at q_2.

These differences in the technological characteristics of production directly influence the effectiveness of regulatory strategies. For example, quotas (or, equivalently, quantity-reducing standards) on agricultural output are not equally effective in the preceding cases. If, as in the first case, the externality output is strictly a function of agricultural output because no possibilities for abatement or input substitution exist, an output quota can be used to achieve socially efficient production. In the second case, however, if possibilities for substitution or abatement exist (externality output is not strictly a function of agricultural output), quotas can be placed on the externality output. Agricultural output will be reduced, but by less than it would be given a quota on agricultural output (Just, Hueth, and Schmitz, 1982).

Standards on output levels are appropriate only to problems in which the actions of individual producers can be identified, namely *point source* pollution problems. In cases in which the contribution of individual producers cannot be identified (nonpoint problems), standards need to be imposed directly on the technology (inputs). Such an approach obviously affects the input price and quality and results in embodied technical change. (This is the approach that has been followed with pesticides, the subject of the fourth section of this chapter.)

A second regulation strategy, Pigouvian taxes, imposes the marginal external costs on producers to force them to equate marginal private costs to marginal social costs. In point source cases with possibilities for abatement and assuming a strict production model, the *ad valorem* tax on agricultural output is computed as the vertical distance (MSC − MPC) at a given output. Producers are thus taxed the equivalent of MEC on agricultural output, which is reduced from q_0 to q_2. Output price increases to p_2. Imposition of a tax on agricultural output in this case produces the same price/output combination as does the use of standards.

If externality production is stochastic (dependent, for example, upon weather), taxes and standards must be managed differently and the results will be asymmetric. A standard can only be imposed on planned output in point source problems or applied to the abatement technology (for example, "closed systems" in pesticide application). Taxes can be placed on production *ex ante* in nonpoint cases or on the externality itself in point source problems. Subsidies on nonexternality-generating activities will likewise differ from taxes on externalities depending on the probability of occurrence (Just and Zilberman, 1979). In general, the establishment of standards requires more information on individual technologies. In such a case, taxes are probably preferable, but both taxes and standards require information about the process generating the externality if they are to be set optimally (Baumol and Oates, 1975).

The above discussion illustrates the difficulties of measuring even the short-run productivity effects of alternative regulatory mechanisms and so far has ignored the decidedly intertemporal nature of most of these problems. In the long run, the goal of environmental and resource policy should be to achieve the dual objectives of economic growth consistent with environmental protection and socially efficient use of natural resources. If regulation is successful, technology should be biased toward environmentally less damaging factors; it will be "saving" in environmentally damaging factors and "using" in environmentally enhancing or neutral factors. Selecting the right regulatory strategy to accomplish this is difficult. Under a standards-based approach, if productivity growth rates decline and production costs increase without commensurate declines in external effects, alternative policy instruments, such as collective action to manage resources or taxes to encourage adoption of less environmentally damaging technology, should be considered. For example, increasing water costs have spurred a shift to water-saving irrigation technologies, such as sprinklers and drip irrigation. Although capital intensive, these technologies result in more efficient water use and can increase water productivity. Policy should be able to provide incentives to encourage technologies that halt depletion rates of scarce supplies more rapidly than a free market solution. Applying taxes to pesticides or subsidies to integrated pest management technology may accelerate the rate of adoption of less environmentally damaging technology and reduce the pressure on resistance development from continued pesticide use.

MODELING AGRICULTURAL EXTERNALITIES

Measuring Productivity

The conventional aggregate total factor or partial productivity measures suffer deficiencies in the explanation of technical change under regulation for several reasons. First, both gains from technology and losses from externalities should be considered in productivity measures. Secondly, because regulation (by way of imposing performance standards on inputs) changes the quality of inputs, technical change should be modeled as embodied technical change, as contrasted to disembodied change measured in TFP residual approaches.

Partial productivity measures may be useful in addressing particular natural resource issues. Measuring the average productivity of land, for example, can identify a change in soil productivity over time. Crosson (1983) employed partial productivity analysis to identify the negative productivity impacts of soil loss during a period of overall yield growth. Conceivably, an average productivity measure could be used to assess the productivity of pesticides under regulation.

However, interpretation of such an index should be done cautiously, as changes in the average product of pesticides or any other input over time may also reflect changes in related inputs, outputs, or technology. Observed land productivity decreases may appear to be smaller than they actually are from adoption of other land-saving technology, such as fertilizers or heavy equipment. Furthermore, the effects of regulation on input quality should be separated from changes in input quality over time that can occur with potentially depletable inputs. For example, the measured productivity of pesticides may evidence decline, but there are at least two potential sources of this decline: regulation may have resulted in less productive materials or the emergence of pest resistance may have lowered productivity.

Obviously, neither of these conventional measures of productivity captures the dynamic adjustments made in input mix, adoption of new processes or output mix that result from regulation or natural resource constraints. Under regulation, technological change occurs through the introduction of new processes (cleanup technology) and changes in the quality of existing inputs. Thus, complete information on the structure of technology and production relationships, including measures of damage, possibilities for input substitution, and changes in output mix, are required if productivity measurement is to be useful in explaining the net contribution of technology to economic growth.

In order to capture this information, production analysis should be performed at a disaggregated level that permits the contribution from intermediate inputs, including the biological and physical inputs, to be identified. Meeting environmental regulations can involve an array of subtle changes that differ by sector and commodity. Therefore, the adjustment process must be modeled at the disaggregate level if certain questions are to be addressed.

Traditional production parameters, such as output elasticities, provide information as to the importance of externality-generating inputs in production (Dasgupta and Heal, 1979) and identify the consequences if regulation or natural resource constraints were to eliminate them from the production possibilities set. Measures of the elasticities of substitution between externality- and nonexternality-generating inputs (herbicides versus cultivation) are essential to identify possibilities for input substitution from the available technology set. Scale parameters can help identify ease of substitution possibiliites at alternative output levels. With this basic production information it is then possible to determine the effects of policy and natural resource constraints on the sectoral measures of agricultural productivity and bias in technical change.

For most situations, duality theory can be employed to estimate these production relationships with greater ease than direct or primal approaches (see chapter 2). However, there are several limitations to the dual approach for welfare analysis when externalities are incorporated. In the presence of unpriced externalities, marginal value products do not equal social marginal opportunity costs of inputs and the underlying assumptions employed in index or duality approaches to productivity measurement are not met. Even though primal measures of production suffer from the estimation difficulties, it is likely that direct estimation of production relationships can incorporate more of the complexities that arise from externalities in agricultural production than dual approaches, providing a more complete framework for analysis of productivity and technical change.

Production Models for Incorporating Externalities

Models that are candidates for incorporating externality effects should reflect a significant degree of disaggregation, as well as have the capability for integrating the biological, physical, and economic processes. In much of the analysis presented in this volume, productivity measurement has been largely at an aggregate level, identifying broad national trends in output and input use over time. Productivity analysis that incorporates externalities needs to be disaggregate, specific to a sector or commodity. This is true for both the agricultural and nonagricultural sectors, as noted by Kopp and Smith (1981). Fortunately, microproduction data are often available at the regional or commodity level, facilitating estimation of externality functions from historical production and regulatory data.

Although federal environmental regulations are uniform for all regions, environmental problems are often location-specific. Thus, analyses and policies are likely to be targeted to specific regional externality problems, such as the selenium contamination of the Kesterson Reservoir in California. Evidence indicates that the emergence of pesticide resistance is restricted to selected regions, crops, and even specific insects. The specificity of these issues makes them no less interesting nor important; however, it does have implications for how production relationships are modeled and how policies are designed.

Many environmental and resource externalities that result from agricultural production also involve complex biological and physical processes, such as depletion of underground aquifers (Noel, Gardner, and Moore, 1980; Lemoine, 1984) and resistance of pests to pesticides (Archibald, 1984). To provide essential information on long-run agricultural productivity, economic models need to integrate these physical and biological processes.

Production externalities most often result from use of one or more specific inputs, such as pesticides or fertilizers. This has been noted by Langham, Headley, and Edwards (1972), who addressed externalities caused by agricultural pesticides in the Florida fruit and vegetable sector. These externality-generating inputs have the characteristics of joint inputs, as any positive quantity simultaneously, or jointly, produces the intended agricultural output and the unintended externality.[5] Technically, they are not separable into the amount used to produce the agricultural output and the amount used to produce the externality. For example, the quantity of pesticides used to produce cotton is also the quantity available to produce pesticide resistance. Nor, in many cases, can these inputs be reallocated away from production of the externality to the production of planned output, as in a normal multiproduct production problem. However, as discussed above, it is often technically feasible to substitute inputs or introduce new production processes or inputs (abatement technology) to reduce the level of the externality without reducing the level of planned output. For example, labor can be allocated to monitor pest population levels to time pesticide applications better, reducing the quantity of pesticide input and thus reducing externalities.

The standard single-output or output index framework that assumes separability of inputs and outputs and no joint production is clearly inappropriate to modeling production externalities. Although joint production specifications in which externalities result from technical interdependencies have been employed in externality analysis (Baumol and Oates, 1975), they are inadequate to capture the complexities of externality generation from agricultural production. This type of analysis has often assumed strict joint production relations, that is, the externality is produced in fixed proportion with the crop output (Bator, 1958; Buchanan and Stubblebine, 1962; Baumol and Oates, 1975). It assumes the level of damages has a direct relationship with the quantity of agricultural production. It does not allow for the technical possibilities of abatement of negative externalities through rearrangement of pro-

---

[5] The joint aspects of input use were noted by Meade (1952).

ductive inputs, and provides too little flexibility to describe accurately many externalities from production.

Multiproduct production specifications, on the other hand, can provide too much flexibility to be useful in analyzing externality production. Such formulations imply possibilities for tradeoffs between intended agricultural output and externality output that do not exist. For example, in the multiproduct model of production it is possible, given a fixed quantity of inputs, to vary inversely the quantities of the two outputs produced. However, agricultural producers cannot allocate more of the joint input to the production of output and less to the production of the externality.

Generalized joint production allows an appropriate degree of flexibility and can be applied to modeling externality production. This framework, first developed by Carlson (1939), allows for joint inputs and the possibility of varying the proportion of intended agricultural output to externality output. Theoretical models employing generalized joint production have been applied to externality analysis (Buchanan, 1966; Whitcomb, 1972).[6] Whitcomb (1972) employed generalized joint production to model the supply of externalities in the static case. For example, this production model can be written in implicit form as

$$F_1(Y, R, V_1, X) = 0$$

$$F_2(R, V_2, X) = 0$$

where Y is a vector of agricultural outputs, R is the externality, and F_1 and F_2 are their respective implicit production functions. R is a vector of externalities received by the firm that enters negatively into the production of output Y. V_1 is a vector of ordinary or nonexternality-generating inputs allocated to the production of output Y, and V_2 is the quantity of these inputs allocated to the abatement of externality R. X is the vector of joint inputs contributing simultaneously to agricultural output and the externality, the same quantity appearing in both equations.

With this formulation, it is possible to reduce the level of externality production without increasing the intended output by diverting non-externality-generating inputs. Sudit and Whitcomb (1976) show that in the translog case the rate of transformation between Y and R depends

---

[6] Buchanan (1966), in a cost function framework, defined an externality to be a consequence of producing an intended output and allowed for the possibility of varying the proportion of intended output to externality output. Production function specifications consistent with this were formulated by Henderson and Dano (see Sudit and Whitcomb, 1976).

upon variable inputs X and V_1. Assuming substitution possibilities or the existence of abatement technology, the quantity of externality outputs in the system can be reduced by reallocating nonjoint inputs, thus avoiding negative output effects implicit in other production models. Externality levels are not proportional to agricultural output levels but are a function of the production technology employed.

Additionally, when externalities are produced and modeled in a joint production framework, several important conventional properties and restrictions do not apply *a priori*. Profit-maximizing firms do not take externalities into account in their input decisions, and therefore the marginal products of the environmentally neutral inputs in the externality function F_2 could be negative. This suggests that a flexible representation of the technology is desirable.

The same advantages of flexible functional forms discussed in chapter 2 apply to production analysis in the case of externalities. Functional forms that allow for variable elasticities of factor substitution are preferable. Furthermore, homotheticity assumptions imply independence of the impact in changes in output levels from variations in input combinations, and eliminate the possibility of greater ease of substitution between externality-generating and neutral inputs at alternative output levels. Thus, nonhomothetic functional forms are more attractive for modeling purposes.

Econometric estimation of a generalized joint production system is feasible if actual observations on the externality damage function and allocated abatement inputs are available. Given information required by the regulatory process, it is likely that such data will be available at the county, state, or regional level.

Current models for productivity analysis are limited in two other important ways. First, they are based on the assumption that firms optimize private returns, ignoring the welfare implications of externalities. As discussed in this section, externalities do affect social welfare. Secondly, most are static analyses, ignoring the dynamics of agricultural production processes. As a result, agricultural production relationships involving production externalities are more accurately modeled as a dynamic multiple output/input system to reflect the fact that current decisions affect future output and net social welfare. Building on the generalized joint production concepts, an intertemporal model of production that incorporates the technical relationships between input use and externality production can be modeled as a recursive production system incorporating time directly. When this system is used as the constraint set in a dynamic economic optimization problem, productivity, technical change, and intertemporal effects of various policies on consumer and producer welfare can be investigated, as demonstrated in the next section with an empirical application to cotton and pesticide resistance in California's Imperial Valley.

AN EMPIRICAL EXAMPLE: ACCOUNTING FOR RESISTANCE DEVELOPMENT IN THE MEASUREMENT OF PESTICIDE PRODUCTIVITY

Chemical pesticides have contributed significantly to aggregate agricultural productivity over the past four decades. The returns to U.S. producers have been estimated to be between $2.00 and $4.00 for every dollar invested (Headley, 1968; Doyle, 1985); for developing countries, figures as high as $14.00 per dollar invested have been estimated (Food and Agriculture Organization, 1972). Maintaining the long-run productivity of chemical pesticides is important if these gains are to continue. Currently, the rate at which insects are developing resistance to chemical insecticides is accelerating. At the same time, discovery of new control mechanisms has slowed. The number of resistant insect species nearly doubled during the 1970s from 224 to 428 (Dover, 1985). A 1981 survey classified nearly 60 percent of resistant pests as agricultural insects (Georghiou, 1981). Eight percent of these are now resistant to control by the four main classes of insecticides.

The economic implications of pesticide resistance for agriculture include crop losses from resistant insect damage; higher pest control expenditures, as chemical effectiveness declines; and, ultimately, adjustments in cropping and land use, as control costs continue to increase. The social costs of resistance include an increased chemical load in the environment; greater "selection pressure" on susceptible, nontargeted insect populations; and potentially higher food costs. The opportunity costs to society of resources devoted to development of new control methods and regulatory program costs must be considered as well.

Toward Measuring the Effects of Pesticide Resistance and Regulation on Productivity

Cotton production in California is particularly suitable for investigating the effects of resistance development as the externality to the long-run productivity of chemical pest control technology. The effects of regulation in both reducing use and encouraging adoption of alternative technology can be assessed. Historically, cotton has been the greatest single user of agricultural insecticides, accounting for over 35 percent of total use. Moreover, since 1950, the cumulative per acre application on cotton nationally has been over 200 pounds of active ingredients (National Research Council, 1982). Yields in California, the second largest cotton producing state, have historically been almost twice the national average. Due to the relatively warm climate, chemical insecticides are relied on heavily to achieve these levels. Imperial Valley, a desert region, yields an average 1,350 pounds of lint per acre, well above even state levels, and in 1982 approximately 40 percent of all restricted pesticides used on cotton in the state were applied in the valley. In addition, the desert climate results in as many as five generations of insects per year and the absence of frost in many years increases the probability of resistant insects surviving from year to year, further speeding the development of resistance.

Trends in aggregate cotton yields and pest control inputs for Imperial Valley indicate that average yields per acre declined over the twenty-year period (1960–80) by about 1.6 percent per year while insecticide applications increased by approximately 12 percent per year over the same period. These trends occurred within an increasingly stringent pesticide regulatory system that controlled the type of materials used; licensed producers, pest control advisors, and applicators; required "prescriptions" for each application; and provided for monitoring applications to assure specific regulations had been followed.

Some aspects of IPM technology (scouting, trapping, and some limited use of biological controls) have been adopted, although not consistently nor universally by all growers. However, in 1981 a crisis was reached in cotton production, as pest control costs exceeded $350 per acre in some cases. As a result, cotton producers formed a pest management district in 1982 to impose a mandatory IPM strategy in the hopes that declining yields and increased insecticide use could be halted. The IPM strategy required all producers to use early season biological controls (pheromones) for pink bollworm, prohibiting use of insecticides at this time to preserve beneficial insect populations and minimize the need for chemical control of secondary pests later in the season.

An empirical analysis of production was undertaken to determine the degree to which observed changes in cotton yields could be explained by the declining productivity of insecticides, the degree to which it could be attributed to resistance, and the contribution, if any, of regulation to this decline. Critical to this analysis was an assessment of whether current regulatory mechanisms appropriately consider long-run biological developments and how alternative regulatory policy could encourage adoption of less environmentally damaging technology without negatively affecting agricultural productivity.

As a first step, a Laspeyres index of output and pest control inputs was computed from primary production data obtained from a random sample of cotton producers for the period 1978–82. Data included actual yields, insecticide use by type of material, IPM inputs, insect population levels, and other nonpest control inputs. These partial indices indicated that, for sample growers, yields declined an average of 3 percent per year over the period, organophosphate (OP) use increased at an annual rate of over 20 percent, and synthetic pyrethrins (SP) increased by about 50 percent (due largely to crossresistance to DDT and high temperatures). IPM inputs increased by much less: use of pheromones rose by about 5 percent per year, use of biological controls by 8 percent per year, and use of pest scouting and trapping rose by about 12 percent per year. The index of mean insect infestation levels (tobacco budworm larvae) increased by nearly a third per year. The year the collective pest management program was in effect, insecticide use declined significantly and, as expected, IPM inputs increased dramatically, with no significant decline in yields. Despite the short time series, this partial productivity analysis suggested some decline in insecticide productivity, relatively slow adoption of IPM technology, and increasing insect resistance.

In specifying and estimating the dynamic production model, a two-stage procedure was used. First, a generalized joint production system was estimated econometrically. Second, these parameters were used in a discrete-time dynamic programming solution of a dynamic optimization problem to examine the effects of an alternative regulatory policy on cotton output and resistance levels, bias in technical change, and overall productivity growth. The salient features of this dynamic production framework are presented below. A complete discussion of the analytical framework and estimation techniques is found in Archibald (1984).

The Production System. A model of cotton production consistent with the concepts of generalized joint production discussed above was constructed to measure insecticide productivity incorporating insect resistance as the production externality that results from collective insecticide use by producers within a given region.

Cotton production is modeled as

$$Y_i^c(t) = f[R_i(t), X_i(t), V_i(t), W_i(t)] \qquad i = 1 \ldots N \text{ producers} \qquad (13\text{-}1)$$

where Y_i^c = cotton output in pounds of lint per acre for the ith producer;

R_i = the number of tobacco budworm larvae surviving insecticide treatment;

X_i = a vector of insecticides in dry pounds of active ingredient by major chemical class;

V_i = a vector of IPM inputs including biological controls, scouting, and monitoring; and

W_i = a vector of non-pest control inputs including soil quality, a capital measure, water use, and fertilizer.

Equation (13-1) has the following properties (subscripts have been suppressed): $\partial Y^c/\partial R < 0$; $\partial Y^c/\partial X > 0$; $\partial Y^c/\partial V > 0$; and $\partial Y^c/\partial W > 0$. The cotton production function (13-1) was specified as a linear model following Talpaz and Borosh (1974) and Feder and Regev (1975).

Production of resistance, the collective externality, is modeled as

$$R(t) = \left[R(t-1), \sum_{i=1}^{N} X_i(t), \sum_{i=1}^{N} V_i(t) \right] \qquad (13\text{-}2)$$

where all variables are as defined above and with properties $\partial R(t)/\partial R(t-1) > 0$; $\partial R/\partial \sum_i X_i > 0$; $\partial R/\partial \sum_i V_i \leq 0$.

A probability model was used to specify equation (13-2). The probabilistic specification follows biological models of pesticide "kill functions" or dose-response relationships. The relationship between insect mortality or survival and a given dose of insecticide can only be determined in a probabilistic sense, as not all insects are susceptible (or conversely, tolerant) to equivalent doses of insecticides and their effectiveness cannot be controlled exactly. However, by invoking the central limit theorem, the tolerance of individual insects can be modeled as a normal density function (see Archibald, 1984).

The probit probability model is commonly employed to estimate these dose-response functions, as it rests on the assumption that the tolerance of individual insects is a normally distributed random variable. The probability that any kill is less than or equal to a given level can be estimated and expressed linearly in terms of the dosage of pesticides. Observed probabilities can be fit by regression techniques to pesticide doses to determine how the probability of being killed changes with a change in dose. It thus provides a measure of the conditional probability of surviving a given pesticide dose (see Maddala, 1977).

Resistance can be measured by changes in the slope of the dose-response function. From the probit coefficients, the lethal dose required to kill 50 percent of the population ($LD_{50}s$) can be determined and changes in $LD_{50}s$ used to measure changes in insect resistance. This was the approach followed in estimating equation (13-2). The system of equations (13-1) and (13-2) was then estimated recursively, using the sample production and pest population data from California's Imperial Valley. Model specification and estimation techniques are summarized in appendix A.

Results of this analysis indicated that (a) the probability of surviving insecticide treatment at mean use levels declined by approximately 32 percent, or 6.3 percent per year, between 1978 and 1982 (five years in the desert climate is equivalent to a much longer period in a more temperate climate); (b) the "efficiency" adjusted insecticide levels, calculated from the regression coefficients as the $LD_{50}s$, increased sharply from 0.7 pounds of active organophosphate (OP) ingredient per acre per application to 1.22, and from 0.213 pounds of active synthetic pyrethroid (SP) ingredient per acre per application to 0.334; and (c) the damage in terms of lost yields due to declining effectiveness (from resistant or surviving insects) increased from 181 pounds of cotton lint per acre to 301 pounds per acre over the period. Thus, the analysis indicated that insecticide use was increasing and yields declining due to the development of resistance.

The Dynamic Programming Model. The dynamic programming model presented here uses the results of the production system as constraints on the optimization.[7] Controls, or decisions, on production input levels consisted of nine pest control strategies employing combinations of chemical insecticides, $X_i(t)$, and IPM inputs, $V_i(t)$. These strategies are presented in table 13-1. Each strategy reflects a combination of organophosphate, synthetic pyrethroid, pest scouting and monitoring, and biological controls (pheromones), and reflected the technical substitution possibilities that were observed in the sample.

The state or output variables in the dynamic model are the externality output, insect resistance $R(t)$, measured as those insects surviving control and entering negatively into cotton production; and cotton output $Y^c(t)$, measured as pounds of cotton lint. One hundred possible discrete states of insecticide effectiveness were calculated from the probit model.

Just and coauthors (1982) demonstrate that producer surplus adequately represents welfare from production under a producer-producer externality if information on costs and benefits are available. The production function and the externality function provide this information. Thus, the objective function maximizes discounted producer surplus over the planning period.

---

[7] Dynamic programming is a solution approach to dynamic optimization that exploits the dynamic decision structure directly (Luenberger, 1979). As illustrated in chapter 2, the discrete time dynamic programming problem can be used to represent better the way producers make dynamic input decisions.

TABLE 13-1. PEST CONTROL STRATEGIES USED IN THE DYNAMIC OPTIMIZATION MODEL

	Pounds of active ingredients per acre			
Strategy	Organophosphate (OP)	Synthetic pyrethroids (SP)	Pheromone[a]	IMP index[b]
A	0.45	0.12	0.665	21
B	0.90	0.12	0	15
C	0	0.24	0.665	15
D	0.75	0.48	0	0
E	0.45	0.24	0	18
F	1.20	0.32	0	15
G	0.90	0.24	0	15
H	0.75	0.24	0	15
I	1.80	0.48	0	15

Note: The levels of input use for each combination were based on actual combinations used by cotton producers.
Source: Adapted from S. O. Archibald, "A Dynamic Analysis of Production Externalities: Pesticide Resistance in California Cotton" (Ph.D. dissertation, University of California, Davis.)
[a] Pheromone is a biological control used to control the pink bollworm population.
[b] Integrated pest management (IPM) index includes scouting, traps, and boll cracking.

The dynamic programming model also incorporates uncertainty associated with the pest control technology. At each period t, production depends on the realization of a random variable—the probability of insects surviving a given control strategy—that affects current period returns. To capture this uncertainty, the probit model is embedded in the dynamic programming model such that in each period, the entire probability distribution of net revenues is calculated.

The discrete-time stochastic dynamic programming problem is then composed of system dynamics,

$$R(t+1) = f[R(t), U(t)]$$

the rate at which resistance changes for a given pest control strategy, where $U(t) = [X(t), V(t)]$; an initial condition, $R(0) = R_0$; the control constraints $U(t) \epsilon U$, where U represents the nine control strategies; the terminal value function, $\tau(T)$, which for this problem represents the "user cost," or externality tax, on insecticide use evaluated in terms of the value of lost yields from pesticide resistance resulting from a given pest control strategy summed over the planning period; and the objective function

$$J = \sum_{t=0}^{T-1} \beta_t \left(\sum_{i=0}^{1} p_i r[R(t), Y^c(t), U(t)] \right) + \tau(T) \qquad (13\text{-}3)$$

The objective function is composed of two terms. The first represents the discounted value of the probability distribution of net revenues; the second is the externality tax on insecticide use. In equation (13-3), p_i denotes the probability of net returns from the optional control strategy, $U(t)$ and $r[\cdot]$ denotes the Hicksian producer surplus measure. The rate of change in resistance is governed by a transition matrix whose coefficients are conditional slope changes estimated in the econometric production model. The producer observes the externality state $R(t)$, selects a control strategy $U(t)$ that results in a cotton output level or state $Y^c(t)$ and causes a transition to state $R(t+1)$. The transition to $R(t+1)$ captures the dependency of the current state of resistance on both the previous resistance state and past control decisions.

The productivity effects and implications for technical change of alternative environmental policies can be assessed from the dynamic programming model. Time rates of change in the value of the optimal objective function can be considered a dynamic analogue to the rate of change in total factor productivity (measured as gains or losses in producer surplus) achieved under alternative policies. This measure is net of externalities and incorporates changes in input quality due to regulation or decay of the resource stock. Alternatively, changes in output levels for the crop and the externality can be used to assess productivity under different policy options. The optimal paths of the externality-generating inputs, relative to the optimal path of nonexternality inputs, provide information about the bias in technical change, from which the effects of regulatory policy on the bias can be assessed.

Model Results

The results of several simulations of the dynamic programming model under alternative policy programs are discussed in this section. The objective is to demonstrate how the optimal actions of the producers change in response to regulations governing the use of chemical insecticides and the impacts they have on the long-run productivity of these inputs. The focus is on two policy programs—the current standards-based program and the externality tax program. The current standards-based pesticide regulations are imposed on the model by restricting the control strategies to lie within the set U—that is, to be one of the nine strategies listed in table 13-1—and setting the externality tax to zero. Under the externality tax program, the objective function includes the terminal value function, which reflects the value of yield losses due to pesticide resistance; the feasible controls are also restricted to lie within the set U.

Optimal Pest Control Strategies. The control strategies for the dynamic programming model under the two policy programs are shown in table 13-2. Under the standards-based regulatory program (column (1)), strategies relying primarily on chemical control dominate other less chemical-intensive strategies. Growers follow a strategy that employs high levels of organophosphates and relatively low use of synthetic pyrethroids (strategy B). They do not employ biological controls and use the lowest possible level of pest scouting and monitoring for the first seven periods. They switch to the IPM controls (strategy A) only after a high level of resistance buildup. Under the base case externality tax policy (column (2)), the lowest levels of insecticide available are employed and the highest levels of IPM adopted (strategy A). Producers alternatively use higher levels of organophosphate and synthetic pyrethrins (6 and 8) as resistance emerges, consistent with optimal resistance management strategies. When the uncertainty associated with the effectiveness of the IPM strategies is increased (column (3)), there tends to be a greater reliance on chemical control. When uncertainty over IPM falls, a greater reliance on IPM strategies is evident (column (4)).

The optimal strategy—not shown here—was also sensitive to the price of cotton. At a higher price of cotton more insecticides are employed. This suggests that high price supports for cotton may contribute to higher insecticide use levels as marginal value product increases. Additionally, at lower cotton prices growers appear not to be as responsive to increases in uncertainty of effectiveness of pesticides.

TABLE 13-2. OPTIMAL CONTROL STRATEGIES UNDER ALTERNATIVE POLICY ASSUMPTIONS

Time period	Standards-based policy	Externality tax policy[a]		
	(1)	(2)	(3)	(4)
1	B	A	A	A
2	B	A	A	A
3	B	A	A	A
4	B	A	G	A
5	B	A	G	E
6	B	B	B	B
7	B	A	B	A
8	A	E	E	A
9	A	A	B	A

Notes: Control strategies are keyed to table 13-1. Both time paths are based on the dynamic programming model, which has a 9 percent discount rate, and a cotton output price of $0.80 per pound.

[a] Column (2) shows the base case externality tax results and corresponds to a standard deviation of the probability of the insects being susceptible to pesticide dosage of 0.20. Columns (3) and (4) employ standard deviations of 0.40 and 0.10, respectively.

The model was also run with several discount rates. The base rate of 9 percent per year was varied from 2 to 12 percent. While net revenues were affected, optimal strategies did not change in any significant manner. The model is more sensitive to state transitions (the rate at which resistance develops) that overwhelm effects of variations in discount rates. This strongly underscores the need for resistance rates to be accurately measured if they are to be useful for productivity-related analyses.

These results provide some preliminary evidence that current standards-based regulatory policy, at least in the case of cotton, has not succeeded in reducing total chemical use due to the neglect of resistance as a collective (or common property) externality. Further, the relatively slow rate of adoption of IPM technology can be explained because the private returns associated with it are less than those associated with chemical insecticides. This situation continues until resistance reaches such a high level that producers adopt IPM approaches. These results suggest that current pesticide regulatory policy needs to be reexamined in light of its neglect of resistance. They further suggest that an externality tax can delay resistance, thus maintaining the productivity of insecticides for a longer period. These findings may also be sensitive to the definition of the control strategies and the length of time horizon chosen and thus should be interpreted carefully.

Output Effects. Under the externality tax policy, the rate of change in resistance is estimated to increase by 1.3 percent per year and cotton output per acre declines approximately 0.3 percent per year. By contrast, under the current standards-based regulatory policy, the rate of increase in resistance is 5.5 percent per year and the rate of decline in cotton yields per acre is 0.4 percent per year. What this analysis suggests is that the private economic returns to producers who follow the IPM strategy may be modest relative to a purely chemical strategy, but that the social benefits of following an IPM approach in terms of reduced pesticide resistance are substantial.

Bias in Input Use. Optimal input use in the externality tax solution, as contrasted to the current standards-based regulatory solution, is biased toward less environmentally damaging inputs, as can be seen in table 13-2. Under the standards-based policy, under which producers face increasing resistance but do not pay for future losses in the current period, producers use twice as much organophosphates relative to the usage levels under the externality tax program. Furthermore, no biological controls are used in the standards-based solution. Optimal quantities of pest scouting and monitoring are also lower relative to the externality tax solution. As resistance increases and insecticide costs per acre rise, growers switch to an IPM strategy in periods 8 and 9 (table 13-2). This is consistent with observed behavior among Imperial Valley producers. Only when control costs and yield losses became so high that cotton was being produced at a loss did producers enforce mandatory IPM practices. In contrast, the externality tax policy induced earlier adoption of nonexternality-generating inputs.

Productivity Effects. Not surprisingly, when resistance is accounted for, previous estimates of insecticide productivity are overstated. Using results from the econometric model, when external costs of resistance are excluded, a dollar invested in chemical insecticides returns $3.50, a figure consistent with earlier studies. On the other hand, IPM technology (including chemical use at lower levels) produced a lower short-run return to individual producers of $2.50 per dollar invested, implying that the chemical strategy provides producers with a higher return in this instance. However, from the dynamic model, if producers face an externality tax that internalizes resistance, producer returns to chemical pest control drop to only $1.00 for every dollar invested.

The changes in the output index that reflect both cotton yields (positive output) and pesticide resistance (negative output) for the two policy simulations are provided in table 13-3. Under the externality tax, the joint cotton-resistance index declines at a substantially lower rate than the same index for the standards-based simulation, primarily due to the lower rates of pesticide resistance produced. This index is not directly comparable to conventional indexes of output growth, as it reflects all outputs of the production process and not simply the marketed products. It clearly demonstrates that substantial differences in productivity in-

TABLE 13-3. OUTPUT INDEX OF CROP AND RESISTANCE PRODUCTION UNDER VARIOUS REGULATORY POLICIES

Period	Standards-based regulation	Externality tax
1	100.0	100.0
2	98.9	99.5
3	95.6	98.9
4	94.2	98.3
5	92.8	97.8
6	81.5	97.5
7	81.0	96.8
8	66.3	96.3
9	66.1	96.1

dices would be forthcoming if the output measures of agricultural production processes included both marketed and nonmarketed products.

CONCLUSIONS

A dynamic production system was developed in this chapter and applied to the production of cotton and the externality from collective pesticide use in Imperial Valley, California. Evidence from this study suggests that, under present regulatory policy, rates of insecticide use have increased, agricultural output has declined, and the productivity of approved materials has declined largely due to resistance development. Furthermore, the dynamic analysis indicates that under the current regulatory program technology, relying primarily on chemical control continues to dominate other strategies over the model's time horizon. Increasing amounts of pesticide are used as effectiveness declines.

When the price of pesticides does not reflect the full cost of their use because of spillover effects on the environment, the combination of pesticides and other inputs that is used may not be the least cost combination from a collective point of view. When producers are faced with an externality tax reflecting the opportunity cost of current pesticide use, the alternative less externality-generating IPM technology becomes the preferred strategy. Further, when total costs (production and user costs) are accounted for, estimates of pesticide productivity decline.

Thus, this study provides evidence that current standards-based pesticide policy has contributed to lower agricultural output without reducing total chemical use and shows why adoption of technology that is less damaging to the environment has been slow. The results suggest that for the cotton sector—in which many externalities involve common property resources, result from collective use of inputs, and exhibit characteristics of nonpoint pollution problems—current policy is failing to meet its objectives, and that alternative policies should be considered that provide economic incentives for the adoption of alternative pest management methods, such as IPM.

In interpreting the results from the dynamic optimization model for this sector, the following limitations of the study should be noted. First, the nine control strategies are defined as discrete combinations, that is, they represent fixed proportions of chemicals and IPM inputs. Only to the extent that these combinations reflect the input mixes growers would select under the alternative policy programs are the implications for levels of input use valid. Secondly, the results are sensitive to the level of uncertainty associated with the effectiveness of chemicals in suppressing the pest population. Improvements in the accuracy of the measurement of pesticide resistance would narrow the bounds of input use. In addition, modifications of the linear specification of the production function for cotton to reflect a more flexible functional form might minimize errors associated with model specification. Finally, it is not clear how the terminal conditions affect results, given the fairly short time horizon for the simulation, although the productive lifetime for pesticides expected by chemical manufacturers is within this bound.

Given those qualifications, the case study presented here suggests that dynamic models can provide useful information for analyzing the effects of regulations on agricultural productivity and technical change. This is a particularly important research tool when natural resources are employed in production and when the longer-run productivity effects of technology must be taken into account for proper analysis of alternative policies.

APPENDIX 13-A.
MODEL SPECIFICATION

Pesticide Effectiveness

Under the assumption that the proportion of insects susceptible to pesticides follows the standard normal cumulative distribution function (CDF), the probit transformation (see Zellner and Lee, 1965) can be used to obtain

$$F^{-1}(p_i) = F^{-1}(P_i) + \frac{e_i}{Z(P_i)} = X'B + \frac{e_i}{Z(P_i)}$$

where F^{-1} is the inverse of the normal CDF and $F^{-1}(p_i)$ and $F^{-1}(P_i)$ are thus the observed and true "probits," respectively, and $Z(P_i)$ is the value of the standard normal density evaluated at P_i. The regressor vector X is the insecticide dose measured in terms of pounds of active ingredient by major chemical class.

The estimating equation is

$$F^{-1}(p_i) = \beta_0 + \beta_1 CB_i + \beta_2 OP_i + \beta_3 SP_i + \beta_4 DD_i + \gamma_1 80CB_i + \gamma_2 81CB_i \\ + \gamma_3 80OP_i + \gamma_4 81OP_i + \gamma_5 80SP_i + \gamma_6 81SP_i$$

where CB represents carbamates, OP organophosphates, and SP synthetic pyrethroids, and DD represents weather, measured in degree days.

Slope dummies on each class of insecticide (γ_i, $i = 1, \ldots, 6$) were included to determine the change in the slope of the dose mortality curve from 1980 to 1981. These changes in productivity were used to approximate the time rate of change in pesticide effectiveness. From these coefficients, LD_{50} s can be calculated and changes used to measure changes in resistance.

Data on pest populations were from weekly observations on tobacco budworm larvae for 179 fields in Imperial Valley. Pesticide quantities applied to each field were obtained from county pesticide use records.

Cotton Production

The probit model provides estimates of the effectiveness of insecticides in reducing insect populations and of changes in effectiveness over time for different classes of chemical compounds. Based on the recursive estimation of equations (13-1) and (13-2), insects that survive pesticides, $1 - p(k)$, enter the production function producing yield loss. Cotton yield levels result, however, from the effectiveness of the entire production technology as well as damage that results from surviving insects and should be included in analysis of pesticide productivity. The physical and biological production input data obtained from the survey of growers were used to estimate their relationship to cotton lint yields. Details on variable definition and measurement are in Archibald (1984).

Yield loss or damage can generally be specified as an increasing function of pests remaining after applying pesticide control (Feder and Regev, 1975). Following Hall and Norgaard (1973), Talpaz and Borosh (1974), and Feder and Regev (1975), a linear cotton production/damage function was specified

$$Y_i^c = f(R_i, I_i, M_i, LQ_i, W_i, K_i) + e_i$$

where $\quad e_i \sim N(0, \sigma^2)$
$\quad\quad\quad E(e_i e_j) = 0 \quad i \neq j$
$\quad\quad\quad\quad i = 1 \ldots 179$ cotton fields
$\quad\quad\quad\quad Y_i^c$ = yields in terms of actual pounds of cotton lint per acre
$\quad\quad\quad\quad R_i = 1 - P(i)$, the number of tobacco budworm larvae surviving X_i
$\quad\quad\quad\quad I_i$ = measure of purchased IPM services
$\quad\quad\quad\quad M_i$ = management skills in equivalent years of formal education by grower
$\quad\quad\quad\quad LQ_i$ = a land quality variable in terms of a Storie index weighted by proportionate soil types in each field and

adjusted to reflect improvements due to drainage investments
W_i = water use in terms of actual acre feet per acre
K_i = a measure of capital availability in terms of a current ratio for each firm.

REFERENCES

Archibald, S. O. 1984. "A Dynamic Analysis of Production Externalities: Pesticide Resistance in California Cotton" (Ph.D. dissertation, University of California, Davis).

Bator, F. M. 1958. "The Anatomy of Market Failure," *The Quarterly Journal of Economics* vol. 72, no. 3, pp. 351–379.

Baumol, W. J., and W. C. Oates. 1975. *The Theory of Environmental Policy* (Englewood Cliffs, N.J., Prentice Hall).

Brattsten, L. B., C. W. Holyoke, Jr., J. R. Leeper, and K. F. Raffa. 1986. "Insecticide Resistance: Challenge to Pest Management and Basic Research," *Science* vol. 231, no. 4743, pp. 1255–1260.

Buchanan, J. M. 1966. "Joint Supply, Externalities and Optimality," *Economica* vol. 33, no. 132, pp. 404–415.

———, and W. C. Stubblebine. 1962. "Externality," *Economica* vol. 29, no. 116, pp. 371–384.

Carlson, S. 1939 (reprinted 1974). *The Pure Theory of Production* (London, published by writers connected with the Institute for Social Sciences of Stockholm University).

Council for Agricultural Science and Technology. 1983. *The Resistance of Agricultural Pests to Control Measures*, Report No. 97 (Washington, D.C., CAST).

Crosson, Pierre. 1983. *Productivity Effects of Cropland Erosion in the United States*, Research paper (Washington, D.C., Resources for the Future).

Dasgupta, P. S., and G. M. Heal. 1979. *Economic Theory and Exhaustible Resources* (London, Cambridge University Press).

Dover, M. 1985. "Getting Off the Pesticide Treadmill," *Technology Review* vol. 88, no. 8, pp. 52–65.

Doyle, J. 1985. "Biotechnology Research and Agricultural Stability," *Issues in Science and Technology* vol. 2, no. 1, pp. 111–124.

Farrell, K. 1981. "Productivity in U.S. Agriculture," ESS Staff Report No. AGE 55810422 (Washington, D.C., U.S. Department of Agriculture, Economics and Statistics Service).

Feder, G., and U. Regev. 1975. "Biological Interactions and Environmental Effects in the Economics of Pest Control," *Journal of Environmental Economics and Management* vol. 2, no. 2, pp. 75–91.

Food and Agriculture Organization. 1972. *Pesticides in the Modern World*, Symposium organized by members of the Cooperative Programme of Agro-Allied

Industries with the United Nations Food and Agriculture Organization and other UN organizations (Rome, FAO).

Georghiou, G. P. 1981. *The Occurrence of Resistance to Pesticides in Arthropods* (Rome, Food and Agriculture Organization).

Greene, W. 1982. *LIMDEP* (Washington, D.C., National Economic Research Associates).

Hall, D. C., and R. B. Norgaard. 1973. "On the Timing and Application of Pesticides," *American Journal of Agricultural Economics* vol. 55, no. 2, pp. 198–201.

Headley, J. C. 1968. "Estimating the Productivity of Pesticides," *American Journal of Agricultural Economics* vol. 50, no. 1, pp. 13–23.

Just, R. E., and D. Zilberman. 1979. "Asymmetry of Taxes and Subsidies in Regulating Stochastic Mishap," *Quarterly Journal of Economics* vol. 93, no. 1, pp. 139–148.

———, D. L. Hueth, and A. Schmitz. 1982. *Applied Welfare Economics and Public Policy* (Englewood Cliffs, N.J., Prentice Hall).

Kendrick, J. W. 1980. "Survey of the Factors Contributing to the Decline in U.S. Productivity Growth," *Proceedings of a Conference Sponsored by the Federal Reserve Bank of Boston*, Edgartown, Mass.

Kopp, R. J., and V. K. Smith. 1981. "Productivity Measurement and Environmental Regulations: An Engineering-Econometric Analysis," in R. Cowing and R. E. Stevenson, eds., *Productivity Measurement in Regulated Industries* (New York, Academic Press).

Langham, M., J. Headley, and W. Edwards. 1972. "Agricultural Pesticides: Productivity and Externality," in A. Kneese and B. Bower, eds., *Environmental Quality Analysis* (Baltimore, Md., The Johns Hopkins Press for Resources for the Future).

Lemoine, P. H. 1984. "Water Resources Management in the Salinas Valley: Integration of Economics and Hydrology in a Closed Control Model" (Ph.D. dissertation, Stanford University).

Luenberger, D. G. 1979. *Introduction to Dynamic Systems* (New York, N.Y., Wiley).

Maddala, G. S. 1977. *Statistical Methods in Econometrics*, 2nd ed. (New York, N.Y., McGraw-Hill).

Meade, J. E. 1952. "External Economies and Diseconomies in a Competitive Situation," *The Economic Journal* vol. 62, pp. 54–67.

National Research Council, Commission on Natural Resources. 1982. *Impacts of Emerging Trends on Fish and Wildlife Habitat* (Washington, D.C., National Academy Press).

Noel, J. E., B. D. Gardner, and C. V. Moore. 1980. "Optimal Conjunctive Water Management," *American Journal of Agricultural Economics* vol. 62, no. 3, pp. 489–498.

Pittman, R. W. 1983. "Multilateral Productivity Comparisons with Undesirable Outputs," *The Economic Journal* vol. 93, no. 372, pp. 883–891.

Portney, P. R. 1981. "The Macro Economic Impacts of Environmental Regulation," *National Resources Journal* vol. 21, no. 3, pp. 459–488.

Sudit, E. F., and D. K. Whitcomb. 1976. "Externality Production Functions," in S. A. Y. Lin, ed., *Theory of Economic Externalities* (New York, N.Y., Academic Press).

Talpaz, H., and I. Borosh. 1974. "Strategies for Pesticide Use: Frequency and Applications," *American Journal of Agricultural Economics* vol. 56, no. 4, pp. 769–775.

Whitcomb, D. 1972. *Externalities and Welfare* (New York, N.Y., Columbia University Press).

Zellner, A., and T. H. Lee. 1965. "Joint Estimates of Relationships Involving Discrete Random Variables," *Econometrica* vol. 33, no. 2, pp. 382–394.

INDEX

AAEA. *See* American Agricultural Economics Association
Abel, M. E., 67
Abramovitz, M., 98
Ackello-Ogutu, C., 334
AES. *See* Allen elasticity of substitution
Aggregation
  Agricultural production: analysis, 67, 336–338, 341–344, 357–358; functions, 18–24
  Agricultural productivity, 1, 2, 6; consistent, 21; evaluation of, 145–146; functions, 50–55; recommendations for improved research on, 185–186
  Approach to technological change: cost functions, 35, 51; neoclassical, 24–25, 26, 51, 53; production functions, 34–35, 316–320; profit function, 28, 29–32, 35–36
  Econometric production models: Antle, 112–113; Binswanger, 109–110, 119; Brown-Christensen, 110, 117, 123; Chan-Mountain, 111–112; Hazilla-Kopp, 110, 119; Lopez, 111, 123; Ray, 110–111; USDA, 106–107
  Inputs, 103–107
Agricultural development
  Constraints on, 248
  U.S.-Japanese, 5, 255–261
  See also Induced innovation theory
Agricultural production
  Aggregation analysis, 67, 336–338, 341–344, 357–358
  Causality, 333, 344; to measure public sector investments, 354; nonstructural approach to, 346; structural approach to, 345–346
  Dynamic nature of, 2, 332, 333, 362; externalities effects, 367; intraseasonal output dynamics and, 334–335; production models, 335–336, 351–352, 353, 362–363
  Externalities, 367; environmental, 366, 367; inputs, 375; internalization of, 369; taxes on, 371–372
  Growth, 255–261, 366
  Price data, 125
  Stochastic nature of, 72, 338–339; and technological change, 339–341
  Technological change and, 3, 67, 98, 113, 189, 366; allocation of gains from, 195–196; biases in, 99, 109, 114, 115–117, 176, 180–182, 196–200; market equilibrium effects of, 202–206; price effects, 190, 195, 200–206
  See also Cost of production, agriculture; Dynamic models
Agricultural productivity
  Competitive market equilibrium approach to, 191–196
  Conventional measure of, 372–373
  Decomposition analysis, 289, 307–312; research specialization dimension, 290, 292; simultaneity problems in, 313; spatial dimension, 293; time dimension, 301–304; variables, 304–307
  Dynamics of, 2; first generation partial adjustment, 67–68; second generation dual cost and profit function, 68; third generation cost-adjustment, 68–72.
  Environment and, 13, 101, 347, 378–380
  Factors influencing, 3, 99, 347
  Indexes: partial, 97, 138, 139, 372–373; TFP, 97–98, 101, 105–109, 138, 139
  Inputs, 3, 101; data for, 124–133, 139–144; elasticity of substitution, 97–98, 117–119; factor demand elasticities, 119–123; growth rates, 103–104; joint versus nonjoint, 229–230, 231
  Land prices and, 190, 195–196
  Methods for measuring, 96–97
  Output: data for, 125–128; indexes of aggregation, 97–98, 101; growth rates, 103; technological change and prices of, 195–196, 200–202
  Post-*1970* measurements, 100–109

395

Agricultural productivity (*Cont.*)
 Pre-*1970* measurements, 97–100
 Revenue shares, 102
 See also Dynamic models; Intertemporal and interspatial agricultural productivity; Research, agricultural productivity; Static equilibrium models; Total factor productivity
Agricultural Stabilization and Conservation Service, 113
Agriculture, Department of
 Data of: evaluation of, 101, 107, 143–153; proposed improvements in, 3, 10–11, 98
 TEP index, 101, 106–107, 138, 139–140
Agriculture and Consumer Protection Act of *1973*, 218
Ahmad, S., 249n, 250
Aitah, A. S., 229n
Akino, M., 229n
Allen elasticity of substitution (AES), 22–23, 285
Ameniya, T., 230, 237
American Agricultural Economics Association (AAEA), on USDA statistics, 101, 107, 143–153
American Feed Manufacturers Association, 142
Antle, John M., 4, 6, 9, 10, 17, 18, 66, 67, 69, 112–113, 229n, 332–333, 334, 339n, 354, 356, 358, 361
Archibald, Sandra, 6–7, 18, 67, 366, 375, 380, 381, 390
Argentina, sectoral growth, 328, 329
Arrow, K. J., 19, 229n
Australia, input substitution, 248

Ball, Eldon, 152, 203
Barker, R., 330
Barton, G. T., 97, 98, 139
Baumol, W. J., 375
Ben-Zion, U., 9
Berndt, E. R., 23, 59, 67, 82, 101, 119, 150, 160, 185n, 186, 209n, 229n, 333, 336
Bertsekas, D. P., 69
Bias, technological change, 3, 8, 36–37
 Defined, 38–39
 Dual measures of, 40–44, 47, 48, 357

 Factor, 47–48, 99, 109, 114, 115–117, 176, 180–182, 196–200, 269–279, 280–281
 Lagged price, 360–362
 Primal measures of, 44, 45, 48
Binswanger, H. P., 66, 109–110, 119, 124, 229n, 249n, 250, 267, 285, 361
Blackorby, C., 20, 37, 185n
Borosh, I., 381, 390
Boussard, J. M., 67
Box-Cox flexible functions, 101, 109, 119
Boyce, J. K., 297n
Brattsten, L. B., 368
Bronfenbrenner, M., 255n
Brown, M., 255n
Brown, R. S., 68, 82, 110, 116, 117, 123, 144, 146, 148, 149, 150, 151, 173, 182
Buchanan, J. M., 375, 376
Bureau of Labor Statistics (BLS), 144
Burgess, D. F., 159, 185n

California, pesticide resistance in cotton production, 13, 367, 378–382, 388, 389–391
Canada, agricultural productivity, 96
 Aggregate econometric production models for, 109, 111
 Cost function model, 166, 168
 Factor demand elasticity, 124
 Input substitution, 119, 248
 Technological change in, 113, 116
 Translog production model, 116
Capalbo, Susan M., 2–3, 4, 9, 17, 96, 101, 113, 116, 159, 160, 189, 204, 229, 350
Capalbo, Vo, and Wade (CVW) TFP indexes, 101, 106–108, 125–129
Capital inputs, 3
 Animal, 129
 Data for, 148–149
 Expenditures, 219
 Factor demand elasticities of, 119, 123, 124
 Labor input ratio to, 317, 325–326
 Land, 129, 219
 Machinery and mechanical power, 141–142
 Measurement of service flow of, 209n
 Real estate, 141

INDEX

Carlson, S., 376
Carter, H. O., 19
Cavallo, D., 328, 329
Caves, D. W., 57, 62, 100, 101, 168
Census Bureau, 144, 149
Census of Agriculture, 141, 142
Census of Manufacturing, 142
CES (constant elasticity of substitution) function, 19, 115
Chalfant, J. A., 119
Chambers, R. G., 185n
Chan, M. W. L., 111–112, 116
Chandler, C., 98
Chaudhri, D. P., 354n
Chemical inputs, 34, 142. *See also* Fertilizer input
Chenery, B. H., 19, 229n
Chile, sectoral growth, 329
Chow, G. C., 69
Christensen, L. R., 19, 23, 34, 57, 59, 62, 68, 74, 82, 100, 101, 102, 110, 116, 117, 123, 144, 147, 159n, 168, 173, 182, 229, 322
Cline, P. L., 229n
Cobb, C. W., 229n
Cobb-Douglas
  Production function, 18; agricultural productivity growth based on, 151; geometric index for, 51, 54, 98; globally concave, 76; simplicity, 19; technology set, 31–32; translog, 19–20, 23
  Production model, 74, 115, 333, 345
Coeymans, J., 328, 329
Conlisk, J., 249n
Convex sets theory, 24–27, 30
Cooley, T. F., 345
Cooper, M. R., 97
Corn, productivity analysis, 209, 220–223, 224, 225–226
Cost function models
  Factor demand elasticities and elasticities of substitution, 117, 119, 123
  Leontief total, 166–168
  Translog, 59, 63, 74–76, 164–166; multiple output, 168–171
Cost-of-adjustment model, 333, 336
Cost of production, agriculture
  Functions, 52–53, 74; monotonicity and curvature restrictions on, 76–80; parameter restrictions on, 80–82; restricted cost and profit, 160

Growth rate, 102, 103, 105
  Index, 52
  Minimization of, 25, 53
  Profit and, 26–27
Cotton production, California, 13
  Pesticide resistance in, 367, 378–382, 388, 389–391
Crops, 110
  Fertilizer effect on yield of, 253–254
  Research program on, 304–307
Crosson, Pierre, 368, 373

Dairy Herd Improvement Association, 143
Danin, Y., 325
Danø, S., 376n
Dasgupta, P. S., 373
David, P. A., 259n
Davis, J. S., 289
Day, P. H., 362
Deaton, A. S., 162
de Groof, R. J., 249n
Denison, E. F., 50, 209n
Denny, M., 20, 63, 65, 69, 101, 113, 116, 160, 209, 210, 212–216
Diamond, P., 10
Diewert, W. E., 19, 24, 50, 51, 54–55, 56, 57, 62, 64, 80, 82, 85, 87, 100, 159n, 160, 162, 209, 232, 233
Dillon, J. L., 19
Disaggregation
  Agricultural outputs, 230
  Models, 185
Divisia index, 54, 56, 105, 125, 145
  Bias from, 65
  Defined, 55
  Uses, 100, 101
Doll, J. P., 263n
Domar, E., 50, 98
Doob, J. L., 341
Douglas, P. H., 19, 229n
Dover, M., 368
Drandakis, E. M., 249n
Duality theory, 6, 232, 233
  Applications, 24–25; to aggregate agricultural production, 109–110; to externalities in agricultural production, 374; to measure technological change, 34–36, 100
  Limitations of, 33
  Models, 3; cost-of-adjustment, 68–69,

Duality theory (*Cont.*)
  74; estimation of output supply and input demand functions, 230; functional forms for, 160; for research productivity, 313; test uses, 185
  Usefulness in econometric measurement, 32–33, 350
  See also Cost function models, Profit function models
Durost, D. D., 139
Dynamic models
  Classification of, 67–68
  Cost-of-adjustment, 68–69, 333
  Issues arising from, 69
  Nerlovian partial adjustment, 67–68, 72, 333, 346–347, 352n
  Production, 69–72, 332–336, 351–354, 362–363
  Profit function, 3, 68, 82–85, 117, 119, 123, 173–177
  Rational expectations, 72–73
  Tractability problem in, 73–74

Easter, D. W., 67
Eckstein, Z., 69, 352n
Econometric approach
  To multiple-output agricultural technology, 230, 232–241
  To total factor productivity, 2, 4, 17, 18; compared with growth accounting approach, 62–63; described, 5–7; technological change in, 57–60
Economic Research Service, USDA, 140, 152
  Labor series, 144
  Output series, 143
Economies of scale
  Agricultural productivity, 3, 98–99, 100–101, 111
  Agricultural production, 66
  Constant versus nonconstant, 163–164
  Productivity measures and, 178–180
  See also Agricultural productivity
Education input, 66, 98, 103
Edwards, W., 375
Endogenous switching models
  Econometric, 230–240; drawbacks of, 241; procedure for estimating, 239–240
  Economic, 231–232
  Methodology for, 230–231

Environment
  Effect on productivity of natural, 347
  Externalities affecting, 366, 367–372
  Regulations of: and agricultural productivity, 13, 101, 369–370, 378–380; focus of, 367–368; regional, 374; resource-related, 368–369, 375; taxes and, 371–372
  Research targeted to, 292–295
  Technological effects on, 6, 366, 372
Environmental Protection Agency, 129
Epstein, L., 69, 74, 101
Evenson, Robert E., 5, 8–9, 11, 66, 67, 114, 229n, 289, 297n, 299, 304–307, 307–312, 348n
Externalities, agricultural production, 367
  Environmental, 366, 367, 374
  Internalization of, 369
  Models for: incorporation of effects in, 374–377; multi-product specifications for, 376; partial productivity measures, 372–374
  Regulations and, 367–369
  Specific inputs and, 375

Fabricant, S., 98
Factors of production, 3, 99
  Substitution, 203, 248; elasticities, 22–23, 97–98, 117–119, 182, 285; and metaproduction function, 262–266; technological change versus, 249–250
  See also Inputs, agricultural productivity
Färe, R., 255n
Farm sector
  Characteristics, 11
  Defined, 3, 146–147
Farrell, K., 366n
Feder, G., 67, 381, 390
FEDS. *See* Firm Enterprise Data System
Feed, seed, and livestock input, data on, 129, 142. *See also* Livestock
Fellner, W., 249n
Ferguson, C. E., 25
Fertilizer input
  Cost, 110
  Data on, 129, 142, 149–150
  Production externalities and, 375
  Relationship between land and, 252–253, 259, 261, 265
  Yield from, 253–254

INDEX 399

Firm Enterprise Data System (FEDS), USDA, 4, 140
 Classification of data, 218–219
 Described, 218
 Evaluation, 227
 Model results using, 220–225
 Weather variation measures for, 219–220
Fishe, R. P. H., 230
Fisher, I., 52
Fisher-Tornqvist growth accounting equation, 210, 212, 216, 227
Flint, M. L., 356
Food and Agricultural Organization, 378
Fuller, Wayne, 240n
Functional forms, in production analysis, 18–19, 159, 184. *See also* Production theory
Fuss, M., 17n, 20, 63, 65, 101, 159n, 171n, 185n, 209, 210, 212–216

Gabler, E. C., 330
Gardner, Bruce, 139, 144, 145, 149, 153, 190, 191n, 375
Gollop, F. M., 147
Grabowski, R., 255n
Granger, C. W. J., 339, 341, 346
Green, J., 26
Griliches, Zvi, 11, 66, 67, 98–99, 100, 138, 144, 147–148, 149, 178, 209n, 229n, 249–250, 262n, 289
Grove, E. W., 149
Growth accounting approach, 2, 4, 17
 Compared with econometric approach, 62–63
 Described, 50
 For intertemporal and interspatial productivity, 211; first order models, 211–213; second order models, 213–216, 225–226
 Measurement of, 18, 51–57

Hall, D. C., 390
Hall, R. E., 46, 80, 102
Halter, A. N., 19
Hanemann, W. M., 230
Hanoch, G., 19
Hansen, L. P., 69, 74
Hardin, L. S., 19
Hatchett, S. A., 74, 334
Hayami, Yujiro, 5, 8, 9, 17n, 18, 66, 67, 96, 99, 189, 229n, 247, 253n, 347–349
Hazell, P., 330
Hazilla, Michael, 4, 18, 68, 77n, 76, 78, 110, 116, 119, 182, 208
Headley, J., 375
Heady, E. O., 19
Heal, G. M., 373
Hechman, J., 230, 234
Heller, W. P., 26
Hellinghausen, R., 324, 328
Henderson, J. M., 376n
Hessian matrix
 Cost function: modified, 78–79; sub-, 81; symmetric, 77, 80
 Diagonal elements, 173, 178
 Growth accounting model, 214, 216
Hicks, J. R., 1, 18, 22, 36, 37, 38, 45, 66, 99, 110, 112, 113, 163, 166–168, 171, 248–249, 325, 347
Hoch, I., 345n
Hocking, J. G., 19
Hockman, E., 230
Hotelling's lemma, 24, 29, 31, 232
Houthaker, H. S., 159n
Hueth, D. L., 369n
Huffman, Wallace, 4, 67, 114, 229
Hughes, D. W., 144
Hulten, C. R., 44, 55

Index number theory, 18, 51, 54–55, 56, 57, 62, 85–87, 100, 209
 Exact, 87–88, 100
 Procedures, 50–51
 Production functions and, 54–55
 Superlative quantity, 54, 100
 TFP measurement and, 55–57
 See also Growth accounting approach
Induced innovation theory, for agricultural development, 1, 2, 5, 18, 66, 67, 99, 109–110, 124, 189, 247, 250, 347–349
 Aggregate production model to test, 352
 Dynamic structural production model to test, 356–362
 Hayami-Ruttan model: dynamic nature of, 347–350; factor substitution in, 251–252; metaproduction function, 253–255, 260; production levels, 247–248; research production function, 250

Induced innovation theory, for (*Cont.*)
  Price effects and, 354–356
  Test of, 266–271, 281–284
  Theory of firm framework for, 248–255
Inputs, agricultural productivity
  Causal relations, 345–347
  Data for, 140–143, 144, 218–219
  Efficiency in using, 97
  Factor demand elasticities, 23, 117, 119–122; compensated versus uncompensated, 123
  Growth rate, 103–104
  Indexes, 98, 101, 285
  Purchased, 219
  Quality changes, 3, 34, 147–148
  Substitution among, 22, 97–98, 117–119, 203–204, 285
Intertemporal and interspatial agricultural productivity, 4, 18
  Defined, 208–209
  Five-state analysis, 209–210, 220–225
  Measurement of total factor, 63–65, 208, 209, 210, 212–216
  Models: data for, 218–219; evaluation of, 226–227; growth accounting, 211–216; translog function, 216–217
Investment, agricultural
  Cost-of-adjustment model for, 68–69
  Innovation and interrelated, 348
  For new productivity techniques, 328, 330
  Public sector, 352–354
  Research on, 312, 353–354, 362–363

Jacobs, R. L., 345n, 346
Jamison, D. T., 67, 229n
Jansson, L., 255n
Japan
  Agricultural development, 5, 255
  Agricultural production: factor biases in, 271, 273, 276, 279; factor shares, 273, 278, 285, 347; factor substitution in, 264–266; factor use in, 258–261; growth, 256–258
Johnson, D. Gale, 329n
Johnson, G. L., 19
Johnson, N. L., 235
Jointness, technological, 23–24, 80, 229–230
Jorgenson, D. W., 17n, 98, 102, 147, 159n, 178, 209n

Translog production function, 19, 74, 322
Judd, M. A., 297n
Just, R. E., 369n, 372

Kako, T., 267n
Kendrick, J. W., 17n, 50, 367n
Kennedy, C., 17n, 96, 249n
Khaled, M., 101, 119, 160, 185n
Kislev, Y., 66, 67, 262n, 273, 299, 348n
KLEMA prices, 218, 223
Kmenta, J., 240
Kohli, U. R., 81
Konüs implicit quantity index, 52–53
Kopp, Raymond, 4, 18, 68, 77n, 78, 101, 110, 116, 119, 182, 208, 367n, 374
  Monotonicity and curvature restrictions, 76, 110
Kotz, S., 235
Krenz, R. D., 209
Kuznets, Simon, 50
Kydland, F. E., 333

Labor input, agricultural
  Capital input ratio to, 317, 325–326
  Cost, 110
  Data for, 129, 140–141, 219
  Demand for, 114
  Family, 103, 194
  Growth rate, 103–104
  Productivity index, 107–109
  Quality index, 103, 105
  Relationship between draft power and, 260–261
  Substitution for other factor inputs, 203, 204–205
Lambert, L. D., 139
Land input, agricultural
  Data for, 129, 219
  Prices, 190, 195–196, 205–206; food prices versus, 200–202
  Productivity index, 107–108
  Relationship of fertilizer and, 252–253, 259, 261, 265
  Technological change effects on, 197–200, 248
Langham, M. J., 375
Laspeyres indexing procedure, 3, 51, 53, 54, 98, 145, 153
Lass, D. A., 230

INDEX

Latimer, R., 304n
Lau, L. J., 19, 46, 67, 74, 84, 114, 159n, 160, 229, 232, 322, 350
Lave, L. B., 98
Leamer, E. E., 345n, 346
Le Chatelier principle, 160, 182
Lee, L., 34, 240
Lemoine, P. H., 375
Leontief, W., 19, 53, 111, 113, 119, 164, 198, 202
LeRoy, S. F., 345
Less developed countries, agricultural production, 255
Lin, K., 240
Livestock, 129, 142
 Inventories, 143
 Research on, 304–307
Lockheed, M. E., 67
Long, J. B., 333
Loomis, R. A., 97, 98, 139
Lopez, Ramon, 9, 111, 113, 114, 116, 123, 124, 164, 166, 182, 185n, 189, 195
Lovell, A. K., 37, 38
Lu, Y. C., 151
Lucas, Robert E., Jr., 68–69, 72, 73, 333
Lurie, P. M., 230

McFadden, D., 17n, 20, 26, 27, 159n, 185n
Machinery and mechanical power, data on, 141–142
Maddala, G. S., 230, 234, 239, 240, 381
Magnus, Jan R., 61
Malmquist, S., 52, 57, 100
Marginal rate of substitution (MRTS), 21, 22
 Neutral and biased technological change in terms of, 36–38
Market equilibrium
 Competitive, 190, 191
 Technological change effects on, 4, 9, 189, 195; factor bias of, 196–200; model to estimate, 191–195, 202–204; price effects and, 200–206
Mehra, S., 330
Milk price support program, 355–356, 363
Minhas, B. S., 19, 229n
Minkowski theorem, 26, 28
Moore, C. V., 375
Morrison, C. J., 67, 186, 333, 336
Mountain, D. C., 111–112, 116

Mowery, D., 249n
MRTS. *See* Marginal rate of substitution
Muller, R. A., 230
Multiple output models
 Econometric, 232–241
 Economic, 231–232
 Purpose, 230
Mundlak, Yair, 6, 9, 20, 23, 316, 324, 325, 328, 329, 344, 345n

Nadiri, Ishaq M., 96
Nash, E. K., 230, 233
National Lime Institute, 142
National Oceanic and Atmospheric Administration, 219
National Research Council, 369, 379
Nelson, F. P., 240n
Nelson, G. L., 144, 251
Nerlove, M. L., partial adjustment models, 67, 72, 333, 346–347, 352n
Newbold, P., 339, 341
Nghiep, L. T., 267n, 273
Noel, J. E., 375
Nordhaus, W. D., 249n
Norgaard, R. B., 390
Norton, G., 67, 289

Oates, W. C., 375
Outputs
 Aggregate, 97–98, 101, 107
 Biases, 47–48
 Data for, 125–128, 143
 Disaggregation of, 230
 Factors influencing, 48–49
 Indexing of, 97–98, 101
 Models for multiple, 230–241
 Quality changes in, 34
 Rate of growth, 103
 Taxes on, 371–372
 Zero-value limits, 4, 232–234

Paarlberg, D., 304n
Paasche index, 3, 54, 98
Paris, Q., 334
Parkan, C., 162
Partial adjustment models, 67, 72, 333, 346–347, 352n
Partial factor productivity (PFP), 97, 144
Parzan, E., 140
Payment-in-kind program, 355–356

Penson, J. B., 144
Pesaran, M. H., 162
Pesticides
   And agricultural productivity, 13
   Data on, 129, 142
   Dynamic model for control strategies with, 382–387, 388–389
   Estimates of effectiveness, 390–391
   Externalities caused by, 375
   Regulation of, 368–369
Peterson, W., 17n, 96, 99, 262n, 273
Petzel, T. E., 67
PFP. *See* Partial factor productivity
Phelps, E. S., 249n
Pindyck, R. S., 74
Pinto, C., 24
Pitt, M., 34
Pittman, R. W., 101
Plosser, C. I., 333
Pollack, A., 52
Pope, R., 185, 230, 233
Portney, P. R., 367n
Prescott, E. C., 333
Prices
   Factor: causal relations between, 347; cost functions, 74–82; elasticities, 182–183; technological change and, 190, 195, 200–206
   Indexes: implicit, 54–55, 86–87; quantity, 85–86
   Primal theory, 3–4
   Estimation of production function, 24–25, 26, 230, 231, 254, 374
   Measurement of factor biases, 48
   Measurement of technological change, 36, 43
   Test uses, 185
Primont, D., 20, 185n
Production technology
   Convex sets approach to, 25–26, 30
   Defined, 24–25
   Profit function and, 27
   Structure, 32
Production theory, 2, 17
   Explanation of output level changes, 48
   Functions: aggregate, 34–35, 50–55, 316–320; forms, 18–19, 159, 184; homogeneous-homothetic, 20–21; intrafirm, 255n; metaproduction, 253–255; separability property, 21–22; transformation, 23–24; translog, 23, 59, 63, 74

Technological jointness in, 23–24, 80, 229–230
Productivity, 1, 2, 4, 17, 18. *See also* Agricultural productivity; Econometric approach; Growth accounting approach; Index number theory; Total factor productivity
Profit function
   Cobb-Douglas, 32
   Cost and revenue functions and, 26–27
   Maximization, 25
   Normalized quadratic, 233
   Restricted, 27, 160
   Technology and, 35–36
Profit function models, 3, 68
   Factor demand elasticities and elasticities of substitution, 117, 119, 123
   Restricted, 173–177
   Second generation, 68
   Translog, 82; monotonicity and curvature conditions on, 83–84; restricted for Hicks-neutral technical change, 85; restricted for nonjointness, 84–85

Rao, K. V. Subba and, P. P., 67
Rational expectations concept, 73
Ray, S. C., 110–111, 116
Real estate, data on agricultural, 141
Reger, U., 381, 390
Regulation, environmental
   Effect on input and output markets, 11
   Effect on technological changes, 366–367
   Focus of, 367–368
   Pesticides, 368–369
   Productivity effects, 13, 101, 369–371, 378–380
   Regional, 374
   Resource-related, 368–369, 375
   Taxes and, 371–372
Research, agricultural productivity, 1, 2
   Borrowable, 300–301
   Crop and livestock, 304–307
   Data base for, 139–144; evaluation, 144–151; problems with 7–10, 150–151; recommendations for improving, 3, 10–11, 152–154
   Decomposition study, 307–312
   Empirical, 332
   Investment policy for, 312, 353–354, 362–363

INDEX

Measurement, 66, 114, 304–307
Productivity of, 5, 8–9, 289–290, 362–363
Proposals for future, 329–330
Specialization in, 290–292
Specificity of location for, 292–300
Time dimension for, 301–304
Revankar, N. S., 19
Rosenberg, N., 249n, 253n
Rotemberg, J. J., 74
Roumasset, J., 67
Rudin, W., 25
Russell, R. R., 20, 185n
Ruttan, Vernon, 5, 8, 9, 18, 66, 67, 98, 99, 189, 247, 249n, 253n, 285, 347–349

Saez, R. R., 146, 230
Salter, W. E. G., 249, 250
Samuelson, P. A., 52, 159, 249n
Sargent, T. J., 69, 72
Sato, K., 267n
Scherer, F. M., 249n
Schmitz, A., 369n
Schmookler, J., 249, 250, 251
Schuh, G. E., 66
Schultz, T. W., 11, 67, 98, 289
Shephard's lemma, 24, 29, 31, 74, 75, 87, 217
Shoemaker, R., 203
Shumway, C. Richard, 3, 10, 101, 114, 138, 153, 229–230, 233
Silberberg, E., 191
Singh, I., 67
Single-aggregate output index, 4, 229–230
Sivan, D., 255n
Smith, V. K., 101, 367n, 374
Solow, R. M., 19, 21, 98, 178, 209, 229n
Soybeans, productivity analysis, 209, 224, 225–226
SRS. *See* Statistical Reporting Service
Static equilibrium models, 2, 7
Bias measures, 40–48, 357
Characteristics, 160–161
Criteria for "best," 162, 177–178, 185
Evaluation of, 184–186
Full: cost function, 164–171; factor price elasticities, 182–183; production function, 162–164
Monotonicity and curvature requirements for, 162, 177–178

Partial: elasticities, 176; restricted cost, 171–173; restricted profit, 173, 177
Productivity measures, 178–180
Statistical decomposition models, 8–9, 307–312
Statistical Reporting Service (SRS), USDA, 140, 148, 149, 153
Stevenson, R. E., 9
Stoker, T. M., 341
Stout, T., 99
Stubblebine, W. C., 375
Sudit, E. F., 376
SUR (seemingly unrelated repression) model, 60–61, 171n
Swamy, S., 52
Swanson, J. A., 101, 168

Talpaz, H., 381, 390
Taxes, as regulation strategy, 7, 371–372
Technological change, 2, 3
Allocation of gains from, 195–198
Based on "obvious compelling need," 253n
Biological, 248, 252, 265n
Defined, 33, 317
Disembodied versus embodied, 33–34
Explanation of, 66–67
Factor substitution and, 249–250, 279
Gains from, externalities in analysis of, 366–367
Growth rate, 113
Input substitution for, 358–360
Land prices and, 190, 195
Market equilibrium effects of, 189–196, 202–206
Measurements: econometric analysis of aggregate, 333, 341–344; equilibrium framework for, 4, 9, 160; growth accounting framework for, 209, 211; importance of improving, 226–227; multiple-output case for, 44–48; neoclassical model for, 339–340; single-product case for, 34–44; static models for, 2, 7, 332, 357
Neutral, 37, 38, 45, 56, 57–59, 63, 66, 81–82, 85, 99
Price effects, 200–202
Productivity change and, 57–60
See also Bias, technological change; Induced innovation theory

Technology
  Abatement, 370–371
  Endogenous, 322, 325–327
  Input-output jointness, 23–24, 80, 229–230
  Techniques for improving productivity of: aggregate data for, 320–323; constraints on, 328–329; estimation, 323–325; investment for, 328, 330; prices and, 327; production function, 316–320; variables, 328
  See also Production technology; Technological change
Teigen, L. D., 152
Texas, aggregate agricultural productivity model, 114
TFP. *See* Total factor productivity
Thirlwall, A., 17n, 96
Thirtle C. G., 115, 249n
Thursby, M. C., 37, 38
Tollini, H., 66
Tornqvist-Theil index, 51, 56
  Implicit price, 54–55, 86–87
  Quantity price, 85–86
  Productivity, 100
  Superlative, 100
Total factor productivity (TFP), 3, 7, 13
  Analysis of intertemporal and interspatial, 63–65, 208–209
  Defined, 17
  Econometric approach to, 2, 4, 17, 18; compared to growth accounting approach, 62–63; described, 57; technological change and, 57–60
  Economies of scale and, 178–180
  Exact number approach to, 87–88
  Growth accounting approach to, 2, 4, 17, 18; compared to econometric approach, 62–63; described, 50; index number measures for, 18, 51, 57, 97–98, 101, 105–109, 124–125
  Growth rate, 107
  Measures of, 106, 180–182
Translog model
  Cost function, 59, 63, 74–76, 82, 110, 117, 164–166, 168, 173; intertemporal and interspatial, 216–217; multiple-output, 168–171
  Evaluation, 184–185
  Production function, 19, 23, 57, 74, 82, 162–164, 322
  Profit function, 82–85, 112–113
Trost, R. P., 230, 240

Uzawa, H., 19, 79

van de Klundert, T., 249n
van den Bosch, R., 356
Vo, Trang T., 2–3, 96, 160, 189, 204, 229, 350
von Oppen, M., 67

Wade, J. C., 101
Wales, T. J., 237
Walters, A. A., 17n
Wan, H. Y., Jr., 249n
Ward, M. P., 345n, 346
Watkins, G. C., 67, 186, 333, 336
Waverman, L., 101, 171n
Weather, measure of variations in, 219–220
Weaver, Robert D., 230
Welch, F., 67, 304
Whitcomb, D., 376
Widdows, R., 263n
Williams, W. A., 34
Winkelman, D., 330
Wood, D. O., 150, 229n
Woodland, A. D., 229n, 237

Yamada, Saburo, 285
Yatchew, A. J., 69
Yotopoulos, P., 229n

Zellner, A., 19, 171n, 332, 344, 346n
Zilberman, D., 230, 372

DATE DUE

AUG 15 2002 BY			
GAYLORD			PRINTED IN U.S.A.